AF516407

Ecology, Technology and Economy

Ecology, Technology and Economy

Continuity and Change among the Fisherfolk of Kerala

P.R.G. Mathur

Indira Gandhi Rashtriya Manav Sangrahalaya, Bhopal
(National Museum of Mankind)

RAWAT PUBLICATIONS

Jaipur • New Delhi • Bangalore • Mumbai • Hyderabad • Guwahati

ISBN 81-316-0197-8

Published by

Indira Gandhi Rashtriya Manav Sangrahalaya
Shamla Hills, Bhopal-462013 (Madhya Pradesh)
email: directorigrms@gmail.com

Prem Rawat for **Rawat Publications**
Satyam Apts., Sector 3, Jawahar Nagar, Jaipur 302 004 (India)
Phone: 0141 265 1748 / 7006 Fax: 0141 265 1748
E-mail: info@rawatbooks.com
Website: www.rawatbooks.com

New Delhi Office
4858/24, Ansari Road, Daryaganj, New Delhi 110 002
Phone: 011-23263290

Also at Bangalore, Mumbai, Hyderabad and Guwahati

Typeset by Rawat Computers, Jaipur
Printed at Chaman Enterprises, New Delhi

Dedicated
to my teacher

Prof. Surajit Chandra Sinha
Former Vice Chancellor, Vishwabharati

Contents

Foreword

India has a long coastline and marine fishing has been primary means of livelihood for a large number of people in India, yet the social scientists in general and anthropologists in particular have not paid adequate attention to study them. This is all the more unfortunate as marine fishing in modern days is going through a major technological revolution which is bound to make profound impact on society and culture of the people who are directly or indirectly involved in fishing industry. It has far-reaching potential. Anyway, one must appreciate the foresight of the Anthropological Survey of India that during 1967–72, while formulating new research programme for the Survey, research on the fishing communities was included in its agenda. Under that programme and under the guidance of Dr. Surajit Sinha, Dr. P.R.G. Mathur did pioneering work on the Mappila fisherfolk of Kerala in 1970 which earned him a Doctor of Philosophy degree from Calcutta University. In 1977 he published his work on Mappila fisherfolk. Earlier, Arne Martin Klausen, a Norwegian, had done a study on Kerala fishermen basically analysing the impact of the various technological-economical stimuli introduced by the Indo-Norwegian project for the development of fisheries in India. While Klausen's study focused on the classical question as to why there was unequal response by different communities to the development stimuli, Dr. Mathur's study was focused on a single community in classical anthropological mould providing rich details on Mappila life, their knowledge about marine environment, their technology, economic organisation, trade and market, their rituals and belief in supernatural power. Dr. Mathur's detailed ethnography illustrated how Mappila's economy was embedded in their social and cultural matrix. It is more than three and half decades have elapsed since Dr. Mathur did his fieldwork among the Mappila and more than four decades when Klausen did his fieldwork among the fishermen. The observations and conclusions derived by them have become extremely relevant for developing a meaningful discourse on change particularly in the wake of profound technological and political changes in the wider arena.

After doing field work among the Mappila between 1970–72, Dr. Mathur did some fieldwork among some Hindu fisherfolk in 1980–82 and then again in 2006 among some other fishing communities, which is the basis of his present study. Both in terms of time depth and spatial spread of his studies, he is in a unique position to provide an insight as to what is happening to the fisherfolk in Kerala.

In the present study the objective was to cover maximum number of fishing communities in Kerala and provide ethnographic details about them, and also the kind of adaptation they are making under the changed circumstances. Obviously, his previous studies on fisherfolk provide rich and substantial background and experience for the present study.

In order to cover maximum number of fishing communities in the shortest possible time for his present study, Dr. Mathur made some significant methodological innovations. For example, he demarcated three major regions of fisherfolk in Kerala. From each region he selected three to five fishing communities. The idea was to capture significant variations among the communities with reference to their fishing operations, their traditional knowledge, trade, rituals and the development.

No doubt, there have been many changes since Dr. Mathur's did his first fieldwork among the fisherfolk of which introduction of sophisticated technology in fishing operation is the most significant. It has so to say removed the limitations imposed by nature. Now trawlers can sweep the ocean clean of any object moving or crawling. Echosounder or fish finders can pick up even the weakest signals generated by fish even at the bottom of the ocean. GPS and constant weather monitoring system have certainly considerably reduced hazards of marine fishing. These, along with rapid changes in communication, transport, refrigeration facilities, have not only enormously increased the yield, but also have expanded the market and profitability of the enterprise. All these certainly have generated enormous wealth and power at least for a few and a lot of opportunities for fishermen in general. Interestingly, all these profound changes did not prompt a large-scale economic diversification. May be marine fishing still has tremendous potentialities though serious concerns of regeneration of resources have arisen which led authorities to impose certain restrictions on fishing and use of certain technology during certain seasons. How far they are effective in a system where there is enough room for negotiations and 'adjustments' is an open question but long-term ecological impact of the new technologies, intense competition and effort to maximise the profit are going to be very serious and also on the communities and people who cannot afford the astronomical cost of modern technology. Dr. Mathur has shown that the role of banks and the middlemen has increased enormously as a result of that stranglehold on weaker sections and real workers have become tighter. One may rightly ask where the level playing field is for them: the world is becoming flatter or pyramidical?

However, it is interesting to note that fisherfolk society and culture have shown remarkable resilience. The caste and community identities have remained intact along with respective cultural, ritual practices and with values associated with status and hierarchy. The gender discrimination too continues. It is legitimate to point out that the existence of them along with revolutionary changes in the technology and globalisation are intriguing particularly in the context of Kerala which is considered to be a model state from development perspective. His study once again points out that the caste question in India in general and in Kerala in particular is extremely complex. The jostling among those who are in the caste system and even with those who had gone out of the Hindu society for higher status is interesting. This jostling continues to make the concept of 'purity' relevant and also the path that they take to raise the status in the caste hierarchy though there have been some new interpretations of symbols they adopt. Hopefully, Dr. Mathur's study will inspire scholars to design research studies to go into these questions pointedly and indicate the nature of changes.

P.K. Misra

President
Anthropological Association
Mysore

From the Desk of the Director

It is indeed a great pleasure for us to publish *Ecology, Technology and Economy: Continuity and Change among the Fisherfolk of Kerala* by Dr. P.R.G. Mathur, who has been working in this field for the last four decades. Nourished in the 'classical tradition' of anthropology in terms of intensive fieldwork, Dr. Mathur is known for his research on Mappilla fisherfolk of Kerala. In the present work, Dr. Mathur has not only emphasised on the interaction between man, environment, technology, society, culture and change among the Mappilla fisherfolk, but also has compared these with those of the Hindu and Christian fisherfolks of Kerala. This is, in a way, provides both synchronic and diachronic perspectives to the fisherfolk studies.

Our museum has been conducting a number of activities in documenting and disseminating various aspects of folk cultures of India. Moreover, houses of fisherfolks of Orissa and Kerala have been displayed in our open air exhibition on Coastal Villages. Thus, Dr. Mathur's work is very much within the ambit of IGRMS. By emphasizing both continuity and change among the fisherfolks of Kerala, Mathur's work very much coincides with the perspective of our museum in terms of portrayal of folk cultures of India not as a romantisised changeless entity, but rather where change is an inherent component.

While I cannot but admire Dr. Mathur for the tenacity of his fieldwork in the ripe age, sincere thanks are also due to the Rawat Publications, Jaipur, our co-publisher for the support in publishing it in a short time.

May, 2008

Kishor K. Basa

Director

Indira Gandhi Rashtriya Manav Sangrahalaya

Bhopal

Preface

In this book an attempt has been made to study the fisherfolk of Kerala both diachronically and synchronically using comparative research among the Mappila, Hindu and Latin Christian fisherfolk. The micro-macro method has been employed, which might also be called anthropo-historic. In a work of this kind, which attempts to make both micro and macro studies of the most backward and widespread communities, certain omissions and inadequacies are bound to be present, even if one could provide a broad frame of enquiry into their ecology, technology and economy.

The first phase of the fieldwork was conducted in about 36 fishing villages (Kadappuram) from June to October 2006. The second phase of the fieldwork was undertaken in the first quarter of 2007. I have tried to update the data collected in 1970–72 and 1980–82 on the Mappila fishermen and Hindu fisherfolk respectively.

I acknowledge with gratitude my profound indebtedness to Prof. K.K. Basa, Director IGRMS, Bhopal, for his kindness in sanctioning the project under the title 'Ecology, Technology and Economy: Continuity and Change among the Fisherfolk of Kerala'. The subject matter of this report is based almost entirely on my book *Mappila Fisherfolk of Kerala* (Mathur, 1978), and on my report, 'Ecological and Cultural Heritage of Hindu Fishing Communities in India', which I submitted to the Director, IGRMS, for publication. I am particularly indebted to

Prof. Basa for his cooperation and advice as well as constructive criticism of the report. Prof. Basa has always been a source of inspiration and encouragement.

I would like to record my indebtedness to Prof. P.K. Misra, an internationally reputed Anthropologist who has written the foreword to my book.

I must also acknowledge my debt to the Indira Gandhi Rashtriya Manav Sangrahalaya without whose financial support this report could not have been presented.

I am thankful to Dr. M. Remesh, Centre for Research in Indigenous Knowledge, Science and Culture, Calicut for preparing the Glossary.

Finally I am thankful to Rawat Publications for the utmost care it has taken in publishing this book.

I am thankful to Mr. V. Sudarsan, Research Assistant, for helping me in collecting data about the 36 fishing villages of Kerala selected. I must also thank Ms. Greeshma for her secretarial assistance.

P.R.G. Mathur

Secretary

Ananthakrishna Iyer International Centre
for Anthropological Studies (AICAS)
Palakkad, Kerala

Glossary of Technical Terms

Adi Nir	Internal current
Adiyil Nilkkunna Min	Demersal fish
Aduthavan	A servicing caste of the Hindu fishing communities engaged for cropping of hair, shaving and ritual cutting of hair *etc.*
Akasa Gandharvan	An ethereal goblin is believed to possess supernatural powers.
Alku Alam	The length of the man's out stretched arms across his chest horizontally
Amaram	Stern of boat
Amarappadi	Crossbar near the stem
Aniyam	Prow or boat's head
Arayan	A subdivision of the Hindu fisherfolk found all along the coastal areas, mainly south of Nattika Panchayath, Thrissur district, Kerala. They are also called Kadalarayans in order to make a distinction between Mala Arayans, a hill tribe found in Kottayam and Idukki district
Arayavathi	A servicing caste of the Hindu fishing communities engaged for cropping of hair, shaving and ritual cutting of hair *etc.*
Avoli	Pomfret, an important pelagic fish
Ayikkora	Seers
Ayila	Mackerel
Ayilachala Vala	A special kind of net for catching mackerels and sardines
Ayila Vala	A special kind of net for catching mackerels
Ayiram Cunda	Long lines
Badari Bait	A kind of bait performed in honour of 313 Badariyangal (the participants in the battle of Badar).
Bait	Chanting of poems in honour of great Muslim saints by singing elegies.

Benthic	On or near the continental shelf
Beppu	Long lines
Beppu Toni	Dugout canoes
Bharathar	A Hindu fishing community engaged in backwater fishing along with other occupation like ferrying, coir making, lime making etc.
Bovis-Mogeyar	A Hindu fisherfolk community found along the coastline north of Trikaripur, Madappally, Kadangodu and Edakad of northern Kerala.
Camban	Horse Mackeral
Cambankadal Tiramala	The longest of the waves which occurs in almost all months at depth of two fathoms.
Cangala Cunda	Chain lines with big shank hooks
Cattamaram	An indigenous fishing craft employed along the East Coast from Orissa to Kanyakumari in Tamil Nadu, Trivandrum and Quilon districts of Kerala.
Cellu	See Piyyu
Chala Vala	A special kind of net for catching sardines
Chemban Muthappan	An ancestor spirit worshiped by Hindu fisherman belonged to the Arayan caste.
Chemmin	Prawns and shrimp
Cheriya Tiramala	Short waves which grows quickly breaks down fast.
Christian fisherfolk	Fisher folk belonged to Christian Community
Chukkan Kalli	Nearest compartment to the stern
Chukkankaran	Steersman
Cori	It is a living organism which serves as the food of several species of fish.
Coyala	It is a kind of whirlpool which occurs at a depth of two fathoms in the month of June, July, August and September.
Edakalli	Place for the third member of the crew.
Etta	Catfish
Gulikan	Son of Saturn who believed to be the causative agent for illness to the crew of the fishing unit.
Hindu fisherfolk	Fisher folk belonged to Hindu Community
Ida Kadal Tiramala	One of the important fishing grounds, the waves breaks out at depth of height of three men (Munnalkku alam).
Idakadal	Part of sea lies up to the depth of three to eight fathoms
Idanir Min	The fish that found in mid water.
Idaniril Nilkunna Min	Mid water fish
Ilakkipidikkunna Nethal Vala	A nylon white bait net consists of several piece of nets which used for catching Vella (white) Nethal
Kaccan Kattu	The wind from north west to south east
Kadal Kuttal	Growing of waves
Kadalamma	Mother goddess of sea.
Kadamkani	Rope net.
Kadikkunna Cori	It is of three colours, namely, red, white and green, generally found when the current flows from south to north or south west

	to north east which resembles hairs in the human chest which cause an itching and burning sensation when contact.
Kadutha	The spirit of the dead Nair sorcerer of the village.
Kaiccunda	Hand lines
Kal Cori	See Nila Cori
Kalli	Compartments in plank built fibre boats.
Kamba	Rope
Kanakkunokkal	Predictions based on astrological charts.
Kanamin	Tunnys, an important Polappu Min
Kanava	Cat fish
Kandakarnan	Defence minister of goddess Bhadrakali
Kara Nir	Easterly current
Karakadal	Part of sea lies up to the depth of eight fathoms(coastal water)
Karakadal Tiramala	Tiramalas, which appear on the coastal waters and sweep over the shore.
Karakattu	Easterly wind
Karavoli	Black Pomfrets
Karimin	A species of Kora
Karin Talayan	Black ribbon fish
Karinakku	Black tongue.
Karinkannu	Evil eye
Karutha Avoli	See Karavoli
Karutha Vavu	New Moon
Kattu	Wind
Kavuthiyan	A servicing caste of the Hindu fishing communities engaged for cropping of hair, shaving and ritual cutting of hair *etc.*
Kayal	Back waters
Kettuvallam	Large plank built boats consist of nine to eighteen planks and can carry 40 to 50 fishermen.
Khadeer	A servicing caste of the Hindu fishing communities engaged for cropping of hair, shaving and ritual cutting of hair *etc.*
Khattam Otikkal	The reading of the full text of Holy Quran by one or more Maulavis(Muslim priest) in honour of Prophets and divine personalities.
Kilappi Cori	It is floating zooplankton with floating habit and white in colour. It is abundantly found in the month of Vrischikam and Dhanu.
Kora	Jew fish
Korunna Vala	A nylon white bait net consists of several piece of nets which used for catching Kari(black) Nethal
Kotha	The spirit of the departed woman sorcerer who belonged to Pulaya community and believed to be the sister of Marutha.
Kudiyiruthal	The consecration of stones representing the ancestors in the village temples.
Kutticathan	A poltergeist sprit propitiated by both Hindu and Mappila fisherfolk.
Kuttipuja	A ceremony performed by the carpenter on finishing a house or boat.

Maalaka	Angel
Maarus	The length of the man's out stretched arms across his chest horizontally (One Maaru is equal to six feet)
Maha Sivarathri	A festival observed in commemoration of the day which Lord Shiva is believed to have saved the world by drinking Kalakuta, the deadly poison obtained by churning of the milk of Ocean.
Mala Pannin	Sea ponal
Manangu	Anchovy, a kind of Polappu Min
Mangoos Maulud	A ritual performed by Mappila fishermen in honour of the Prophet to mark the launching of new nets and boats.
Mani Cori	It is as cold as ice, serves as the main food of Avoli and appears in the month of Kumbam and Minam.
Mannan	A servicing caste of the Hindu fishing communities engaged for washing cloths, performing rituals *etc*.
Mantal	Sole
Mantravadi	Healer, sorcerer and ritual specialist.
Mappila fisherfolk	Fisherfolk belonged to Muslim Community
Marutha	The spirit of the Pulaya woman sorcerer of the village.
Mathi	Sardine
Mathikollivala	A special kind of net for catching sardines
Maulud	The most important rituals performed by the Mappila fisherfolk, in which the birth and life history of the prophet and his predecessors and descendants or other saints are recited.
Menir	Surface current
Mepantiya Min	Pelagic fish
Mogavirar	A Hindu fisherfolk community found in the northern extreme of the Cannanore district, Kerala.
Mohiuddin Ratheeb	A ritual conducted on the 11th day of every lunar month at midnight for getting good things.
Mokayan	A Hindu fisherfolk community found in Cannanore district and northern parts of Kozhikode district, Kerala.
Mukkuvan	A Hindu fisherfolk community found all along the coastline north of the Nattika Panchayath in the Thrissur district, Kerala.
Mullan	Silver Belly
Mumbakallikkaran	A fisherman who sites on the Aniyam Kalli
Munnodi Appan	The spirit of the dead ancestor propitiated by Araya fisherman.
Muriyam Potti	Sometimes Ida Kadal Tiramala breaks out a furlong away from the coast where the depth is comparatively less but it appears unexpectedly.
Murttisevayullavar	People who possess malevolent spirits by inheritance.
Mutterikkal	A magical rite performed by Mappila fisherfolk to ward off the evil eye.
Nalalku Alam	Four Maarus
Namannu	It is kind of phytoplankton which is generally found in the bottom of the sea, particularly beyond the depth of ten fathom.
Necca Veli	Neap tide
Nemmin	Seers
Nethal	Whitebait, a pelagic fish and is caught from Karakadal

Nethal Vala	White bait net.
Nila Cori	It is blush in colour and has the shape of an egg or a ball. Their legs are like those of Octopus and appear in the month of Kumbam and Minam.
Nilavu	Moonlight
Nir	Water current
Nittielakkal	A technique employed using nylon net which used depends upon the flow of the wind.
Nittivilakkal	A technique employed for catching mackerel according to the direction of the wind and the migration of fish.
Nullayan	One of the Hindu fishing community found exclusively in Thiruvananthapuram district of Kerala and Kanyakumari district of Tamil Nadu.
Odam	Dugout canoes
Olivunokkal	Predictions based on astrological charts.
Olukkuvala	A kind of floating gill net made up of hemp which used for catching Ayıkkora and Nemmin
Onam	A national festival of Kerala celebrated in the month of Chingam.
Oncuvala	A nylon white bait net consists of several piece of nets which used for catching Vella (white) Nethal.
Otta Cunda	A modern technology of hook and line fishing.
Ottathadi Vallam	Dugout canoes made up of the trunk of huge trees.
Ozukkuvala	Gill net
Padi Kadal Tiramala	Tiramala, which break out in the fishing ground
Padikadal	Part of sea located at the west of Idakadal at the depth of 2 fathoms
Paitakora	See Kora
Pallimin	A species of Kora
Pandi	Low portion of the boat
Panniakkal	A servicing caste of the Hindu fishing communities engaged for cropping of hair, shaving and ritual cutting of hair *etc.*
Parassinikadavu Muthappan	An incarnation of god Shiva belongs to Tiyya family.
Paravan	A servicing caste of the Hindu fishing communities engaged for washing cloths, performing rituals etc.
Pattu	The reading of the full text of Holy Quran by one or more Maulavis (Muslim priest) in honour of Prophets and divine Muslim saints.
Peedam Vekkal	Placing a tripod in the shrine in honour of the dead
Pelagic	All surface water
Petiyil pedunna Min	Demersal fish
Peyitham Vala	A nylon drag boat seine net which is in extensive use in Malabar area
Peyunna Nethal Vala	A nylon white bait net consists of several piece of nets which used for catching Kari(black) Nethal

Piyyu	The Piyyu is living organism which floats according to the direction of the current. It also lives in puttu.
Polappu Min	The fish which have their habitat near the surface water and appear in shoals.
Polappu Min	Pelagic fishes like Sardine and Mackerel
Ponal	A kind of phytoplanktons abundantly found on the Malabar cost in the month of Chingam, Kumbam and Minam
Ponkala	An annual festival performed by Hindu fisherfolk in honour of Kadalamma.
Pottanmar	Body guards of goddess Bhadrakaļi
Potti Cori	See Kadikkunna Cori
Pretham	Ghost of a person who commits suicide.
Punkara	A kind of phytoplankton, that serves as the food of pelagic fish is Punkara. It is the spores of sea weed which flows along with the currents, particularly in the month of Makaram, Kumbham and Minam
Puram Kadal Tiramala	The waves that occur in the part of sea lies beyond 10 fathoms and within the distance of 100 fathoms from the continental shelf.
Puram Kattu	The wind blowing during the South West monsoon from the west to east (Westerly wind).
Puram Nir	Westerly current
Puramkadal	Part of sea lies beyond 10 fathoms and within the distance of 100 fathoms from the continental shelf
Putiyappila Kora	A species of Kora (goat fish)
Puttu	It is one of the important zooplankton, refers to a mound thrown up by worms on the rocky bottom of the sea.
Rafai Ratheeb	An expensive ritual performed generally by very rich fishermen once year for getting good things and avoiding diseases.
Rani Vala	A kind of synthetic net used in traditional wooden boat.
Ratheeb	A special ritual exclusively performed in honour of Mohiuddin Shaik and Rafi Shaik for favours received by Mappila fisherman.
Sevamurttiyullavar	People who possess malevolent spirits by inheritance.
Sravu	Shark
Sura	Cuada, a kind of Polappu Min
Suranat Kaimal	An ancestor spirit worshiped by Hindu fisherman belonged to the Nair caste.
Talayan	Ribbon fish
Talepedunna Min	The fish that live in the bottom of the sea.
Tali Vali	Waves that sweep the shore and fall backwards as if the water is released from a reservoir.
Tammankallikaran	The oarsman who occupies in Tammankalli.
Tangu Vala	A kind of purse seine net generally operated for catching mackerel and sardine.
Tanni Cori	It has short legs. It is generated abundantly during the month of June, July and August. It does not cause an itching sensation to the body.
Tazhepedunna Min	Demersal fish

Tekkan Kattu	Southern wind
Tekkanvada	When the southern wind blows, the southern current also flows from south to north in the same direction.
Tekkan Nir	Southern current
Tembrayil Kattu	South-westerly wind, generally blows in the month of Idavam and Mithunam
Tembrayil Nir	South-westerly current
Tengara Kattu	South- Easterly wind
Tengara Kattu	South- easterly wind
Tengara Nir	South- Easterly current
Thottam Kazhikkal	A ritual performed by employing cabalistic figures using coloured powder.
Tirandi	Skates and rays
Tiras	Waves
Toni	Plank built boat.
Tottappu	When Tali Valli occurs, the area of the sea to its east called as Tottappu
Tulavatta	Tembrayil Kattu blown in the month of Tulam
Tuyi	Light of the current
Tuyikku Pokal	Fishing expeditions at night with the help of Tuyi.
Ucca Veli	Spring tide
Ulivekkal	The occasion in which a carpenter first starts his work on a trunk of a tree meant for building a new boat.
Ulkadal	The mouth of a large river (estuary)
Unnikka Muthappan	An ancestor spirit worshiped by Hindu fisherman belonged to the Arayan caste.
Vadakkan Kattu	Northerly wind
Vadakkan Nir	Northerly current
Vadavartha Kattu	The wind blows from north-west to south-east on the Malabar Coast.
Valaccilakkal	A technique used by the Mappila of Tanur for fishing mackerel which depends upon wind and current.
Valachilakkuna Vala	A kind of boat seine net generally operated for catching mackerel and sardine.
Valakorkkal	The joining of the pieces of nets
Valan	A Hindu backwater fisherman community found in Cannanore, Thrissur and Ernakulam district, Kerala.
Valiya Vallam	Big fibre glass canoe.
Valinjiar	A servicing caste of the Hindu fishing communities engaged for washing cloths, performing rituals *etc*.
Valippu	Growing of waves
Vallam	Dugout canoes
Vanchinanakkal	An auspicious time and day for launching new boat as instructed by an astrologer.
Vanji	Dugout canoes
Vannan	A servicing caste of the Hindu fishing communities engaged for washing cloths, performing rituals *etc*.

Varana Kadal	See Puram Kadal Tiramala
Varana Kadal Tiramala	The waves which grow in the Puramkadal during the Monsoon.
Vattakkara Kattu	North Easterly wind
Vattakkara Nir	North easterly current
Vayya Kadal Tiramala	It is believed to be the swiftest wave which breaks during the months of June and July.
Vayyakadal	Part of sea located at the depth between five to eight fathoms
Velan	A servicing caste of the Hindu fishing communities engaged for washing cloths, performing rituals *etc.*
Veliyettam	High tides
Veliyirakkam	Low tide
Vella Cori	It is of almost the size of paper weight, with small legs and hand and white in colour. It makes the Tuyi to more shine.
Vella Avoli	Silver Pomfrets
Velladu	Presence of thin films of oil on the surface because of the swimming of the shoals inside the water
Vellam Irakkam	Low tide
Vellam Velli	High tides
Vellamvali	Forward and backward masses of water
Velli Talayan	Silver ribbon fish
Velutha Vavu	Full Moon
Venganavoli	A species of Pomfret
Vishu	A festival of Kerala which considered as an astronomical New Year day.
Yama Kadal	Giant waves that generally fall on the shore from where the waters either move up to 100–50 feet or return to the same point or move backwards to the place from where they fell originally.

Malayalam Months	**English Months**
Chingam	August-September
Kanni	September-October
Thulam	October-November
Vrichikam	November-December
Dhanu	December-January
Makaram	January-February
Kumbham	February-March
Minam	March-April
Medam	April-May
Idavam	May-June
Mithunam	June-July
Kakkitakam	July-August

List of Plates

Introduction

Fishing is one of the most important traditional occupations of mankind. Marine fishermen are mainly dependent upon the sea-life thriving in a particular ecosystem. Sea is the most fertile source of food although it is often hostile and inhospitable to man with its dangerous waves, tides, currents and winds. Over ten million fishermen in India subsist on fishing. The sea is, therefore, an inexhaustible source of wealth, particularly for the fishing communities of Kerala who exploit it for their livelihood. Thus, the whole mode of the fisherfolk's life develops from their constant contact with the sea.

India has a total coastline of 4,677 kms, out of which Kerala's share is nearly 580 kms in length and nearly 115 kms in width. The Malabar coastline is 292.5 kms long. Kerala is bound on the west by the Arabian Sea and on the east by the Western Ghats. The continental shelf along the Kerala coast is the richest fishing ground in the country. Ecologically, the state has three belts – the Highland, the Middle Lateritic Plain and the Coastal Plain. The Kerala coast is indented by numerous lagoons, creeks, etc., which have provided facilities for the construction of seven fishing harbours in recent years. The most important fishing harbours are Mappila Bay (Cannanore), Beypore (Kozhikode), Neendakara (Quilon), and Cochin port.

The long coastline of Kerala has brought a number of ethnic groups in direct contact with the fishing communities that include Hindu fisherfolk, Mappila fishermen and Latin Christian fishermen. The Mappila fishermen who inhabit the Malabar and Cochin coastal villages are reputed to be experts in fishing in the changing environmental conditions of the sea. They possess very valuable traditional ecological knowledge of the sea like the different species of fish and their habitat, the various planktons, and of the breeding seasons of the different species. The Hindu fisherfolk are found on the coastal villages of Malabar, Cochin and Travancore; the Latin Christian fisherfolk are distributed in the districts of Quilon and Trivandrum. The fisherfolk of Kerala depend upon the Arabian Sea for their livelihood. The technology of fishing of these communities is related to the ecological conditions of the particular region which they inhabit. These conditions govern to a great extent the distribution of the communities, their settlement pattern, the nature of their economy, their rituals, and their cosmological ideas. The fishing technology and economy of the Kerala fisherfolk are embedded in and enmeshed with the ecological conditions, their rituals and their cosmological ideas (Mathur, 1978).

The fisherfolk of Kerala follow a life pattern, which is moulded by their marine environment. Their cultural ecology and economic organisation can also be seen to be the direct result of their marine environment. The repercussion of the marine environment on fishing communities is one of the most outstanding instances of the intricate relationship that exists between man and ecology. The fisherfolk's indigenous knowledge of the ecology is their main asset, providing the means for their livelihood, and aid in times of dangers at sea. Since the fisherfolk get their livelihood from the sea, they believe in several superstitions, which have resulted from their utter dependence on the ecosystem. Superstitions are the means to overcome their fear and cover the uncertainties in their lives. Therefore it may be said that the fisherfolk of Kerala are steeped in supernatural beliefs and practices.

The Problem

The present study is based on fieldwork undertaken in Kerala among the Mappila fisherfolk between 1970 and 1972, among Hindu fisherfolk between 1980 and 1982 and other fishing communities between June and October, 2006.

This study has updated our previous fieldwork among the Mappilas and Hindu fisherfolks of the state. Writing on the Mappila fisherfolk of Kerala, Surajit Sinha observes: "The fishing economy of Mappila is inseparable from their kinship structure and belief systems. Dr. Mathur's book brings into relief how the economy is embedded in the social and cultural matrix". He has also pointed out that "the book provides a wealth of information on the fisherman's detailed knowledge of the marine environment and cosmology. They enumerate eleven types of waves, eight kinds of winds and eight kinds of currents and carefully adjust their technical operations, connected to particular kinds of fish, on the basis of a careful assessment of the objective situation. Although the fishermen construct a rich world of symbolism and sentiments in presenting their ideas about the marine environment and cosmology Dr. Mathur's presentation convinces us about the enormous capacity to take rational technological decisions on the basis of detailed objective knowledge of the situation" (Mathur, 1978).

The current study presents a descriptive analysis of the *Ecology, Technology and Economy: Continuity and Change among the Fisherfolk of Kerala*, who inhabit overwhelmingly the coastal regions of Kerala. The Mappila fisherfolk form a distinct community within the Muslim society. These fisherfolk have constant contact with their Hindu and Christian counterparts who live on the contiguous coastal villages of Kerala. This study focuses attention first of all on the ecology, technology and economic organisation of fishing, followed by an analysis of their social and cultural fabric woven around their decisive livelihood activity of fishing. The technology of fishing of the fisherfolk of Kerala cannot be separated from their natural laws and cosmological ideas. Thus peculiar notions of the Mappilas about winds, currents, habitat, behaviour of the fish, waves, tides and phases of the moon guide their technology. In this study we have thoroughly reinvestigated their ideas about the marine ecology and relevant cosmos and have tried to compare them with those of the Latin Christian and Hindu fisherfolk. Apart from undertaking a study of the technology and economic organisation as well as marine environment and climate conditions, an attempt has been made to explain certain aspects of fishing problem like:

1. To what extent do the Mappila fisherfolk share common techno-economy, techniques and organisation as well as ecology and cosmology with the Hindu and Christian fishermen of Kerala.
2. To what extent has the value system of the fishing communities undergone change under the impact of modern technological innovations.
3. How does fishing as an occupation affects the cultural ecology of fishing communities?
4. How do the external conditions of the fishing communities affect the market economy in which they are to operate?
5. What role does technology play in moulding the economic organisation of the fisherfolk?
6. To focus attention on finding out how religion, magic, rituals and ceremonies are interwoven into the technology and economy of the fisherfolk of Kerala.
7. To highlight the customary rules governing the fishing organisation, mobilisation of the labour force, distribution of catches and principle of reciprocity.
8. To find out the fishermen's knowledge of the marine environment, particularly about the species, the volume of catches, the relation with the habitat and behaviour of the fish on the one hand, and technology of fishing and the marine environment on the other.
9. To study the impact of various laws enforced by the Union Government and State Government for the regulation of over-fishing.
10. In sum, the object of the present study is to focus attention on the interrelationship between marine environment, technology, society and culture and changes among the Mappila fisherfolk and compare them with those of the Hindu and Christian fisherfolk.

It is assumed that the Hindu fishing communities of Kerala are now generally lagging behind in adapting to modernisation while the

Christian and Muslim fishermen have been able to forge ahead. How is that some of the non-Hindu fishing communities and even some Hindu fisherfolk are able to adapt modernisation more easily and thereby succeed while others remain stagnant, at times even worsening their lots. This calls for an in-depth understanding of the social and cultural mechanism and the different networks acting and functioning within and outside various communities. Since these factors play important roles in the ability of the various fishing communities to accept or reject modernisation. This study of the ecology, technology and economy of the fisherfolk of Kerala would help us to understand better as to how and why some of them have succeeded in adapting modern, technological innovations and also how development programmes should be planned according to the felt needs of the target fishing population.

The Mappila fisherfolk are converts from Hindu fishing communities, particularly from Mukkuva and Mogeyar. They are found in almost all the coastal villages of former Malabar and Cochin areas. The Mappilas occupy the lowest social position in the hierarchy of the Muslim society of Kerala. There is no intermarriage between the Mappila fisherfolk and the non-fishing Muslims and the Marakkan, a Muslim fishing community of Trivandrum, do not marry with the non-fishing Muslims of Travancore.

The Latin Christians are converts from Bharathar/Paravar communities of Quilon and Trivandrum districts. Bharathar and Paravar, who have been included in the Scheduled Caste list, still follow their traditional occupation of fishing and ferrying boats. Thus, Latin Christians and Muslim fishermen are converts from the Hindu fold and despite the fact they had been converted many centuries ago, several ritual practices and cultural traits still persist among these non-Hindu fisherfolk.

Research Methods

The field method used for this study was the participant observation method. This is the most popular method of anthropological study, which is of a lone researcher spending several months in a community in order to understand the social phenomena in the ways that local community understands them in the local dialect. Indeed, Clifford Geertz, one of the anthropology's most widely known modern figures, describes "being there" as the central tenet of anthropological research (Geertz, 2000). Most anthropologists still argue for the necessity of spending several months in the field in order to develop sufficient familiarity with complex research problems and to develop enough rapport with local communities and research subjects.

Applied anthropologists like Steward (1955) and Rappaport (1984) have argued for the necessity of analysing human and environmental systems in tandem. Their goal was to understand the ways in which local environments shape culture and the ways in which cultural practices, in turn, impact the environment. This focus upon both natural and cultural systems has resulted in productive recent works in which scholars bring a multi-cultural lens to environmental research (Milton, 1996) and analyse environmental values and actions as cultural phenomena themselves (Brosius, 1999).

"The research commitments required of ethnographic methods have been particularly useful in small-scale societies that are not well suited to other social science research methods. Ethnographic methods have been employed in larger communities as well, though the focus is often on a smaller subset within the urban social group. In these diverse contexts, participant observation and ethnography can yield incredibly rich information, but to do so effectively requires a serious investment of time *in situ*" (Sepez et al., 2006: 281).

While conducting the study we encountered large geographic scale of many national programmes, including fishery management plans implemented by the central and state government which was in tension with the time and field work commitment of ethnography. From Kasargod to Kanyakumari on the west coast of Kerala there are a number of communities that participate in fisheries in some way, and over a dozen of these are significantly involved in commercial, recreational and/or subsistence fisheries. The marine communities involved in fishing on the west coast of Kerala are segmented into Mukkuva, Mokayan, Bovis-Mogeyar and Mogavirar of Malabar, Mukkuva and Valans of Cochin and Arayans of Travancore. Apart from the above Hindu fisherfolk there are Latin Christian fishermen of Travancore and Mappila fisherfolk of Malabar engaged in subsistence and commercial fishing. Each of these communities has a different history, ethnic composition, economic organisation, social and cultural tradition.

In order to delineate this tension a bit more, we tried to map out what it might mean to adopt participant observation as the primary method in studying fishing communities of Kerala. In the West Coast fisheries of Kerala, over a dozen communities significantly involved in commercial fishing were selected using a multidimensional model for measuring fisheries involvement. With three months of field work in each community (a truncated period, since the amount of time an anthropological purist might devote to a fieldwork project is at least twelve months), we estimated that a typical participant-observation-based ethnography encompassing these fishing communities would require about a year. If we shorten the allotted time even further from three months to one week, it would still take one year of non-stop fieldwork. At seven days per community, it is still more than ten months.

Obviously, there was a need to adopt the method to the scale of the area requiring study. Given the constraints of two full-time researchers (one at the Malabar area and one at the Travancore area), and unpredictable funding from year to year for research assistants, the remedial options were relatively straightforward:

1. We could use conventional ethnographic methods, but conduct research in fewer communities;
2. We could conduct research in more of the communities, but spend less time in each one;
3. We could alter the methodological approach completely, and conduct research on these communities without actually spending time in them.

Each of the approaches and methods described below that were employed in a patchwork have advantages and disadvantages and they reflect the method of data collection in the context of fisheries management which had to be evolved that was responsive to both geographic and social scales.

For the study we decided to cover only those Muslim, Hindu and Christian communities studied earlier. As mentioned above, we used quantitative fishing information to select half a dozen communities out of each of them that had a significant involvement in commercial and subsistent fisheries. We then employed two slightly different strategies for selecting twelve among them for fieldwork, but both were based on dividing the large area into regions and selecting a small number of representative communities within each. This approach of selecting communities within partitioned regions was similar to that employed by Wilson, et al. (1998) in their social impact assessment of the highly migratory species management plan, which covered a similarly wide geographical area.

The State of Kerala, with its small size and small number of communities involved in fisheries, is divided into three regions, i.e., Malabar, Cochin and Travancore, based on geographic boundaries. We selected three to five communities from each region for intensive studies to represent the area. We focused first on the area most heavily involved in commercial fisheries, i.e., Malabar. The communities visited there were the Mappila fisherfolk, Bovis-Mogeyar, Mogavirar and the Mukkuva. For Cochin and Travancore, we used the state boundaries themselves as regions. Communities were selected for field work within each area based on their involvement in fisheries and their accessibility. Communities on whom field study was conducted included Mukkuva, Mappila fishermen and Valans of Cochin, and Arayan, Latin Christians and Marakkan of Travancore.

According to Johnson (1990), the benefit of this approach is that it partitions the study area into manageable pieces. The problem is that the selected communities are not necessarily representative of other communities in the area. For example, while size is an important general factor for grouping similar communities, other elements may be equally important, such as ethnic composition, facilities and services, socio-economic structure, and ecosystem characteristics. Such case studies are thus limited to being an example rather than being *exemplary* of communities in the state or region. However, in future research, community data could be used in a cluster analysis of all the communities involved in fishing. Such an analysis would present a methodologically defensible means of selecting a smaller number of exemplary communities for fieldwork from among the dozen communities.

In addition to focusing on fewer communities, we spent less time in the selected communities. In Malabar, we spent three to seven days each in Kadappuram, rather than three months or longer. This was because we found that after about two days of key informant interviews and participant observation, the rate of new information began to level off. It is true that more time in the community would have led to a deeper understanding of social dynamics, but to reach that next level of insight would take considerably more time. Three to seven days for each community visit in the Malabar area was just long enough to assess the fisheries infrastructure, complete a few interviews, and acquire a sense of the local ethos.

Another viable alternative method for spending less time in selected communities is the interdisciplinary rapid assessment process or RAP. Beebe defines RAP as "intensive, team-based qualitative inquiry using triangulation, iterative data analysis, and additional data collection to quickly develop a preliminary understanding of a situation from the insider's perspective" (2005: 285–86). While RAP and related methods were originally conceived primarily as a basis for judging the efficacy of rural development projects in agricultural settings, in recent years the techniques have proven quite useful within marine and coastal environments (Lamug, 1994; McCracken, 1990). Pido et al. (1996) developed a methodological framework for conducting rapid assessment research of fisheries management systems. Their approach underscores the need for on-the-ground cooperation between the researchers and local communities (Sepez et al., 2006).

The recommended RAP research time frame is five days to six weeks for the entire process. However, the research requires a team effort, preferably a multidisciplinary effort that also includes at least one "insider". The real savings in RAP is of time, not necessarily money. That makes RAP a good method for social impact assessment and an important tool where insider perspectives are an important target of research.

The third approach we have used to confront the ethnography/geography predicament is the compilation and analysis of social indicators, especially quantitative indicators, which we have used to describe communities involved in fisheries in the coastal villages of Kerala.

Social indicators are sets of variables used to estimate the economic and social conditions in a community at a given point in time. We have used social indicators since the 1970s to estimate various research parameters of interest, such as racial composition, religion, income and employment, and health status, in order to understand large-scale structural change in fishing societies.

Moreover, social indicators have become more nuanced and more accurate through the iterative process of repeated research (Gahin and Paterson, 2001). As a policy-making tool, social indicators have proved useful in forecasting economic change, allocating resources, directing social intervention programs (Ferris, 1988), and assessing sustainability (USDA, 2003).

Anthropologists working in various research organisations have long been interested in the sociocultural aspects of fishing communities and technological innovations. As the State Government is responsible for management of the marine resources, the Supreme Court has directed the State Government to extend the trawling ban up to 15th of August, 2006. The apex court has emphasised the importance of fishery resources, livelihood of fishing communities and national standard of fishery conservation and management. The management of fisheries in India is governed by rules and regulations formulated under the Indian Fisheries Act, 1897 and later under the Marine Fishing Regulation Act, 1978, demarcating the fishing zones. The Government of India in 1977 enacted the Exclusive Economic Zone Act extending the rights to explore, exploit and utilise the living and non-living resources available in 200 n.m. zone from the shore. As development of marine fisheries in the territorial waters extending upto 12 n.m. from the shore is a state subject, different maritime states formulated their own rules and regulations for the management of the resources.

The regulatory measures formulated under the above acts and regulations, by and large, cover prohibition of destruction of resources by explosives and poisonous means and by destructive gears. The other regulatory measures include regulation of fishing in the nursery areas where juveniles are concentrated; indiscriminate fishing or catching of breeder in their migratory phase and leasing/licensing system of fishing rights, particularly in the inland waters. In the marine region, the regulatory measure that has been adopted as an administrative approach to the management of fisheries is the demarcation of fishing zones aimed at mainly safeguarding the interests of small-scale and medium-scale fisheries. The other management solutions discussed and advocated are: (1) regulation of fishing effort for exploiting the resources, particularly the shrimp resource which is a single critical resource and centre of most of the controversies and conflicts in the country; (2) restriction of number of fishing gears which exploit the juvenile phase in the backwaters, estuaries and shallow inshore waters through licensing; (3) mesh size regulation; (4) minimum legal length for capture, and (5) closed seasons and areas. Among these, although, the licensing of fishing gears engaged in the juvenile fishery is in force through regulation as in Kerala, its implementation has not been successful mainly due to socioeconomic constraints, particularly lack of alternative employment opportunities for fishermen. Similarly, mesh size regulation could not be enforced due to multispecies, multigear nature of the fisheries and also for socio-economic reasons.

The crucial problem of conflicts in the monsoon fisheries of Kerala was discussed by several committees and commissions appointed for the purpose. After considerable deliberations and detailed analyses of the pattern of the fisheries, information available on the resources and other related social and economical implications, it was decided to ban trawling in the territorial waters of Kerala during June-August and this has been implemented since 1984 with relaxation in certain areas and with varied duration. One of the Commissions (Kalawar Commission) appointed to examine this issue recommended regulation of effort of 1,145 trawlers during this period with the trawl codend mesh size not less than 35 mm taking into account the area, extent and intensity of fishing. However, the arguments for ban of trawling and extending into total ban of fishing are being continued.

The main objective of regulatory management of fishery resources is to ensure maximum sustainable yield or maximum sustainable economic yield. However, while considering the regulatory management tools, it is essential to consider the socio-economic conditions and employment opportunities of fishermen belonging to different economic and ethnic groups so that these measures could be implemented ensuring maximum benefit after safeguarding their interest even though the total objective of that regulatory measure or the fishery output may suffer. In such a situation, the classical management tools such as limiting the effort, regulation of entry by gear type and closed seasons or areas may not be totally acceptable unless alternative opportunities for their livelihood and basic needs are ensured or created. The failure of the total implementation of the present regulatory measure of ban of trawling to the satisfaction of different sectors appears to centre around this crucial point, although the data available on the resource exploitation and cognate argument of adverse effects of trawling on spawning population and juvenile exploitation do not advocate total ban of trawling in the fishing grounds. Nevertheless in consideration of the conservation of the resource and unrestricted exploitation of juveniles, it has become imperative to stress viable management measures.

As the success of implementation of regulatory measures largely depends on the involvement of fishermen, it is necessary to take this aspect into consideration. It is observed that self-regulation practiced by fishermen themselves as in the case of management of purse seine operation in Karnataka and sharing of day and night fishing between the artisanal and mechanised sectors in Tamil Nadu have been successful in conflict management. Such a conflict management system with the total involvement of fishermen, administrators, politicians and others should work effectively as against the exclusively administrative approach that is being followed now. It is therefore clear that the management of fisheries cannot be considered as a matter concerning administration or biology or resource assessment only, but as an integrated approach taking into account the sociological, economic and development objectives and priorities as well.

The Government of Kerala has prohibited the use of bottom trawl in the sea along the entire coastline of the state within the area specified under G.O. (P) No. 136/g4/PWF&PD dated the 30th November, 1984, published as S.R.O. No. 1496/84 in the Kerala Gazette Extraordinary No. 1005 dated the 3rd December,1984, for a period of 45 days with effect from the 15th day of June 2006 to the 29th day of July 2006 (both days inclusive).

However, at present, several themes have emerged, including dependence on fishing, fishing community resiliency, and external forces of change. Kerala state fisheries management is faced with serious dual challenges of declining resources and rapid changes in marine environments. The communities that depend on these marine resources are caught between efforts to conserve or restore the resources, the social and economic realities of marine development, and the emerging effects of climate change. For many years, Union Government focused on assessing the status of living marine resources (fish and other economically useful species) to manage and conserve these resources to provide food, recreational opportunities, and economic stability in the fishing economy. This was largely an attempt to maintain a balance between what was in the ocean and what was taken through harvest. Only recently has this effort expanded to recognise that many additional forces are at work (e.g., environmental degradation, world ocean climate change, changing social and economic conditions in fishing communities from non-fisheries sources, and competing uses of ocean space) that influence the availability of fishery resources and the communities that depend on them. The present study examines factors that precipitate cultural and economic transformation in fishing dependent communities by looking at the transformation of the relationship between cultural identity and the natural environment as affected by complex forces such as economic globalisation and ecological health. To place these data in context, we first briefly outline the history of efforts within the state of Kerala to address the sociocultural context in fisheries management, and then identify some critical issues in maritime anthropology.

In 1984, during the early years of the State Fishery Department, a sociocultural data necessary for analysing the potential impacts of regulations was rarely available in the formats or geographic coverage needed. Under these circumstances, the phrase "*no data available*" was often written in the relevant sections of required documents. It was under these circumstances that the Government of India appointed a committee in 2005 with the Director of CMFRI as chairman to study the impact of closed season on fisheries.

Colburn et al. (2006), in their article 'Anthropological Applications in the Management of Federally Managed Fisheries', have observed that "conservation and management measures shall, consistent with the conservation requirements of this Act (including the prevention of overfishing and rebuilding of overfished stocks), take into account the importance of fishery resources to fishing communities in order to (a) provide for the sustained participation of such communities, and (b) to the extent practicable, minimise adverse economic impacts on such communities." These new requirements stimulated increased attention to sociocultural analysis in partnership with economic analyses.

The data presented in the study provide a sample of the work that has been undertaken on challenges and approaches to meeting the needs for rigorous social scientific analysis of human impacts from Government of India, fisheries management actions. The data include basic research and social science methodology as well as more specific issue-related research. All examine, in some form, the relationships among community, household, and individual well-being as they related to resiliency in response to regulatory pressures to protect marine resources and to other aspects of marine environmental change – including economic changes originating at regional, national and global levels.

This is at the core of what the state requires to consider how fishing-dependent communities can adapt and sustain their engagement in marine resource harvesting and processing in the face of complex pressures. From a more purely anthropological perspective, the study includes important discussions on the cultural nature of fishing, in particular, an examination of the enduring question that if fishing is among the most physically dangerous occupations in Kerala, and is also economically threatened, why do Kerala fishermen still fish?

We have tried to provide in our book 'Mappila Fisherfolk of Kerala' a look at two very different communities' responses to modern innovations. Our comparative analysis of the structural characteristics of prawn and bottom fishing in northern Kerala uses rational choice theory to evaluate the factors that contribute to quite different levels of support for conservation legislation in the bottom fishery. We have tried to identify several factors and mark the progress on defining conditions that may encourage resource users to develop sustainable practices and codes to maintain their livelihood.

Acheson (2006) provides a look at two very different community responses to external pressures. His comparative analysis of the structural characteristics of lobster and groundfish fishing in northern New England uses rational choice theory to evaluate the underlying factors that contribute to quite different levels of support for conservation legislation in the lobster fishery and the groundfish fishery. He identifies several factors, and makes progress on defining conditions that may encourage resource users to develop sustainable practices and codes to maintain their livelihood.

In our case study we have examined a single species of fishery, namely prawn, that has responded to a regulatory shift from open access to limited access. This classic management strategy can be implemented in a variety of ways including limiting boats permitted, catch size, days fished and area fished. We argued for ethnographic investigations that capture the relationship between fishing dependent cultures and political economy, and the variety of management regimes that fall between theoretical poles such as market-based versus community-based. To predict how communities may adapt to these management measures it is critical to look closely at what actually constitutes a fishing community with its multiple interests, statuses, alliances, and relative standings, and how it responds to shifts in property relations.

While advocating the preservation of fishing communities of Kerala and culture, we have tried to use the cultural model approach to evaluate the degree of concerns between different fishing communities. Cultural knowledge and interests of fishing communities, as well as other stakeholders that are part of the local community, are necessary for preserving the heritage of the fishing communities of Kerala. We have tried to provide an ethnographic description of the social impacts of fishing regulations at individual, household and community levels by examining the closure of long line fishing grounds in Kerala and the residual social impact within the predominantly Latin Christian fishing trawlers. The study reveals historical resilience in the face of extreme social change citing the immigration of fisherman from southern Travancore to the northern part of Malabar. The study illustrates the importance of social ties within multicultural settings and across physical space including international boundaries.

Our study would address the question of what motivates the fisherfolk of Kerala to fish despite great personal risk and sometimes marginal economic rewards. We have tried to examine their psycho-social characteristics and measure their job satisfaction using structured survey instruments. While there are community-level policy implications of this study, its greatest value lies in its contribution to our understanding of the interrelationship between the psychological antecedents that predispose people to prefer to fish for a living over other pursuits, and point out how occupational mobility can result in problematic well-being outcomes. The self-identity of fishermen is characterised by a high degree of adventure and risk seeking behaviour, and a need for independence not found in the self-identities of those in other occupations. We have argued that the Hindu fisherfolk of Kerala may resist change because the hold between their psycho-social profits and the needs of the fishing occupation is particularly tight.

Using ethnographic methods we have tried to evaluate the concern for potential risk identified by the fishing industry. Our study has revealed that there is clear need for refinement and or development of methods for social and cultural analysis of fishing communities. We have tried to address both the issues of scale and selection of methods of analysis for baseline social and cultural information for a large geographical region of Malabar, Cochin and Travancore. Community data from our earlier works ('Mappila Fisherfolk of Kerala', 1978, 'Hindu Fisherfolk of Kerala', 1982) are included as baseline information in the present study. This study considers the need for developing a basic anthropological research program and data base that covers a wide geographic area stretching from Kanyakumari to Kasargod. This very wide area coverage conflicts with traditional anthropological research and methodology that was generally confined to community level ethnographic studies. We have proposed a nested scale analytical framework that overlies the community level information with macro-level considerations. A series of social indicators has been developed making intercommunity comparison possible between the in-depth ethnographic data collected from selected communities. This has also made it possible to more consistently identify fishing related socio-cultural structures that may not have otherwise been identified using community-profiling methods on such a large-scale.

The large-scale community profiling methods outlined by us contrast with a highly focused ethnographic study. We have tried to present a narrative account and analysis of the daily life of marginalised fishermen on the Kerala coast. Our study highlights the personal costs and risks that fishermen face in this transient fishery while uncovering further evidence of the psycho-social characteristics that drive participation in the industry. We have tried to describe a "vicious circle" where fishermen go home to escape the rigours of life at sea, feel uncomfortable there and soon return to the docks or vessels to continue fishing, only to long for home once again. Our account also identifies issues of infrastructure and community stability in the face of real estate development of coastal Kerala that is outside the circle of agency fish stock assessment and managed access to fisheries, but is nonetheless having a profound impact on industry and fisherfolk's lives in several coastal areas.

The selection of chapters in this volume weaves together several common themes. It is now recognised that many fisheries throughout the world are in crisis (FAO, 2005: Part 1, p. 32). As a result, some fishing communities are also in crisis-multiple factors including declining stocks, globalisation effects, coastal development, climate change, and environmental degradation, combining to put extraordinary stress on many of these communities and cultures. The role of maritime anthropology in fisheries management is to evaluate and predict the response of communities to change induced by management efforts and external forces. This role raises a number of key questions related to the culture of Hindu fishing communities, Mappila fisherfolk and Latin Christians. Do the fisherfolk quit fishing or do they adapt to new technological innovations? Can social structure and cultural identity survive either response? To fully understand the response of different communities, households, and individuals it is critical to understand why fishermen continue to follow their traditional occupation and what contributes to the resiliency of some fishing communities while others decline and even disappear.

1

Fisherfolk's Indigenous Knowledge of the Marine Environment

In this chapter, we shall make an attempt to describe and analyse the fishermen's indigenous knowledge of the marine environment. Our main concern is the fisherfolks' place in the ecosystem. An examination of the various aspects of interactions of the organisms – marine environment, planktons, benthos and benthic animals and nektons – is attempted in this chapter.

We shall confine ourselves to a detailed description of the indigenous knowledge of Mappila and Hindu fisherfolk and their understanding of cosmology. This will enable us to probe into the relation between the varieties of winds, currents, tides and cross-currents. and the technology of fishing. Furthermore, we are also interested in systematically describing the ethno-science of the Mappilas and comparing it with the ideas of the Hindu fisherfolk and of the modern scientists, namely, marine biologists. The present fieldwork carried out from June to October 2006 has shown that the Mappila fisherfolk and Hindu fishing communities share common indigenous and ecological knowledge of the marine environment. We have, therefore, reproduced our information on the ecological knowledge of the Mappila fisherfolk and compared it with that of the Hindu fishing communities after updating the data.

Knowledge of Oceanography

A proper knowledge of the ecosystem and behaviour of the marine fish necessitates a comprehensive study of the sea and the fishing grounds. The marine fish of commercial importance are distributed in different areas of the Arabian Sea and the fishermen of Kerala, particularly the Mappilas, possess a wide knowledge of the behaviour of fish, the regions of their occurrence and their migration patterns.

Fishing-grounds

Fishing operations employing traditional methods are generally confined to the shallow waters or coastal belt, which extends to 6–10 miles from the beach. From the economic point of view the continental shelf is the most important area of the sea for all kinds of fish and other marine animals. The continental shelf, i.e., the area of the sea within 100 fathoms depth along the west coast of India except off Kathiawar and Bombay is relatively narrow varying from 15 to 65 miles where it exceeds, at certain places, 190 miles (Mathur, 1978).

Marine biologists divide the marine environment mainly into two zones: (1) benthic (on or near the continental shelf) and (2) pelagic (all surface water). The division of marine environment into benthic and pelagic is very important from the point of fishing. It is in these two zones that the maximum number of species have their habitats.

We shall make an attempt to highlight the distinguishing features of different fishing grounds as conceived by the fisherfolk of Kerala, especially the Mappilas. The Hindu and Mappila fishermen of Kerala classify the fishing grounds on the basis of the depth of the sea. The fisherfolk often say that such and such a Kadal has got such *Maarus* or *Alku Alam* (the length of a man's out-stretched arms across his chest horizontally). The fishermen say that a fathom corresponds to one *Maaru* (6 feet) or *Alku Alam* (*Alam* literally means depth). The depth of any fishing ground is generally measured with the help of a rope, at the end of which a stone is tied as a sinker. The rope is lowered to the bottom of the sea and then pulled up in order to measure it by *Maaru*. The Mappilas of Tanur hold that *Nalalkku Alam* (four *Maarus*) is equal to one mile whereas in Calicut it corresponds to a mile and a half.

The fishing grounds, according to the fisherfolk of Kerala in general and the Mappilas in particular, fall into five zones: (1) *Karakadal*, (2) *Idakadal*, or *Edollamkadal*, (3) *Padikadal* (4) *Vayyakadal*, and (5) *Puramkadal*. We shall examine the significance of each of these fishing grounds, their relation with catches and finally, with the economy of the fishermen.

1. *Karakadal:* The Mappila and Hindu fishermen hold the view that the *Karakadal* lies upto the depth of eight fathoms. They say that almost all species of fish are caught in the *Karakadal* by employing traditional and modern nets.
2. *Idakadal: The Idakadal* lies between the *Karakadal* and *Padikadal* at the depth of three to eight fathoms. The Mappila and Hindu fishermen say that almost all species of fish appear in the *Idakadal* between September and October and they get bumper catches during that period. They also hold the view that mackerel and sardines migrate beyond the *Idakadal* (beyond 10 fathoms) during their spawning in the months of May, June and July. They stay there until such time that their little ones grow up and are able to migrate to the *Idakadal*. Mackerel and sardines are caught abundantly from *Idakadal* in the months of July-August and August-September because they migrate to this fishing ground during this period.
3. *Padikadal:* The *Padikadal* is located to the west of *Idakadal* at a depth of 2 fathoms. This is the most dangerous of all the fishing grounds and the fisherfolk of Kerala say that if a boat capsizes in the *Padikadal* it is impossible for them to salvage it, particularly during monsoons.
4. *Vayyakadal:* The *Vayyakadal* lies below the *Padikadal* and *Puramkadal*. The most important fishes like sardine and mackerel are abundantly caught in the *Vayyakadal* during the months of September, October, November and December. This is also a dangerous fishing ground located at 5–8 fathoms and it is also called the *Chatiyan* or cheater.
5. *Puramkadal:* The *Puramkadal* lies beyond 10 fathoms but is within the distance of 100 fathoms from the continental shelf. About 35 years ago when the fishermen reached the *Puramkadal*

they could not see their village on the coast. But nowadays both the Hindu and Mappila fishermen use GPS for locating their village when they set out on fishing expeditions to the *Puramkadal* in the months of August, September and October. The fishermen say that within the continental shelf upto a depth of 100 fathoms the *Puramkadal* is the most important area of the sea from the economic point of view. The fisherfolk of Kerala go out for fishing to the *Puramkadal* with their inboard boats using modern fibre boats at the depth of 50 fathoms.

Food, Habitat and Behaviour of Fish

The fisherfolk of Kerala possess extensive knowledge of different species of fish, their habitat, various plankton which serve as their food, and their breeding seasons. They are familiar with the spawning period of almost all kinds of fish because they correlate their catch with the availability of plankton and spawning. It is during the spawning period that the fisherfolk of Kerala expect bumper catches. The duration of the spawning period differ from one species to another. The fisherfolk of Kerala hold that the season of spawning is linked with the abundance of plankton, i.e., plankton usually precedes spawning. The catches of the fish also depend upon their migration. Thus, the advent of particular species of fish in different times is related to their habitat, food habits, spawning and the availability of plankton in the sea.

Taking these factors into consideration, the State Government imposed a ban on trawling during the spawning period from June 15th to July 31st, which was extended up to August 15th by the order of the Supreme Court on July 7, 2006.

The fisherfolk of Kerala in general, and the Mappilas and Latin Christians in particular, possess a fairly good knowledge of various kinds of plankton. The marine biologists have classified the plankton into two varieties, i.e., phytoplankton and zooplankton. Similarly, both the Hindu and Mappila fisherfolk classify the plankton into two groups: (1) plankton connected with floating plants, seaweeds and diatoms, etc., and (2) plankton associated with floating living organisms.

Ponal

The most important phytoplanktons, according to the Mappila fisherfolk, are the *Ponal* or *Mala Pannin* and *Punkara*. The Mappilas say that the *Ponal* is abundantly found on the Malabar coast in the months of *Chingam*, *Kumbham* and *Minam*. The advent of the *Ponal* is associated with the discharge of rain rivers into the sea in the months of Edavam and Midunam. The *Ponal* is in plenty on the Kerala coast immediately after the south-west monsoon. When the muddy rainwater enters the sea, it does not dissolve quickly. It is brought to the surface layer and is transformed into small *Taris* (bubbles) according to the direction of the current. The transformation of rain water into Taris is called *Ponal* or *Mala Pannin*. The Taris are about the size of a paddy seed and are of whitish colour. They are also called phytoplankton as they have no life. The *Ponal* is the most important food of *Karutha Avoli* (black pomfret). The *Ponal* also serves as the food of *Mathi* (sardine) and *Ayila* (mackerel), which are hauled abundantly in the months of *Chingam* and *Kanni* when they are available in plenty on the Malabar coast.

Punkara

Another important phytoplankton that serves as the food of pelagic fish is *Punkara*, which is abundantly found from Cochin to Trivandrum. The Hindu fisherfolk, particularly the Arayans of Cochin, say that there is a plant called *Pala* (seaweed) in the sea. During the flowering period of this plant a kind of powder falls into the sea from its blossoms. This powder is called *Punkara*, which has the fragrance of Attar (rose water). The Mappila and Hindu fishermen believe that the *Punkara* flows along with the currents, particularly in the months of *Makaram, Kumbham* and *Minam*. But, if the wind blows against the current, the *Punkara* gathers lengthwise in a place. The fish are immediately attracted and eat the *Punkara*. They then swim in the direction of the fragrance of the *Punkara*. The *Punkara* is relished by *Avoli* (pomfret), *Tirandi* (skates and rays), *Etta* (cat fish) and *Palamin*. Both Mappila and Hindu fishermen believe that the availability of *Punkara* is responsible for the development of larvae (eggs). The more the fish eat the *Punkara*, the more the increase in the production of eggs.

Namannu

Namannu is another important phytoplankton, which is generally found in the bottom of the sea, particularly beyond a depth of 10 fathoms. This is a kind of mud that glitters like gold. *Thirandi* (skates and rays) and *Kakka* (shell-fish) are fond of *Namannu*, which has the colour of *Tavidu* (rice-bran). There are three species of skates, viz., (1) common (*Metirandi*), (2) long-nosed (*Irumukkan*), and (3) White (*Pulli Tirandi*). *Metirandi, Pulli Tirandi* and *Irumukkan Tirandi* are the most important species of skates and rays found on the Malabar coast. The *Pulli Tirandi* grows to a giant size after eating *Namannu* and a sea mud called *Calikudi*. The Metirandi live on shell-fish and *Namannu*. The *Irumuukkan Tirandi* does not eat sea-mud but small fish and *Namannu*. These important skates and rays are caught in the *Puramkadal* in the months of *Makaram* and *Kumbham* favoured by the *Karakattu* (easterly wind). The skates and rays are also occasionally caught from the Karakadal.

The Hindu and Mappila fishermen classify the zooplankton into three main groups: (i) *Cori*, (ii) *Puttu* and (iii) *Piyyu*.

Cori

The *Cori* is a living organism which serves as the food of several species of fish, leading to better catches. The Mappila and Hindu fisherfolk classify *Cori* into six types: (a) *Kadikkunna Cori*, (b) *Tanni Cori*, (c) *Vella Cori*, (d) *Nila Cori*, (e) *Mani Cori*, and (f) *Kilappi Cori*.

Kadikkunna Cori

Kadikkunna Cori is of three colours, namely, red, white and green. It is also known as *Potti Cori*. Contact with any of these three types of *Kadikkunna Cori* causes an itching and burning sensation. The Mappila and Hindu fishermen say that *Kadikkunna Cori* is generally found when the current flows from south to north or south-west to north-east. The length of a *Kadikkunna Cori* is like the hair that grows on the human chest. It has also legs. Sometimes when the *Kadikkunna Cori*, particularly the red one (*Cuvappu Cori*) appears, the fishermen find it difficult, if not impossible, to launch their nets. The *Kadikkunna Cori* (all the three types) is an appealing food of bigger fish like *Etta* (cat fish), *Palamin* and *Nemmin* (seer) and *Avoli* (pomfret).

Tanni Cori

The *Tanni Cori* has short legs. The Mappilas say that it does not cause an itching-sensation to the body even if it touches the fishermen. It is the principal food of *Avoli* (pomfret), Parai (king fish), *Mathi* (sardine) and *Ayila* (mackerel). The *Tanni Cori* is generated abundantly during the months of June, July and August when the rivers drain their waters into the sea. The *Tanni Cori* helps the fishermen to get bumper catches of sardine and mackerel.

Vella Cori

The *Vella Cori* is of the colour of marble. According to the Mappila and Hindu fishermen, the *Vella Cori* is of the size of a paperweight and its legs and hands are very small. *Vella Cori* makes the *Tuyi* (light of the current) shine brilliantly. Since, generally the *Vella Cori* emits the light of a 100 W bulb. In spite of this light, however, it is as cold as ice. Sometimes, along with the fish, the *Vella Cori* is also hauled into the boat, but it disappears within the twinkling of an eye. The *Vella Cori* produces a very bad smell but it helps the fishermen to catch more fish as it increases the power of the light of *Tuyi,* particularly on dark nights. The fishermen are able to see the migration of the fish shoal with the help of light produced by the *Vella Cori* and *Tuyi.* The *Vella Cori* is the staple food of the larger fish like the cat fish, pomfret, skates and rays.

Nila Cori

The *Nila Cori* is also known as *Kal Cori*. It is bluish in colour and has the shape of an egg or a ball. Its legs are like those of an octopus. The fish do not consume it. *Nila Cori* appears in the sea in the months of *Kumbham* and *Minam* when there is *Kaccan Kattu* (the wind from north-west to south-east), *Vadakkan Nir* (northerly current) and the north-western current. The mere touch of the *Nila Cori* is sufficient to cause fever and pain, particularly *Kazhalakoluttal* (swelling chiefly in the groin), which keeps the fishermen bed-ridden for a long period.

Mani Cori

The *Mani Cori*, which is also as cold as ice, serves as the main food of *Avoli* (pomfret). The Mappilas say they get bumper catch of pomfret when there is plenty of *Mani Cori*. The *Mani Cori* is generally found in the months of *Kumbham* and *Minam*. Its presence causes the appearance of pomfret in plenty; hence, the fishermen get bumper catches.

Kilappi Cori

The *Kilappi Cori* is another important zooplankton. It is white in colour and is a floating and living organism. It is also an important food of *Avoli* (pomfret). It flows according to the movement of the current and is abundantly found on the Kerala coast in the months of *Vrischikam* and *Dhanu*. The Mappilas say that the *Kilappi Cori* is as big as a *Lolakku* (earring). The catch of pomfret in November, December and January, depends upon the abundance of *Kilappi Cori* in the sea.

Puttu

The *Puttu* (ant-hill) is the most important zooplankton known to the Hindu and Muslim fishermen. It refers to a mound thrown up by worms on the rocky bottom of the sea. The worms inside the *Puttu* as well as the luminous soil thrown up by them serve as the main source of food of demersal fishes like *Etta* (cat fish), *Tirandi* (skates and rays) and *Kora* (jew-fish). The worms of the *Puttu* do not ordinarily float along with the currents. When there is an upwelling in the sea during the south-west monsoon the *Puttu* is uprooted and thrown into the surface. Pelagic fish like mackerel and sardine find this plankton a delectable dish. The Mappilas say that they get bumper catches of sardine and mackerel during the south-west monsoon due to the appearance of *Puttu* on the sea.

Piyyu

The *Piyyu* or *Cellu* is a living organism which floats according to the direction of the current. The *Piyyu* is the most important food for *Etta*. It also lives in a *Puttu*, which is very broad. Sometimes, the *Piyyu* is collected and dried in the sun for manuring tobacco and vegetables. The Mappila hook-and-line fishermen also use *Piyyu* as prey for catching *Sravu* (shark), *Etta* (cat fish), *Nemmin* and *Palamin*.

Habitat and Behaviour of the Fishes

Marine biologists divide fish into two main groups, viz., demersal and pelagic, on the basis of their habitat. The demersal are those fish, which live near, or on the sea bottom feeding on the benthos (flora and fauna of the sea bottom), whereas the pelagic fish have their habitat on the surface of the open sea, swimming in shoals, feeding on plankton and small flies. The Mappila and Hindu fisherfolk however divide the fish into three groups: (1) *Talepedunna Min* (the fish that live in the bottom of the sea); (2) *Polappu Min* (the fish which have their habitat near the surface water and appear in shoals); and (3) *Idanir Min* (mid-water fishes). The *Taleppedunna Min* and *Polappu Min* of the Mappila and Hindu fishermen appear to correspond with the demersal and pelagic fish of marine biologists. However, the Mappilas and Hindu have another group of fish that live in mid-water, called *Idanir Min*. Most of the demersal fish, particularly prawns, shrimps, soles, sharks, cat fish, skate and rays etc., are caught on the Malabar coast by hook-and-line, trawl nets and boat-seine nets. Among the pelagic fish, herring, sardine and mackerel are significant from the economic point of the Mappila fishermen. The maximum catch of fish along the Kerala coast is during the months of October, November and December whereas the catch is minimum in April, May and June. Despite the adaptation of nylon nets and mechanisation of boats the traditional fishermen, particularly the Mappila, depend upon their indigenous knowledge for catching different species of fish.

Demersal Fish

Chemmin (Prawn and Shrimps)

Among the demersal fish the *Chemmin* (prawns and shrimps) occupies the most important place in the economy of the Mappila fishermen. The prawn and shrimp fisheries in Kerala rank among the most important commercial species of the state. The different species of prawns and shrimps, which are caught in large quantities, inhabit various fishing grounds of the sea.

The Mappila and Hindu fishermen say that the *Chemmin* have their habitation in *Karakadal* (coastal water), *Ulkadal* (estuary) and *Kayal* (backwater). These fishermen hold that the *Chemmin* prefer to live in the muddy bottom. The principal food of the prawns and shrimps consists of detritus of plants and animals which is found in abundance in *Patam* or *Cheru Chakara*. The Mappilas say that the *Patam* or *Cheru* appears in the sea when the south-west monsoon breaks. The Hindu fisherfolk, particularly the Arayans of Cochin and Travancore, hold identical views on the subject but they use the term *Chakara* for the mudbank.

Breeding

Both Mappila and Hindu fishermen believe that the *Chemmin* (prawn) are generally bred in *Cheru* (mud) during the south-west monsoon. They also hold that the *Chemmin* are bred in *Namannu* (a kind of plankton which serves as food of the fish) and in calm sea water. The *Mutta Tari* (larvae) glistens below the stomach near the legs. When the breeding season commences, the *Chemmin* rubs its belly on the mud of the *Namannu*, (spawning-bed). Sometimes this *urakkal* (rubbing) is done on shores or rocky places of the sea bottom. When such *urakkal* takes place the larvae is delivered from the stomach to the sea and they move according to the flow of the current.

The Arayans hold that the prawns and shrimps are bred in dirty sea-water. When the sea water gets spoiled, it transforms into a number of bubbles, which the fishermen call *Katta*. There are a number of black spots in the shape of mustard seeds inside each *Katta*. When these black spots are broken, the fishermen say, the prawns and shrimps are bred or hatched (*Chemmin mutta podinnu*). These offsprings of prawns and shrimps are called *Puvalan* and when they grow larger in size, they are known as *Naran Chemmin* or *Vella Chemmin*.

The beliefs of the Hindu fishermen about the habitat and breeding of the prawns do not differ from their Mappila counterparts.

Both the communities employ similar techniques in catching the prawns. But, unlike the Hindu fishermen, some Mappilas believe that the prawns and shrimps are bred by the miraculous powers of Allah.

The prawns and shrimps are generally caught by the sardine boat-seine nets, sole-nets and trawl-nets. The Hindu fisherfolk who confine their fishing to the backwaters use stake-nets, Chinese-nets, and drag-nets for catching the prawns and shrimps. The stake-nets are employed by the Valan community in the *kayal* (backwaters) of Cochin by setting against the tides which drive the prawns. Consequently, they are hauled at the turn of the tide.

Mantal (Sole)

The Mappila fishermen believe that the *Mantal* (sole) belong to the group of demersal fish. The *Mantal*, are therefore, bottom-feeders. The season for the sole-fishing commences immediately after the south-west monsoon and the peak season extends up to the month of *Tulam.* The breeding season of the sole is generally from *Kanni* to *Makaram*. As a general rule, the sole appear on the Malabar coast as soon as the *Chakara* (mudbank) or *Cheru* are formed during the south-west monsoon and the temperature and salinity of water increase. They appear in large *Polappu* on surface water and in midwater in *Chingam*. The sole disappear from the inshore waters when the south-west monsoon lashes the land and disturbances take place in the sea-bottom on account of the heavy downpour and storm which washes away the *Chakara* (mudbank). The sole are generally caught within a depth of 8 to 10 fathoms with the help of *Peyitham Vala* (surrounding net). Sinkers are added to the net to make a clean sweep over the sea bottom for fishing the *Mantal*. The introduction of trawl nets in recent years has of course increased the catch of the sole. In fact the Mappila fishermen keep ready their gears in tiptop condition to catch the sole when the season commences.

Sravu (Shark)

The shark is one of the important demersal fishes of the Kerala coast. Sharks were caught throughout the year on the Kerala coast from *Karkkitakam* to *Minam* at 20–30 fathoms with the help of the *Tirandivala* as well as the *Sravu* Vala in the past. The hook-and-line fishermen used to employ their traditional hook-and-lines for catching different species of sharks beyond the depths of 15–20 fathoms. Nowadays, modern hook-and-lines, nylon nets and fibre-covered plank-built boats are used for catching sharks. According to the Mappila fishermen, the *Kalla Sravu* (Tiger Sharks) migrate to the rivers and backwaters for eating carcasses. The main food of the shark is sardine and mackerel. Some sharks travel to the *Puramkadal* as far as 120 kilometres where they spawn in the months of *Chingam* and *Kanni.* The spawning of the sharks takes place within four fathoms of depth of water. Although, nowadays, the *Sravu* are caught by the nylon-nets set for other species of fish, they often destroy the gear. Mackerel and sardine are the main prey used by the fishermen for capturing the *Sravu*. Marine biologists have classified the *Sravu* into 22 species. The Mappilas do not lag behind the marine biologists in their indigenous knowledge of the different species of sharks. They classify sharks into 15 kinds. The Arayans of Cochin have limited their classification of sharks into nine species. The behaviour of the sharks of different species varies from one another according to the Mappila and Hindu fisherfolk (Mathur, 1978).

Tirandi (Skates and Rays)

The skates and rays are important demersal fish. They are generally caught from 10 to 15 *Maarus* depth of water in season and out of season except in the month of June when the sea is very violent. The spawning of the skates and rays takes place in the months of *Chingam* and *Kanni.* Both the Hindu and Mappila fishermen believe that the breeding of the skates and rays takes place in the *Puramkadal* during the monsoon. The skates and rays migrate from one area of the sea to another in search of food. They feed on *Namannu, Chali* and shell fish. The abodes of the skates and rays in the bottom sea are rocky places, *Tadakams* (natural dykes) and caves.

Etta (Cat-Fish)

The cat-fish belongs to the demersal category. It is widely distributed on the coast of Kerala, particularly on the shores of Malabar. The Mappila fishermen handle them very carefully as they have poisonous fins.

Both the Hindu and Mappila fishermen maintain that though the cat-fish is demersal they come up as *Polappu Min* in the month of *Chingam* and *Kanni.* Some of the species of cat-fish also appear on surface water during the south-west monsoon. During this season they migrate to the *Puramkadal* and hide at a depth of 50 fathoms.

All cat-fishes eat prawns and shrimps, *Kora* and other small fish. But *Pottankuri* feed on snakes, *Nandu* (Crab), *Mullen* (silver-belly) and other small fish. It may be mentioned here that as soon as the rain waters disappear from the sea, the cat-fish appear on the surface. They are usually caught from the bottom sea during the months of *Vrischikam* and *Minam.*

The spawning period of the cat-fish is generally from *Idavam* to *Mithunam*. The female *Ettas,* particularly the *Tailamkuri,* keep their eggs in their mouth and travel in the sea in the months of *Chingam* and *Kanni* till they are finally hatched in the month of *Tulam.* They do not take any food until their eggs are hatched. Once the eggs are hatched, the young cat-fishes are scattered.

The *Vellatta* come near the shore and eat the night soil. They prey upon small fish like *Karikkadi,* sole, shell-fish and eat the zooplankton, *Piyyu.* They feed mainly on shell-fish in the months of *Kumbham* and *Minam. Piyyu* is baited by the Beppukkar for catching *Navetta*. Most other *Ettas* are caught by long lines baited with smaller fish and sometimes with *Piyyu.* The cat-fish species are very well distributed all along the Kerala coast but the best catches come from Malabar. Nowadays they are caught from the coastal waters of Kerala with the help of trawl nets. To sum up, the cat-fish, as distinguished from other species, is of great economic importance not only to the Mappilas but also to the Hindu and Christian fisherfolk of Kerala.

Kora (Jew-Fish)

Both the Hindu and Mappila fishermen divide the *Kora* (Jew-fish) into three species, viz., (1) *Pallimin,* (2) *Putiyappila Kora* (goat-fish), and (3) *Karimin* or *Paitakora*. They are considered demersal fish and are caught in abundance when the *Chakara* appears during the monsoon, particularly in the months of *Chingam* and *Kanni.* There was a bumper catch of Jew-fish in September in the fishing villages of Puthiyappa and Beypore. Their breeding takes place in the months of

Kumbham and *Minam* and therefore they start appearing in the bottom sea during the months of *Idavam* and *Mithunam.*

Polappu Min (Pelagic Fish)

The two important *Polappu Min* (pelagic fish) caught in Kerala are sardine and mackerel. The nets and fishing methods used in catching the *Polappu Min* are different from those employed for landing the species feeding on substances in the bottom of the sea while fishing the Polappu, a unit generally engages itself for catching only one species at a time.

Mathi (Sardine)

Among the different species of *Mathi, Nalla Mathi* (oil sardine) is the most important commercial fish and is widely used for food, either fresh or cured. The introduction of nylon nets has helped the Mappila and Hindu fisherfolk to meet the growing demands of the oil sardine. This sardine inhabits within 8 to 12 kms of the coastal waters.

The Mappila and Hindu fishermen hold that the oil sardine migrate from inshore waters just before their breeding. This migration takes place in the months of *Idavam, Mithunam* and *Chingam.* The season for fishing the *Nalla Mathi* commences from *Chingam,* immediately after the onset of the south-west monsoon, and ends in *Minam.* The peak season for the oil sardine is from *Chingam* and *Dhanu.* The fishermen of Kerala in general and the Mappilas in particular know well the behaviour of the oil sardine, particularly its migration, breeding and food habits. This migration takes place during the months of June, July and August while the spawning period is from August to September. The eggs are seen in the plankton collected from the fishing grounds. The mortality rate after breeding is very high, particularly among the females. The fishermen of Tanur, Puthiyappa and Blangad told us that they got bumper catches of oil sardine during July, August and September, 2006.

Food of the Nalla Mathi

The oil sardine feeds on pelagic organisms and occasionally resorts to bottom feeding. It is believed that the availability of their favourite food influences their migration.

The Mappila and Hindu fishermen believe that the *Nalla Mathi* feed on *Ponal* (one of the phytoplanktons) during the period of south-west monsoon. The fishermen say that when the rain water enters the sea at the onset of the monsoon, it does not dissolve in the sea. Instead it remains on the surface. This water is red in colour. Its freshets transform into *Ponal* which flows along with the current. The odour of Ponal is like the flower of *Payini* tree (Acanthus ilicifolius). The *Ponal* appears more in the month of *Chingam* immediately after the monsoon. This plankton is available in plenty on the surface water as seaweeds or the green enveloping the sea, in the months of *Kumbham, Minam, Medam* and *Idavam*. The oil sardine appears on the surface water during this period to eat *Ponal* enabling the fishermen to get bumper catches. After *Chingam* and *Kanni,* the *Ponal* disappears from the sea, after which the sardine feed on mud. Thus they migrate to the bottom of the sea in the months of *Dhanu* and *Makaram*.

The shoals of oil sardine in the sea bottom are detected by divining the *Mariyam* (bubbles) that come up to the surface. The reddish Mariyam are released from the gas bladders of the oil sardine. Besides locating the Mariyam, the fishermen also detect the bottom shoal by smelling a strange odour in the sea. This smell is attributed to the presence of the sardine in the bottom of the sea.

The Mappila and Hindu fishermen believe that apart from the availability of food, the sardine go to the bottom of the sea due to changes in the current (*Nir*). They say that when the *Tengara Nir* (south-easterly current), *Tembrayil Nir* (south-westerly current) and *Tekkan Nir* (southern current) flow violently (*Kudutal Varakal*), the sardine go down to the bottom of the sea. In such a situation the Mappila and Hindu fishermen labour hard but often fail to get good catches.

The Mappila fishermen appear to be very familiar with the shoal behaviour of sardine, particularly that of oil sardines. They adjust their fishing operations according to the size of the shoal and the behaviour of the sardine. The Mappila and Hindu fishermen, like marine biologists, classify the shoals of sardine into two kinds: (1) *Mukal Polappu* (top shoal), and (2) *Talatte Polappu* (bottom shoal). These fishermen say that they can detect the Mukal Polappu by hearing the sound of the *Mathi*. Their movements on the surface are of four kinds: (1) *Kadannu Cadal* (leaping), (2) *Olam Adiccu Cadal* (rippling), (3) *Sabdam Untakki Cadal* (pattering), and (4) *Melpottu Cadal* (flipping). It is very interesting to point out here that the knowledge of the Mappila and Hindu fishermen of the shoal behaviour of sardine is almost similar to that of marine biologists.

The Mappila fishermen say that during the nights of the dark phase of the moon, when the *Tuyi* is moderate they can detect the shoal of sardine. Most species of fish can see the net if the *Tuyi* is brilliant and hence they run away, resulting in a poor catch.

It may be mentioned that darkness is preferred for catching sardine during the night. The Mappilas go out in the sea to a depth of 5 to 8 fathoms with torches made of coconut leaves. This bonfire attracts the sardine in larger numbers and they are easily caught. The sardine, particularly the immature ones, cannot see the net when torches are lighted.

Ayila (Mackerel)

The *Ayila* (mackerel) are called pelagic fish or simply *Polappu*. The catch of the mackerel is related to their feeding habits and spawning seasons. The mackerel subsist on phytoplankton and zooplankton. The Mappila and Hindu fishermen hold the view that mackerels being plankton-feeders, their catch is related to the occurrence of this substance in the sea. The appearance of the plankton in the Arabian Sea varies from season to season. The fishing season of the mackerel in the southern parts of Kerala is from *Chingam* to *Makaram* while in Malabar it commences in September and lasts upto April. The fishermen attribute the bumper catches of mackerel to the production of the phytoplankton *Ponal* immediately after the monsoon. Like the sardine, mackerel also migrate to the *Puramkadal* for spawning in the months of *Medam* and *Idavam*. The tiny mackerels return to the Karakadal in the months of *Chingam* and *Kanni.*

Ayikkora and Nemmin (Seers)

Since the *Ayikkora* and *Nemmin* (seers) travel in the midwater as well as in the surface water, they are called midwater fish. They feed on small fish, particularly mackerel and sardine. The seer is an important pelagic fish on the coast of Malabar. It is caught from the inshore regions all along the Malabar coast when it appears, on the surface water from *Kanni* to *Medam*. In Malabar the technique of hook-and-line fishing is mainly used for catching the seers. Small fish like sardines and prawns are baited for catching them. Sometimes nylon seine-nets are also employed for hauling them. In addition, both Hindu and Mappila fishermen sometimes employ the Olukkuvala (gill-nets) in the months of *Knani, Tulam* and *Vrischikam*. Sometimes a seer jumps up to a furlong over the surface water holding the mackerel or sardine in its mouth. But while jumping they often fall into the Olukkuvala (gill-net). The yarn used for making the gill-net is of eighth or ninth counts. The breeding of the seer is an interesting process. They hatch eggs by floating their larvae from their stomachs into the water. This floating larvae of the seer is called *Panhinu*. The seer concentrate at the bottom of the sea in the months of *Kumbham* and *Minam*. At that time the Mappilas increase the length of the floatline (*kavi kayar*) of the seine net by five Marus to catch the seers hiding there. This lowering down of the net is called *Valaekkal*. Once the float-rope is extended by five Marus the net goes down to a depth of 10 Marus into the muddy bottom.

***Avoli* (Pomfret)**

The Mappilas classify the *Avoli* (pomfret) into three species: *Vella Avoli* (silver pomfrets), *Karavoli* or *Karutha Avoli* (black pomfret) and *Venganavoli*. The pomfret is an important pelagic fish. The spawning season of the *Avoli* lasts for two months from *Chingam* to *Kanni*. They are caught in the *Puramkadal* in these months. The catch of pomfret is related to the production of the *Ponal*, their principal food. They come up to the surface to eat the *Ponal* and are caught abundantly with the help of the *Avolivala* or *Mathikolli* nylon net. The remarkable trait of the *Venganavoli* is that they do not appear in *Polappu* (shoal) but migrate individually. The Arayans of Cochin call *Karavoli Macean*. According to them the pomfret subsist on *Ponal, Punkkara* and *Chali*. They believe that the plankton *Punkkara*, is responsible for the production of a large quantity of larvae of the pomfret. The pomfret in Malabar is caught in *Idanir* and *Mukal Nir* by using the nylon *Mathikolli Vala*.

Kanamin (Tunnys)

The *Kanamin* (Tunnys) is another important *Polappu Min* caught by the Mappila fisherfolk in the months of *Chingam* and *Kanni*. They are notorious net-breakers with their sword-like tail fins. In olden times the cotton nets were, not infrequently, destroyed by the Tunnys. They feed on water and *Nethal* (white bait), and frequent the Kerala coast in small shoals. The fishing season for the Tunnys begins from *Kanni* and lasts up to *Makaram*. They are caught in the *Puramkadal* by employing hooks-and-lines. The hatching of their eggs is called *Panhine Olukkal*.

Nethal (Whitebait)

The *Nethal* (whitebait) is a pelagic fish and is caught from the *Karakadal* in the months of *Chingam* and *Kanni*, but in other months they are hauled from the *Puramkadal*. The *Nethal Vala* is the main gear used to catch the fish. The breeding season of the whitebait extends over a period of eight months from September to May. The spawning of the whitebait is called *Panhine Olukkal*. The *Ponal* and water are the main food of the whitebait.

Talayan (Ribbon Fish)

The Mappila and Hindu fisherfolk divide the *Talayans* (ribbon fish) into two species (1) *Velli Talayan* (silver ribbon fish), and (2) *KarinTalayan* (black ribbon fish). The ribbon fishes are caught in abundance in the months of *Chingam, Kanni* and *Tulam* with the help of the *Vakkuvala* in the past. They have the same technique of hatching eggs as Panhine Olukkal. The *Ponal* and small fish are the main food of the ribbon fish. Almost all species of fish except *Vella Avoli* and *Karavoli* are afraid of the ribbon fish because they are capable of biting them. According to the Mappila and Hindu fisherfolk, the *Velli Talayan* glitters like gold and its fins are of golden colour. On the other hand, the *Kari Talayan* is a very black ribbon fish.

Sura (Cuda)

The *Sura* (Cuda) belongs to the *Polappu Min* group. The Mappila and Hindu fishermen (Mukkuva) claim that the Cuda is one of the Tunny species like the *Kanamin*. They are mostly caught in the *Puramkadal* and bred in the months of *Chingam* and *Kanni*. They feed on water, mud and *Punkara*.

Manangu (Anchovy)

The *Manangu* (anchovy) also belong to the *Polappu Min*. Their food consists of plankton, shrimp, dirt and mud. More than 90 per cent of the anchovies are caught from the west coast of India. The fishing of the anchovies lasts for five months from August to December. The anchovies migrate to the *Karakadal* from the *Puramkadal* immediately after the south-west monsoon. However, the Mappila fisherfolk do not take much interest in fishing the anchovies as it involves hard labour but gives poor returns. The Hindu fisherfolk of the villages of Beypore and Puthiyappa catch the anchovies with the help of boat-seine nets.

Camban (Horse Mackerel)

The *Camban* (Horse mackerel) is a significant pelagic fish. This fish inhabit the *Puramkadal*. Its food consists of small fish like the sardine, mackerel, *Vellri* and *Koluva (Sura)*. It is bred in the monsoon and the spawning takes place in the deep waters of the *Puramkadal*.

Mullan (Silver Belly)

The fisherfolk of Kerala say that the *Mullan* (Silver belly) is both a pelagic and a mid-water fish. It is caught in the *Idanir* (midwater) and *Mukal Nir* (surface water). Mud, water and dirt constitute the food of the Silver belly. The peak season for the catch of the Silver belly is in the months of *Chingam, Kanni* and *Tulam*. Their breeding takes place in the *Puramkadal*.

With the introduction of inboard bigger boats in which there is a crew of 40–50 fishermen, fishing is done beyond 50 fathoms of the *Puramkadal*. Both the Mappila and Hindu fisherfolk have their traditional knowledge of plankton and behaviour of the fish with which they catch them according to their spawning period. In fact, they correlate their catch with the availability of Plankton and spawning.

Thus, we note that the Mappila and Hindu fisherfolk, particularly the Mukkuvas, divide the fish into three categories: (1) *Polappu Min* or *Mepantiya Min* (Pelagic fish), (2) *Tazhepedunna Min* or *Patiyil Pedunna Min* or *Adiyil Nilkunna Min* (demersal fish), and (3) *Idaniril Nilkunna Min* (midwater fish). Most species of fish appear as pelagic fish on the surface water in the months of *Chingam, Kanni* and *Tulam.* The fish that inhabit the *Patiyil* or *Adiyil* (bottom) are called demersal fish. During the south-west monsoon the fish come up to the surface of the water because of the cold. The bottom of the sea becomes cold during the summer months and so the majority of fish migrate to the bottom of the sea at that time. The midwater fish are caught throughout the year. The Mappila and Hindu fishermen say that in the month of *Tulam,* all species of fish live in the *Idanir*. When the fish come up to the Idanir the fishermen can smell fish oil. Once they receive this odour they launch their nets. The Mappila and Hindu fishermen hold that the midwater fish are found in the fishing ground situated 20 miles away from the coast. The presence of a thin film of oil on the surface because of the swimming of the shoals inside the water is called *Velladu*. The appearance of the *Velladu* heralds bumper catches.

It will be noticed that both the Hindu (Mukkuva) and Mappila fishermen hold almost identical views on the different fishing grounds, various species of fish and their sub-species and of their ecosystem including their behaviour. The fisherfolk adjust their technology and fishing activities on the basis of objective knowledge and inductive reasoning, according to the availability of the different species of fish and their interrelation with other organisms like the plankton. The fishermen's catch and their economy is related to the landing of various kinds of fish, which in turn is influenced by the availability of food, their habitat, seasons and climate. The fishermen are fully aware of these objective conditions.

The Hindu fishermen of Kerala (Mukkuva), like the Mappila fisherfolk, have acquired enormous indigenous knowledge of the network of relation between the different organisms and their marine environment. Besides the location of the various fishing grounds, availability of the right kind of plankton is the important factor in determining the fisherfolk's quantum of catches and their relation with the ecosystem. This is noticed in the adjustment of their technological knowhow in different fishing grounds. Their catch is usually affected by certain developments in the sea. The above analysis conclusively shows that the presence of certain planktons helps bumper catches while their total absence results in very poor hauls. The availability of the plankton (food) plays an important role in the success or failure of the catch of different species of fish. The knowledge of the fisherfolk about breeding and spawning seasons of different species of fish and their behaviour shows that the traditional fishermen of Kerala, both Hindu and Mappila, do not lack in zoological knowledge. There is very little a priori deductive arbitrariness in their knowledge of the environment that is relevant to their technology.

Climatic Conditions and Cosmology

The Hindu and Mappila fishermen are aware that the sea fish of colder waters are of better quality. That is why they are caught in greater abundance than the ones in warmer seas. During the south-west monsoon, from July to August, the maximum number of certain species likes prawns, sardines and mackerels are caught in Kerala. The low night temperature during the months of November, December and January is also considered by the fisherfolk of Kerala to be conducive to bumper catches of fish. Like depth and movement of water, the temperature of water also directly affects the variety and abundance of fish, which call for different techniques. From October, the temperature increases slowly till April, which is the hottest month in the year. The catch is negligible during the months of April and May in Malabar. Mid-summer is generally welcomed by the Mappila and Hindu fishermen, as they are able to dry and salt their catches before they perish.

The fisherfolk of Kerala in general, and the Mukkuva and Mappila fishermen of Malabar in particular, modify their technology according to the season. They use the maximum number of floats for the nylon gill-nets and boat-seine nets and lighter sinkers during the south-west monsoon, particularly from June to August. On the other hand, in summer months, only 50 per cent of the floats of a net are used. Similarly, the weight of each sinker is increased from 2 kg to 4 kg. In other words, in summer heavier sinkers are used by the traditional fishermen.

Both during the summer and the rainy seasons the length of the float ropes are adjusted. For instance, the length of the float ropes used in the rainy season varies from two feet to four feet from the cod end to the mouth of the nets, particularly that of boat-seine sardine nets. This shows that only shorter float ropes are used in the rainy season. Contrary to this, during the summer months, particularly from January to May, the length of the float ropes ranges from 6 to 12 feet from the cod end to the mouth of the nets.

When the length of the float ropes and the weight of the sinkers are increased the fishermen can lower the net to the bottom of the sea for catching demersal fish. Most shoals of fish have their habitat in the muddy bottom during the summer, while they live on surface and midwaters in the rainy season. The fishermen very often say that the fish in the monsoon months live in the '*Edanir*' (mid-water). As has already been stated, all the floats of a net are used during the rainy season while their number is generally reduced to one half in the summer, particularly from January to May. Above all, during the summer, two floats of the head rope of the *Mathikolli* nets are removed before shooting them. The Mappilas do not make these technological innovations when the current is very strong. They use all the floats and sinkers both in the summer and the rainy season when the flow of the currents is very strong. In other words, they adjust their technology according to the velocity of the current, regardless of the season.

2

The Fishermen's Ecological Knowledge of Tiramala, Kattu, Nir and Nilavu

Tiramalas (Waves)

Both the Mappila and Hindu fishermen classify the Tiramala (waves) into eleven types, which are as follows:

Karakadal Tiramala

The *Tiramalas*, which appear on the coastal waters and sweep over the *Kara* (shore) are called *Karakadal Tiramalas*. The *Karakadal Tiramalas* generally occur in the months of June, July and August. The Mappilas say that when there is *Chakara* or *Patam* the *Karakadal Tiramalas* do not occur. They believe that even their fibre covered plank built boats are liable to be capsized during the occurrence of the *Karakadal Tiramala* at a depth of *Munnalkku Alam* (three grown-up men's height). Sometimes, the boats of one or two fishing units collide with each other in the *Karakadal Tiramalas*, which result in the drowning of the crew. This knowledge is shared by the Hindu fishermen of the villages of Blangad and Puthiyappa.

Cambankadal Tiramala

The Mappila and Hindu fishermen hold that gusts create short and long waves. The longest of the waves are called *Cambankadal*. They occur in almost all months at a depth of two fathoms. Sometimes, these waves have a length of several miles. When the *Cambankadal Tiramalas* occur in the *Puramkadal*, the fisherfolk do not go out for fishing, particularly in the months of June and July. Whenever the wind and current move in the opposite directions very swiftly and violently, the *Cambankadal* breaks out.

Cheriya Tiramala

The Mappila and Hindu fishermen hold that when the winds and currents originate from the same direction the sea is generally calm. However, the *Cheriya Tiramala* (short wave) breaks out. The *Cheriya Tiramala* grows quickly and breaks down fast whereas the *Cambankadal* (largest wave) grows more slowly and disappears gradually. The fishermen are able to predict the depth of water by seeing the speed of the *Tiramalas* that lash out obliquely on the shore. They hold that the velocity of the *Tiramala* is generally dependent upon the depth of the sea.

Yama Kadal

The *Yama Kadal* refers to the giant waves that generally fall on the shore and from where the waters either move up to 100–50 feet and return to the same point or move backwards to the place from where they fell originally. Sometimes they move up to a distance of a furlong from the place where they fell first. The *Yama Kadal* is responsible for the capsizing of many a fishing boat. The Mukkuvas of Blangad call this wave *Engatari*. But the Hindu and Muslim fishermen say that the *Yama Kadal* is very violent during the New Moon and Full Moon. The fishermen are generally unable to undertake fishing expeditions when the *Yama Kadal* lashes the sea.

Varana Kadal Tiramala

The waves which grow in the *Puramkadal* during the monsoon are called *Varana Kadal Tiramala*. When the *Varana Kadal* breaks out the traditional fishermen do not go out for fishing.

Vayya Kadal Tiramala

The *Vayya Kadal,* which breaks out during the months of June and July is very unpredictable. It is believed to be the swiftest wave. The Mappilas often say that the *Vayya Kadal* is a *Catiyan* (cheater) as it grows very unexpectedly. In fact, once a boat is capsized in a *Vayya Kadal,* there is little hope of survival of the crew.

Padi Kadal Tiramala

The *Padi Kadal* is one of the most important *Tiramalas* that occur in Malabar. It originates at a distance of 200 to 300 feet from the coast. When the *Padi Kadal Tiramala* grows in the sea fishing becomes impossible as the boats are liable to be capsized and the crew seldom escape death. It is in the *Padi Kadal* fishing ground that *Coyal Vellam* (the coldest water, 10 times colder than ice) occurs. If a person is caught in the *Coyal Vellam* his body becomes frozen and he dies instantly. About 35 years ago a Mappila fisherman of Tanur went out to Kuttayi, 5 miles away, for fishing. His boat capsized in the *Padi Kadal* and he was drowned in the most dangerous *Coyal* water. His body could not be traced.

Thus, the *Tiramalas,* which break out in the fishing ground, viz., *Padi Kadal,* is called *Padi Kadal Tiramala.* This wave occurs at a distance of half a furlong from the coast in June, July and August.

Ida Kadal Tiramala

The *Ida Kadal* is one of the important fishing grounds. The waves occurring in this fishery are called the *Ida Kadal Tiramala.* The *Ida Kadal Tiramala* breaks out at a depth of *Munnalkku Alam* (height of three men). The Mappilas say that boats do not generally capsize when the waves called *Ida Kadal* occur. On the other hand, the same *Ida Kadal Tiramala* that breaks out during the monsoon at a depth of *Munnalkku Alam* causes capsizing of fishing boats. Sometimes the *Ida Kadal Tiramala* breaks out a furlong away from the coast where the depth is comparatively less but it appears very unexpectedly, particularly in the month of *Idavam.* Then the *Ida Kadal* is called *Muriyam Potti.* The occurrence of this wave prevents an easy launching of boats.

Puram Kadal Tiramala

The *Tiramalas* or waves that occur in the *Puramkadal* fishing ground are called *Puramkadal Tiramala* by the Mappilas. The Arayans of Cochin call them *Varana Kadal.* These waves break out generally beyond 8 to 10 fathoms of depth in the months of May, June, July and August. During this period the fishermen do not go out for fishing. Their boats are liable to be capsized if they undertake fishing expeditions when the *Puramkadal Tiramalas* lash. Although the main fishing ground of the hook-and-line fishermen is the *Puramkadal,* they very seldom go out for fishing when these waves occur with violence. It is during this bleak period that the fishermen are obliged to pawn the ornaments of their womenfolk for meeting their daily expenses and also borrow large sums from the traders against security of their future catches.

Coyala

The *Coyala* is a kind of whirlpool. It occurs at a depth of two fathoms in the months of June, July, August and September. The *Coyala* is the most violent wave, which grows very swiftly and it moves upwards very violently. Even when it falls its velocity is very high. The Mappila fishermen believe that it is created by the Sea Satan. The *Coyala* is 5 to 10 times colder than ice. Once a fisherman falls into the *Coyala* his entire body becomes frozen. In olden days when trained allopathic doctors were not available in the village several Mappilas were reported to have drowned in the *Coyala.* Nowadays whenever a fisherman is drowned in the *Coyala,* he is pulled up to the boat or to the shore with the help of the rope and he is immediately taken to the nearest hospital or physician. In several cases, fishermen have been saved from the jaws of death in the *Coyala* by giving timely and proper medical aid. The Hindu and Mappila fishermen believe that the *Coyala* is caused by the mischievous spirit, Gulikan. The custom among the Mappilas is to remove their girdle and float it along with the current when their fishing boats capsize in the *Coyala.* The *Coyala* generally grows and falls when there is gusty winds and rains.

Tali Vali

There are some *Tiramalas* that sweep the shore and fall backwards as if the water is released from a dam. The Mappilas call such *Tiramalas*

Tali Vali. When the *Tali Vali* occurs, the area of the sea to its east is called *Tottappu*. When the *Tottapppu* grows the water moves backwards and when it disappears, the *Coyala* occurs. Here the *Coyala* is ten times colder than ice. If a fisherman slips into this *Coyala* he can seldom escape from the clutches of death. Sometimes, when there is thunder and showers accompanied by gusty winds, the fishermen are unable to locate the *Coyala*, in which case they are liable to be drowned along with their fishing boats.

In the *Kara Kadal* (shallow water) where the major fishing operations take place the nature of the waves often changes. By seeing the movement of the waves the Mappilas row their boats and launch their nets in such a manner that they are not capsized on the waves. In other words, they adjust their technological knowhow according to the pattern of the wave during their fishing operations. Despite the experience and familiarity with the waves, sometimes, particularly during the monsoon, mishaps do occur resulting in damage for their boats and death of the members of the crew. However, their mastery over, and perfect knowledge of the nature of the waves, help them in adjusting their fishing operations when the sea is very rough.

Thus, we note that the sea is never calm. Even in calm summer days ripples and waves appear on its surface. Wind is the source of most of the waves. The waves are caused by the disturbance that takes place on the surface or in the depths below the sea when the winds blow. In other words, a wind blowing over the sea causes various kinds of waves. Once the waves are created they travel in the direction established by wind. When the waves move upward, particles of water move up and down, back and forth. On the open sea, waves travel across the surface, as though the water inside it is moving forward.

We have already noticed from the above description that the waves travel in various speeds depending on their length. The largest wave travels for the longest period. As the wind continues to blow over them, the waves grow higher and higher and move faster, until they move as fast as the wind. On the other hand, the smaller waves grow quickly becoming quite steep in order to break down and disappear. As long as the wind blows there may be numerous kinds of waves. Sometimes the waves move faster than the wind and they break down on the surface water by the pull of the wind. The native notions regarding winds are given below:

Kattus (Winds)

The speed of the *Kattu* (winds) on the coast of Malabar is generally light and moderate. During the south-west monsoon in the months of May, June, August and September, the current of the wind is generally strong. During the months of *Dhanu* to *Medam*, the winds are stronger, especially in the afternoon. The Mappila and Hindu fishermen (Mukkuva) say there are eight kinds of winds.

The Mappilas believe that the winds are created by the *Maalaka* (angel) Mikayil. They say that their catch is very much related to the types of *Kattu* in a particular season or month or a day. It is interesting to note here that both Hindu and Mappila fishermen have similar notions about the nature of the winds and their relations with the catch of different species of fish. The importance of each kind of wind is examined below with reference to the techno-economy of the fishermen of Kerala, particularly the Mappilas.

Puram Kattu (Westerly Wind)

The *Kattu* blowing during the south-west monsoon from the west to the east is known as *Puram Kattu* (westerly wind) among the Mappilas of Kerala. The *Puram Kattu* during the monsoon makes the waves and the sea violent. It generally starts blowing from *Makaram* to *Medam* and continues throughout the monsoon. The Mukkuvars of Blangad hold that the *Puram Kattu* blows in the months of February, March, April and May in an indiscriminate manner. The Mappilas say when there is *Puram Kattu, Etta* (cat fish), *Chemmin* (prawns), *Mantal* (soles), *Kora* (Jew fish), Karikkadi, etc., migrate from the *Puramkadal* to the coastal waters at a depth between six and seven fathoms. In fact, the fishermen get good catches when the *Puram Kattu* blows.

Vadavartha Kattu (North-westerly Wind)

The wind that blows from north-west to south-east on the Malabar coast is called *Vadavartha Kattu* (north-westerly wind). It occurs more often in the months of February, March and April. When the *Vadavartha Kattu* blows violently it is called *Kaccan Kattu*. The *Etta* migrate from *Puramkadal* to the Kara Kadal at a depth between 7 and 10 fathoms when the wind blows, in order to enjoy the breeze. During this period the fishermen get bumper catches of the *Etta*. The Mukkuvas of Blangad, however, claim that the *Vadavartha Kattu* is the most common and important wind that occurs in their village, generally in February, March and April. In some years the wind blows in the months of *Kanni* and *Tulam* as well. During the occurrence of the *Vadavartha Kattu* and *Vadavartha Nir* (north-westerly current) the fishermen expect good catches, particularly small fishes like *Karikadi, Mantal* and *Kora*.

Vadakkan Kattu (Northerly Wind)

The *Vadakkan Kattu* (northerly wind) blows over Tanur more frequently in the months of *Chingam* and *Kanni,* whereas in Blangad, the same wind blows in the months of *Makaram* and *Kumbham*. When the *Vadakkan Kattu* blows violently it is called *Mancan Kattu* by the Mukkuvas of Blangad. The striking feature of the *Mancan Kattu* is that when it blows the fishermen cannot go out for fishing as it generally causes capsizing of boats. On the other hand, the *Vadakkan Kattu* is regarded as a calm and pleasant breeze, which occurs generally up to 10 a.m. After that it turns into *Mancan Kattu.* However, the Hindu and Mappila fishermen say that the occurrence of the *Vadakkan Kattu* or *Mancan Kattu* is the indication of a bumper catch in the months to come. A Hindu fisherman of Blangad told us that although "we see that the *Vadakkan Kattu* blows from north to south, it gives us a warning to adjust our technological knowhow as if it were the *Vadavartha Kattu* (north-westerly wind)". The Mappilas say that when the *Vadakkan Kattu,* and *Vadakkan Nir* occur in the months of *Mithunam, Karkkitakam, Chingam* and *Kanni,* the sea becomes calm. They expect good catches of *Mantal, Kora, Camban* (Ribbon Fish) and *Talayan,* when there is heavy showers brought by the *Vadakkan Kattu* and the *Vadakkan Nir.* The fishermen believe that when there is *Vadakkan Kattu* and *Tekkan Nir* (southern current), the sea becomes violent.

Vattakkara Kattu (North-easterly Wind)

The *Vattakkara Kattu* (north-easterly wind) blows during the monsoon, particularly in the months of *Idavam* and *Mithunam*. This encourages the fish to migrate to the *Puramkadal*. In fact, the fishermen face untold difficulties when the *Vattakkara Kattu* blows. The *Vattakkara Kattu* blows almost throughout the year. When there is simultaneous *Vattakkara Kattu* and *Vattakkara Nir* (north-easterly currents) the *Beppukar* and *Valakkar* get good catches because both the wind and current occur from the same direction, i.e., from the north-east. The *Vattakkara Kattu* generally blows up to 10 a.m. from the north-east, after which it changes its direction to the north. In the evening it blows from *Vadavartha* (north-western side). Thus the *Vattakkara Kattu* turns into *Vadavarta Kattu* (north-westerly wind) in the evening and continues to blow up to 10 p.m. This helps the fishermen to catch the fish at a depth of 10 fathoms.

Tekkan Kattu (Southern Wind)

The *Tekkan Kattu* (southern wind) occurs mostly in the months of *Tulam* and *Vrischikam,* but there is no certainty about its occurrence. Sometimes it blows in season and out of season and on all days, causing *Valippu* or *Kadal Kuttal* (growing of waves). When the *Tekkan Kattu* blows, an important wave called *Neduverru Valippu* grows. This occurs when 10 to 15 waves (*Tiras*) join together and appear on the surface, simply to disappear within a short time. Then the sea becomes calm for at least 10 minutes. Sometimes, the *Neduverru Valippu* occurs for two or three days at a stretch. When the *Tekkan Kattu* blows, particularly in the months of *Tulam* and *Vrischikam,* the fish, especially *Camban, Aduva, Kora, Talayan, Mullan,* etc., migrate to the *Puramkadal*. In other words, when there is the *Tekkan Kattu,* the fishermen undertake fishing expeditions in the distant seas. Sometimes, when the *Tekkan Kattu* blows, the *Tekkan Nir* (southern current) also flows from south to north in the same direction. The fishermen call this phenomenon *Tekkanvada*. The occurrence of the *Tekkanvada* helps the fish to migrate to the *Puramkadal*. However, when the *Tekkanvada* is very violent, particularly in the months of *Chingam, Kanni,* and *Tulam,* the fishermen cannot launch their nets and boats. In case the *Nir* is calm (*Patam*) and the *Tekkan Kattu* is wild, the fish migrate according to the flow of the current. The mackerel and sardine do not, however, migrate to the *Puramkadal* when the *Tekkanvada* takes place because they do not travel against the current and wind. In such an event the Mappilas as well as Hindu fishermen employ the *Tattipeyyal* technique to catch the fish with the help of the *Mathikolli* nylon net. The Hindu fishermen of Blangad hold that when the *Tekkan Kattu* and *Vadakkan Nir* act in unison, rain and epidemic are likely to follow.

Tembrayil Kattu (South-westerly Wind)

The *Tembrayil Kattu* (south-westerly wind) generally blows in the months of *Idavam* and *Mithunam*. The fish migrate to the *Puram Kadal* when this wind blows. It is difficult to get good catches at this time. When the *Tembrayil Kattu* and the *Tembrayil Nir* (south-westerly current) occur together the fishermen do not get any catch at all. When the *Tembrayil Kattu* blows in the months of *Tulam*, *Vrischikam* and *Dhanu*, the hook-and-line fishermen also do not get any catch. If the *Tekkan Nir* (southerly current) flows along with this wind the traditional fishermen get good catches in the *Kara Kadal*.

Tengara Kattu (South-easterly Wind)

The *Tengara Kattu* (south-easterly wind) blows from south-east to north-west, particularly during north-eastern monsoon, in the months of *Kanni, Tulam* and *Vrischikam*. The Mappilas say that when the *Tengara Kattu* blows there is little likelihood of the north-eastern monsoon. Once this *Kattu* occurs, the clouds disappear; hence, there is no chance of rain. The Mappilas go out for fishing on such occasions both at night and in the day. When the *Tengara Kattu* blows very violently after the 15th of *Vrischikam*, it is called the *Kara Kattu* (easterly winds). The Mappila and Hindu fishermen of Blangad are unable to employ the Olukkuvala (gill-nets) during this period and they consequently migrate to Tanur and other villages in the north for undertaking fishing expeditions. When the *Tengara Kattu* blows in the months of *Tulam, Vrischikam, Dhanu* and *Makaram* in Tanur, the fish, particularly mackerel and sardine, migrate to the *Kara Kadal* and they are caught in abundance by the Mappilas.

Karakattu (Easterly Wind)

The *Karakattu* blows from east to west. When the *Karakattu* (easterly wind) blows, mackerel and sardine migrate from the *Puram Kadal* to the shore, and the water flows to the west. This westward flow of water is called *Vellam Valiyal*. The fish then migrate to the eastern seashore, and the fishermen get a bumper catch. Since these fish travel against the current they are caught in the months of *Vrischikam, Dhanu* and *Makaram*.

The most important wind that blows over the Malabar coast in the months of *Vrischikam, Dhanu* and *Makaram* is the *Karakattu*. During this period the hook-and-line fishermen restrict their fishing to between 15 and 25 fathoms. The traditional fishermen too do not go beyond this limit. The Beppukkar use three masts when they go out for fishing against the winds while the Valakkar employ only a boat without any mast. When there is a strong *Karakattu*, the Beppukkar say that the central line is likely to break off. In fact they have to wait for 2 to 3 hours just to ascertain whether any fish is entangled or not. They pull up the central line and examine the net only when the wind becomes calm. The traditional fishermen shoot their nets, particularly the mackerel net, in the east-westerly direction when the *Karakattu* blows, whereas they launch their boats against the wind.

During the south-west monsoon the *Kaccan Kattu* (strong *Vadavartha Kattu*) blows over the Malabar region. The traditional fishermen confine their fishing during this period (in the months of June, July and August) between 6 and 8 fathoms. They take one to one-and-a-half hours to cover this distance due to the high velocity of the wind blowing against their direction as well as the roughness of the sea. They hardly stay more than one-and-a-half hours in the sea for fishing owing to strong winds and violent tides. When the *Kaccan Kattu* blows they normally take only 45 minutes for covering the same distance on their return. If the *Kaccan Kattu* is very strong the return journey is accomplished within 15 to 20 minutes. The traditional fishermen use their masts when they return to the shore with their catch if the *Kaccan Kattu* is blowing. The wind that blows over the Malabar region in the morning hours is from south-easterly and north-easterly directions in November and December. The weak westerly wind that starts blowing after 10 a.m. is called *Puram Kattu* (westerly wind). Mackerel, cat-fish, etc., come up to the surface of the water to enjoy the wind when the *Kaccan Kattu* blows. On the contrary, when the *Karakattu* blows these fish go to the bottom mud. The fishermen, however, locate them by seeing the muddy bubbles coming up from the sea. These bubbles are called *Mariyam*. When the mackerel and sardine move in the muddy bottom the *Mariyam* appear on the surface water. The fishermen launch their nets on seeing these bubbles for catching *Ayila* and *Mathi* when the Karakattu blows in the months of *Vrischikam, Dhanu, Makaram* and *Kumbham*.

The winds are uncertain and unpredictable in the months of August, September, October and November. The Mappilas say that during this period the *Karakattu* (easterly wind), *Kaccan Kattu, Tekkan Kattu* (southerly wind) and *Vadakkan Kattu* (northerly wind) blow indiscriminately. The fishermen take more time for conducting their fishing operations during these months as well as for their return. *Tulavatta (Tembrayil Kattu* blown in the month of *Tulam)* also blows over Malabar, particularly in Tanur. The Mappilas assert that the advent of Tulavatta is an indication of the appearance of all species of fish in the months to come, though they do not go out for fishing in the month of *Tulam* when the wind is violent.

Winds and Launching

The mackerel is caught by the *Ayila Chalavala* by employing one of the two major techniques called *Nittivilakkal*. This technique is employed according to the direction of the wind and the migration of the fish. When the *Karakattu* blows from east to west and the fish travel from north to south, the net is launched in the south from east to west in the same direction as the wind. When the *Karakattu* blows, the fifth piece (*Murivala*) of the mackerel net is first launched in the east and moved towards the west until the remaining five pieces are launched. The boat is launched in such a way that its *Aniyam* (prow or boat's head) facing the east is against the wind. Once all the piece-nets (webbing) are launched, the *Kadessu* is tied to one end of the *Kamba* (rope) while the other end of the rope is tied to the crossbar very near the Aniyam. But if there is no change in the direction of the wind, the rope is tied to the *Amarappadi* (crossbar near the stem-*Amaram*). The *Toni* (boat) swerves according to the direction of the wind. A bamboo pole is tied to the *Pandi* (low portion of the boat) to stop it from tilting.

The technique of the launching of the *Olukkuvala* (gill-net) is the same as that of the mackerel net. When the *Karakattu* blows the fishermen haul the net and put it into the boat. If they fail to do so all the floats and piece-nets get entangled, preventing them from launching it again. In that case, the net is taken to the shore for separating the floats and the pieces. When the wind blows against the net, it should be hauled immediately. However, this net is used to catch *Polappu Min* (Pelagic fish) in the months of *Chingam* and *Kanni* with the help of minimum shore-sinkers (50 per cent) and minimum length of float-rope.

As most of the fishing grounds of the Mappila and Hindu fishermen are located on the western front, the northerly and the southerly winds do not appreciably affect their catches. The Mappilas adjust their technological knowhow and equipment according to the direction of the winds in the Malabar region. In other words, they adjust the technology of fishing to suit the natural environment. Thus, we note that the Mappilas and the Hindus adjust their fishing activities in accordance with the direction of the wind that blows over the Malabar region.

Nir (Currents)

The Mappila fishermen of Tanur and the Hindu fisherfolk of Blangad have a fair knowledge of the *Nir* (current), which flow in their fishing grounds. The Mappilas define *Nir* as the forward and backward movement of masses of water. They also call it *Vellamvali*. But they say that the velocity of the *Nir* is caused by the angel 'Mikayil'. On the other hand, marine biologists hold that the currents are the continued forward movement of mass of water. They are caused by a combination of factors such as the prevailing winds supported by the revolution of earth, the difference in the salinity of the seas and oceans as well as their temperature and the consequent variation in density. The movement of the current depends upon the direction of the wind. Like marine biologists, both the Hindu and the Mappila fishermen hold that the occurrence of a particular *Nir* depends upon the blowing of the winds.

We shall therefore make an attempt here to examine the different kinds of *Nir* and their relation to the winds, technology and catches. According to the Mappilas and Mukkuva Hindu fishermen there are eight kinds of *Nir* flowing in Tanur and Blangad from different directions. They are called (1) *Vadakkan Nir* (northerly current), (2) *Vattakkara Nir* (north-easterly current), (3) *Kara Nir* (easterly current), (4) *Tennara Nir* (south-easterly current), (5) *Tekkan Nir* (southern current), (6) *Tembrayil Nir* (south-westerly current), (7) *Porayi* or *Pura Nir* (westerly current) and (8) *Vadavartha Nir* (north-westerly current).

Vadakkan Nir (Northerly Current)

The *Vadakkan Nir* (northerly current) refers to those currents which flow from north to south. It occurs in Tanur and Blangad throughout the year. The maximum occurrence of this *Nir* is from *Kanni* to *Minam*. But, in some years, the *Vadakkan Nir* occurs in the months of *Medam, Idavam, Mithunam* and *Chingam* while in others this particular northerly current does not appear in the remaining six or seven months. The *Vadakkan Nir* is one of the important currents flowing in this region and most species of fish, particularly mackerel, sardines, prawns and mullets are caught during the movements of this *Nir.* On the whole, the Mappila and Mukkuva fishermen, particularly the Valakkar, welcome this *Nir,* as they get bumper catches during its flow. Unlike the Valakkar, the Beppukar do not wholly welcome this *Nir.* The Beppukkar say this *Nir* causes the breakdown of their hooks and lines and overturn of the fishing boats. A steersman belonging to the hook-and-line fishing unit told us that had the *Vadakkan Nir* been a little more stronger and faster "we would have got bumper catches". But he explained that it should not be very fast and violent or very slow. The fishermen should extend their float ropes when this particular current flows.

Along with the occurrence of the *Vadakkan Nir* there should be different winds and flow of other currents simultaneously in order to get good catches.

The Mappila and Mukkuva fishermen hold the view that the sea becomes calm and free from the violence of the waves when the *Vadakkan Nir* flows. They also correlate the movement of the clouds to the occurrence of the currents. For instance, they say that when the *Vadakkan Nir* flows, the clouds move from the north to the south. In other words, the clouds move according to the direction of the *Nir.*

Vadakkan Nir and Vadakkan Kattu

The Mappila and Mukkuva Hindu fishermen expect bumper catches when the *Vadakkan Nir* (northerly current) and *Vadakkan Kattu* (northerly wind) occur simultaneously. They say that several species of fish appear on the surface waters as *Polappu* (shoal) when the *Vadakkan Nir* and *Vadakkan Kattu* occur simultaneously. Among the important pelagic fish which appear in shoal are mackerel, sardine, *Chamban*, catfish, *Mullan* (silver belly), *Chemmin* (prawns), *Manannu* (anchovies), *Avoli* (pomfret) and *Cuda* (Tunny). These fish come up to enjoy the pleasant breeze of the *Vadakkan Kattu* and the most favourable current is the *Vadakkan Nir*. The occurrence of the *Vadakkan Nir* and the *Tengara Nir* (south-easterly current) simultaneously as undercurrent induce the fish to go to the muddy bottom and this results in a very poor catch. Thus, the *Tengara Nir* is regarded as the most unfavourable current for the fish.

Vadakkan Nir and Vattakkara Kattu

Both the Mappilas and the Mukkuvas believe that it is very easy to get good catches when the *Vadakkan Nir* (northerly current) and *Vattakkara Kattu* (north-easterly wind) occur simultaneously. The *Vattakkara Kattu* generally blows throughout the year in the mornings up to 10 a.m. After 10 a.m. this wind blows directly from the north and then it is called *Vadakkan Kattu* (northerly wind). Sometimes, in the evening, it blows from *Vadavarta* and continues to blow up to 10 p.m. when it is called *Vadavartha Kattu* (north-westerly wind). However, the fishermen go out for fishing up to a depth of 10 fathoms when *Vattakkara Kattu* blows and *Vadakkan Nir* flows. When the light of the *Tuyi* increases the fish notice the net and run away.

Vadakkan Nir and Kara Kattu

As a result of the simultaneous flow of the *Vadakkan Nir* (northerly current) and the *Kara Kattu* (easterly wind) the fishermen get good catches as the fish remain in the *Edanir* (midwater) making bubbles. The fishermen launch their nets at the sight of these bubbles on the surface of the sea. They say that the fish hide in the midwater only when the *Vadakkan Nir* and *Kara Kattu* occur simultaneously. During this time the sea becomes calm because of the *Vadakkan Nir* and *Kara Kattu* which make the sea very calm.

Vadakkan Nir and Tengara Kattu

When there is the concerted action of the *Vadakkan Nir* (northerly current) and *Tengara Kattu* (south-easterly wind), the *Olannal* of *Tiramalas* (velocity of the waves) increases. This makes the pelagic fish to migrate to the bottom of the sea and hide there. Even, the demersal fish like cat-fish and shark appear on the surface water as *Polappu* in the month of *Chingam*, but if the *Vadakkan Nir* flows and the *Tengara Kattu* blows they go down to the bottom sea. In that event the float ropes are generally extended to catch the bottom feeders. This increases the manual labour of the Valakkar and hence they do not welcome the occurrence of the *Vadakkan Nir* and the *Tengara Kattu* at the same time. On the other hand, the Beppukkar welcome the advent of the *Vadakkan Nir* and the *Tengara Kattu* simultaneously because they say they get good catches.

Vadakkan Nir and Tekkan Kattu

The traditional fishermen have to labour hard when there is the united action of the *Vadakkan Nir* (northerly current) and the *Tekkan Kattu* (southerly wind). However, this phenomenon brings good catches to the hook-and-line fishermen because the fish migrate to the muddy bottoms. Thus the violent *Tekkan Kattu* and the *Vadakkan Nir* are very unfavourable for the traditional fishermen, as they do not get good catches. Although the Beppukar have discarded their traditional hook-and-line fishing and have taken to modern techniques they too depend on the occurrence of current and wind for catching fish.

Vadakkan Nir and Tembrayil Kattu

The traditional fishermen do not get good catches when the *Vadakkan Nir* (northerly current) flows and the *Tembrayil Kattu* (south-westerly wind) blows simultaneously, as the fish migrate to the sea-bottom. But, the hook-and-line fishermen get good catches when this is the case. However, if the traditional fishermen want to get good catches they have to put in more labour than usual because it takes more time for launching the net.

Vadakkan Nir and Puram Kattu

The *Vadakkan Nir* (northerly current) and the *Puram Kattu* (westerly wind) occur throughout the year. The traditional fishermen employ the technique of *Tattipeyyal* for catching mackerel and sardine. The nets are launched in such a manner that fish can enter it through the meshes below the head rope. The fishing hands do not need to labour much.

Vadakkan Nir and Vadavartha Kattu

The fishermen get bumper catches when the *Vadakkan Nir* (northerly current) and the *Vadakkan Kattu* (northerly wind) act in unison. The net is launched on the *Vadavarta* side (north-western side). When they launch the net on the north-western side during the occurrence of the *Vadakkan Nir* and the *Vadavartha Kattu* the anchor (*Irimbu*) is lowered to the bottom by the *Mumbakallikkaran* (a fisherman who sits on the *Aniyam Kalli*). The fishermen get good catches, particularly in the months of *Chingam, Kanni, Kumbham* and *Minam*, when the *Vadakkan Nir* and the *Vadavartha Kattu* act conjointly. Despite the fact that fibre boats fitted with 10 HP motors are used by both the Mappila and Hindu fisherfolk, they correlate the catch with the prevailing current and wind.

Vattakkara Nir

The *Vattakkara Nir* (north-easterly current) flows from north-east to south-west. This current is present throughout the year but it is more noticeable in the months of *Chingam* and *Kanni*. When the *Vattakkara Nir* flows the sea becomes calm and the fish migrate from the *Puramkadal* to the *Karakadal*. The fishermen say even if the sea happens to be choppy the presence of the *Vattakkara Nir* makes it calm. The fish, which migrate to the coastal waters, appear on the surface and are caught in abundance.

This *Nir* is generally welcomed by the Mappila and the Hindu fishermen as it is believed that it tempts the fish to migrate towards the shore. Most of the species of fish are caught when the *Vattakkara Nir* flows. When the current flows from north-east, the fishermen believe that the colour of the water becomes either brown or yellow and it is generally muddy. All the floats of the net are tied while fishing in the muddy water. But, the fishermen use the minimum number of floats and lighter sinkers if they launch the nets according to the direction of the current. However, if the nets are shot against the *Vattakkara Nir,* the number of floats and weight of sinkers are generally increased. The Mappila and the Hindu fisherfolk expect good catches in the advent of the *Vattakkara Nir.* The Mappilas also believe that this *Nir* brings the sea mud to the shore, heralding a bumper catch. Sometimes, when the catch falls short of their expectation, the fishermen make a vow to donate a portion of their haul to their favourite temples and mosques for the occurrence of the *Vattakkara Nir* and the presence of the mud-bank (sea-mud-*Chakara*) in the shore waters. This is sure to result in bumper catches.

Vattakkara Nir and Vadakkan Kattu

The Mappilas expect a good catch when the *Vattakkara Nir* (north-easterly current) is accompanied by the *Vadakkan Kattu* (northerly wind). Both the Mappila and the Hindu fishermen believe that if the undercurrent is the *Tengara Nir* (south-easterly current), the net gets entwined, which prevents the entry of the fish into it. However, when there is the occurrence of the *Vattakkara Nir* and the *Vadakkan Kattu,* the Mappila and the Hindu fishermen employ the *Tattipeyyal* technique for catching sardines.

Vattakkara Nir and Vattakkara Kattu

Both the Hindu and the Mappila fishermen are sure that the *Vattakkara Kattu* (north-easterly wind) and *Vattakkara Nir* (north-easterly current) occur from *Chingam* to *Minam*. Although during this period the other *Nirs,* particularly *Tengara Nir* (south-easterly current) and *Tekkan Nir* (southerly current) also occur, the *Vattakkara Nir* and *Vattakkara Kattu* occur frequently. This makes the *Tuyi* shine, which is most favourable for fishing, particularly at night. When the above current and wind occur the fishermen say that "*Nir Mayakkam Vannu* or *Tuyi Nannayi*" (the current is very favourable as its light is very good for fishing).

Vattakkara Nir and Puram Kattu

It is neither very good nor very bad for fishing when the *Vattakkara Nir* (north-easterly current) and the *Puram Kattu* (westerly wind) occur. The fishermen do not have to labour much for launching and hauling their nets and they get moderate catches.

Vattakkara Nir and Tengara Kattu

The traditional fishermen have to labour hard for launching their nets if they want to get good catches during the occurrence of thc *Vattakkara Nir* (north-easterly current) and the *Tengara Kattu* (south-easterly wind) but they get good returns during the occurrence of these winds and currents.

Vattakkara Nir, Tembrayil Kattu and Vadavartha Kattu

The fishing folk get moderate catches when the *Vattakkara Nir* (north-easterly current) and *Tembrayil Kattu* (south-westerly wind) occur. However, they have to face enormous difficulties in launching their nets and boats during this time. When the *Vattakkara Nir* and *Vadavartha Kattu* (north-westerly wind) are present the velocity of the current is reduced by the *Vadavartha Kattu.* This results in getting poor catches.

Vattakkara Nir and Kara Kattu

The fishermen, particularly those who stick to their tradition, do not get good catches during the occurrence of the *Vattakkara Nir* and the *Kara Kattu* (easterly wind) simultaneously. When the *Puram Kattu* (westerly wind) blows along with the *Vattakkara Nir* the fishermen get good catches. The Mappila and the Hindu fishermen designate the brighter current as *Telinna Nir.* Since it shines brightly, the sardine cannot see the nylon net and do not enter into it.

Kara Nir

The *Kara Nir* (current from the shore waters or easterly current) flows from east to west. There is no catch at all when this *Nir* occurs in the months of *Idavam, Mithunam* and *Karkkitakam*. The *Kara Nir* occurs the most in the months of *Tulam* and *Vrischikam*. When the *Kara Nir* occurs the fish migrate to distant areas of the sea, 8 to 12 miles away from the coastal area and hide in the muddy bottom, particularly upside down. The fishermen find it very difficult to go to such distant areas of the sea, particularly during the monsoon, when the *Puram Kattu* (westerly wind) blows against the *Kara Nir*. In these circumstances most of the species of fish migrate to the *Puramkadal*, and take shelter in the muddy bottoms. The *Kara Nir* is also responsible for increasing the ferocity of the sea (*Kadalkopam*) during the monsoon, particularly in the months of *Mithunam* and *Karkkitakam*. Sometimes, the *Adi Nir* originates from the bottom of the sea and flows in the opposite direction when there is *Kara Nir*. The Mappila fishermen launch their boats at a depth of six fathoms during this period when the sea is not violent. They row the boat according to the direction of the *Kara Nir* and the boat follows the movement of the current. The Mappilas believe that the *Kara Nir* flows only up to a depth of three-and-a-half fathoms. They follow the direction of the current upto three-and-a-half fathoms to cover the six fathoms depth and for the rest of the distance they depend upon the direction of the *Adi Nir,* which flows from west to east as the *Pura Nir*.

Kara Nir and Winds

The fishermen have to labour hard and sometimes they are unable to reach their landing station during the occurrence of the *Kara Nir* (easterly current) and the *Kara Kattu* (easterly wind). On the other hand, when there is *Kara Nir* and the *Vattakkara Kattu* (north-easterly wind), the fishermen can reach the shore without any hardship. Both the Mappila and the Hindu fishermen sometimes get moderately good catches when the *Vattakkara Kattu* blows along with the *Kara Nir*. The presence of any other wind with the *Kara Nir* makes the fishermen draw blank but their safety is not in jeopardy. The *Kara Nir* is not at all favourable for fishing and the fishermen get a poor catch, or sometimes no catch at all.

Tengara Nir

The *Tengara Nir* or *Tengarakkara Nir* flows throughout the year from south-east to north-west, especially in the months of *Idavam,*

Mithunam, Tulam, Vrischikam, Dhanu and *Makaram*. When the *Tengara Nir* (south-easterly current) occurs the fish go down and hide in the muddy bottom. No good catch is expected during this period as the sea is very violent, particularly in the months of June and July. As the shoals move to distant areas of the sea in search of safety and hide in the muddy bottom, the fishermen face hardships in conducting their fishing operations.

Tengara Nir and Tengara Kattu

When the *Tengara Nir* (south-easterly current) and *Tengara Kattu* (south-easterly wind) occur simultaneously in the months of *Vrischikam, Dhanu* and *Makaram* the fish migrate to the *Puramkadal*. The prevalence of the *Tengara Nir* and the *Puram Kattu* (westerly wind) prevents the fishermen from getting a good catch. The same is the fate of the fishermen when the *Tekkan Kattu* (southern wind) blows and the *Tengara Nir* flows, as the fish descend to the bottom and hide there upside down. The blowing of the *Tembrayil Kattu* (south-westerly wind) together with *Tengara Nir* helps the fishermen get a moderate catch, but they face difficulties in catching fish. The action of the *Vadakkan Kattu* (northerly wind) together with the *Tengara Nir* also results in this predicament. Similarly, when the *Tengara Nir* and the *Vadavartha Kattu* (north-westerly wind) interact the fish migrate to the *Puramkadal* and go to the bottom. The shoals also migrate to the *Puramkadal* when the *Vattakkara Kattu* (north-easterly wind) blows while *Tengara Nir* flows. The Mappilas attribute the cause of the failure of the catch to the occurrence of the *Tengara Nir* and the *Tengara Kattu* besides the presence of the violent wave, *Yama Kadal*. When there is *Tengara Nir* and *Puram Kattu* (westerly wind) the fishermen get good catches.

Tekkan Nir

The movement of the *Tekkan Nir* (southern current) is from south to north. It takes place in the months of *Tulam, Vrischikam, Dhanu* and *Makaram*. The Hindu fishermen of Blangad say that the *Tekkan Nir* occurs frequently during the south-west monsoon. This *Nir* makes the sea violent and therefore the Valakkar dare not go out for fishing. Though the fish do not visit the coastal waters, they are caught from the areas where there is *Chakara* or mud-bank. The *Chakara* is generally found in the sea during the monsoon when there is no wave. The freshets and dirt flow to join the sea to form accumulations. Westerly winds carry these accumulations three to five miles away from the shore. The fish take shelter under this mud-bank. Minimum sinkers are used to catch the fish that feed on this mud-bank. The *Chakara* move to different directions when the sea becomes choppy due to wind, current and rain. No current of any sort is welcomed by the Mappilas, particularly the hook-and-line fishermen. These fishermen complain that their lines are liable to be cut off when the *Tekkan Nir* flows. Sometimes the fish cannot be caught by hooks-and-lines in the face of the *Tekkan Nir*. They also frequently face a lot of difficulty in returning to the landing station when the *Tekkan Nir* flows. The *Tekkan Nir* occurs very often in the months of *Tulam* and *Vrischikam*. The fishermen cannot go out for fishing when the *Tekkan Nir* is powerful.

It must be pointed out here that the Mappilas believe that there is a polluted current called *Kedu Nir,* which sometimes occurs along with one of the eight kinds described above. An old informant told us that the main current that always flows at a depth of 50 miles is the *Tekkan Nir* (southerly current). Another aged fisherman confessed that no one dares to go so far up to 50 miles and check up the *Nir*. They do not expect good catches during the period of the *Tekkan Nir*.

Tekkan Nir and Winds

The Mappilas expect good catches of *Ayila* and *Mathi* whenever the *Tekkan Nir* (southerly current) and the *Tekkan Kattu* (southerly wind) occur moderately. But if the *Tekkan Nir* and the *Tekkan Kattu* are very strong, the sea becomes very violent and the fishermen call it '*Kol*'. They do not go out for fishing whenever there is a '*Kol*'. Even if they venture out they seldom get a catch. The '*Kol*', present on account of the *Tekkan Nir* and the *Tekkan Kattu,* drives the fish towards the south, and if the *Nir* increases its velocity (*Vadaval*), the fish float according to its direction. It becomes difficult, if not impossible, for launching the fishing nets in that case and the result is a poor catch.

The traditional fishermen get good catches when there is *Tekkan Nir* and *Vadakkan Kattu*. Though the hook-and-line fishermen do not get good catches in this situation, their safety is not in danger. The concerted action of the *Tekkan Nir* and the *Vattakkara Kattu* (north-easterly wind) helps the fishermen haul bumper crops of mackerel and sardine. On the contrary, when the *Tekkan Nir* and the *Vadavartha Kattu* occur, the net cannot be launched easily because the fishermen say that the fish can see it. This prevents easy catch of the fish. Those fish which are in the bottom do not come up and consequently of the difficulties in launching the net increase. The *Tekkan Nir* and the *Tembrayil Kattu* (south-westerly wind) occur simultaneously in the months of *Tulam, Vrischikam* and *Dhanu* when this happens the hook-and-line fishermen seldom get a good catch but the traditional fishermen are able to get moderate catches in the *Kara Kadal*. Under the concerted action of the *Tekkan Nir* and *Tengara Kattu* (south-easterly wind), the hook-and-line fishermen get poor catches from *Tulam* to *Minam*. They sometimes lose their hooks-and-lines. Although there is some difficulty in launching the net, the traditional fishermen find it worthwhile to conduct their operations as they get good catches.

Tekkan and Vadakkan Nirs

In the months of *Kumbham, Minam* and *Mithunam* the *Tekkan Nir* (southerly current) flows as the upper current (*Menir*) and under it the *Vadakkan Nir* (northerly current) flows as the *Adi Nir*. This adds to the difficulties of fishermen in catching the bottom fish.

Tembrayil Nir

When the *Tembrayil Nir* (south-westerly current) flows from south-west to north-east in the months of *Chingam, Kanni, Tulam* together with *Tembrayil Nir* (south-westerly current), the Mappilas get bumper catches of sardine, mackerel and *Paitakora*. Some of the Mappilas hold that whatever be the wind that blows during the occurrence of this *Nir* the fishermen get good catches. The old Mappilas, however, contradicted this and said that they did not get the expected catch of mackerel and sardines when the *Tembrayil Nir* flowed. They held that the waters of the sea became bluish when this current flowed and the fish were not able to see the net. In order to prevent the fish from seeing the net, whenever there is the flow of the *Tembrayil Nir* it is generally lowered from its original position. The fishermen reduce the number of floats to one-third of the total for lowering the net. Sometimes other currents alsc occur along with the *Tembrayil Nir,* spoiling the fishing activities entirely.

Tembrayil Nir and Kattu

The *Tembrayil Nir* (south-westerly current) occurs more often in the months of *Chingam, Kanni* and *Tulam.* The occurrence of the *Tembrayil Nir* and the *Tembrayil Kattu* (south-westerly wind) together helps the fishermen to get a bumper catch. The Mappila and the Hindu fishermen of Malabar hail this *Nir* and *Kattu* as *'Nammude Ariyude Karrum Nirum'* (Our breadwinner current and wind). On the other hand, when the *Tembrayil Nir* flows and the *Vattakkara Kattu* (north-easterly wind) blows the fish go down to the muddy bottom rather than migrate to the shore waters (*Kara Kadal*). This causes immense difficulty in catching them. But it is easy to catch the fish if the *Tembrayil Nir* flows in conjunction with the *Vadavartha Kattu* (north-westerly wind). When the *Vadavartha Kattu* is coupled with the *Tembrayil Nir,* the fish come up as *Polappu* for enjoying the pleasant breeze, which results in good catches. The *Mathi* relish the *Tembrayil Nir* and the *Tekkan Kattu* (southerly wind) and they come up the surface, when they are caught abundantly. The occurrence of the *Tembrayil Nir* and the *Kara Kattu* together drive the fish to the muddy bottoms. The fishermen get moderately good catches of sardine and mackerel when the *Tembrayil Nir* flows in conjunction with the *Puram Kattu* (westerly wind).

The flow of any other *Nir* along with the *Tembrayil Nir* causes the fishermen to get either a bumper or a poor catch. For instance, when the *Tembrayil Nir* flows as a *Menir* (surface-current), and the *Vattakkara Nir* (north-easterly current) as the *Adi Nir,* the fishermen can catch the fish very easily, particularly in the months of *Chingam* and *Kanni.* Above all, in these two months the fish come up as *Polappu* and they become drowsy on account of the flow of *Tembrayil Nir.* On the other hand, if the *Tembrayil Nir* flows as the *Adi Nir* and the *Vattakkara Nir* (north-easterly current) as the *Menir,* a good catch is expected. Both these *Nirs* are generally liked and enjoyed by the fish. When the *Tembrayil Nir* occurs as the *Adi Nir* and the *Kara Nir* (easterly current) as *Menir,* the fishermen cannot get a catch easily. If the *Tembrayil Nir* flows over the *Kara Nir* as the *Menir,* they get bumper catches; conversely, when the *Tembrayil Nir* flows as *Adi Nir* and *Kara Nir* as *Menir,* they get a good catch. This is because the velocity of the *Adi Nir* makes, the fish giddy so that they are unable to detect the net. If the *Tembrayil Nir* occurs as the surface current and the *Vadavartha Nir* as the under current, the fishermen expect moderate catches.

Puram Nir

The *Puram Nir* (westerly current) flows on the Malabar coast mostly from west to east in the months of *Chingam* and *Kanni,* particularly after noon. When the *Puram Nir* occurs, sometimes *Kakkacci* (a sea-crow) appears on the surface water. The Mappila and the Mukkuva fishermen hail the appearance of the *Puram Nir* and the sea-crow as the harbinger of good catches, particularly the fishes of *Camban, Podimathi* and *Netal.* The *Puram Nir* induces the migration of various species of fish like *Adavu, Netal, Manannu, Mullan, Camban, etc.,* from the *Puramkadal.* This enables the fishermen to get bumper catches from the coastal waters. Thus, the occurrence of the *Puram Nir* is generally welcomed by the Mappila fishermen as they expect a good catch. Most species of fish are caught when the *Puram Nir* flows. The Mappila fishermen believe that all species of fish come out of their muddy bottoms and move to the surface when there is *Puram Nir.* This facilitates the fishermen to catch the fish. There is a saying among the Mappilas that "*Perramma Cathalum Puranir Olukumbol Pulayilla*" (when the westerly current flows there is no observance of death pollution even for one's own mother).

Puram Nir and Kattu

When the *Puram Nir* (westerly current) and the *Puram Kattu* (westerly wind) occur simultaneously the fishermen get good catches. But if the *Puram Kattu* is very strong they have to put in more labour than usual. On the other hand, if the *Kara Kattu* (easterly wind) blows along with the *Puram Nir* the fishermen need not exert themselves labouriously to get good catches. If any other wind blows along with the *Puram Nir* good catches are expected. But it may be mentioned here that if any of these winds blow violently the fishermen have to labour hard. Sometimes the *Kara Kattu* blows very violently along with the above current, which prompts the fish to hide in the bottom of the sea.

Vadavartha Nir

The *Vadavartha Nir* (north-west current) flows from north-west to south-east throughout the year. But it occurs more frequently in the months of *Tulam, vrschikam, Dhanu, Makaram, Kumbham* and *Minam*. In some years it occurs frequently during the monsoon. If any wind blows along with this *Nir* during the monsoon, the sea becomes violent and the fishermen cannot go out for fishing. The *Vadavartha Nir* generally occurs when the strong *Kaccan Kattu* (westerly wind) blows over the region. When this current flows the fish migrate to the depths of the sea and hide in the muddy bottom. As a result, the muddy water comes up from the bottom to the surface, a clear symptom of *Mariyam*. When the *Mariyam* appears the Mappilas resign to their fate since the fish take refuge in the muddy bottom. This means that they will have to exert heavily for catching the fish by extending the length of the float-rope and increasing the weight of the sinkers.

Tuyi, Nir, Kattu and Nilavu

The *Tirandis* (skates and rays) are generally caught from a depth of 2 to 3 fathoms employing the *Tirandi Vala* (skate-and-ray net). The catch of the skates and rays is related to the *Tuyi* (light), *Nir* (current), *Kattu* (wind), and *Nilavu* (moonlight). Bumper catches of the skates and rays are obtained in the month of *Makaram*. During this period when the *Kara Kattu* blows from east to west the skates and rays migrate to the shore waters and they are easily caught. The skates and rays are also caught during the nights when there is moonlight, because this light fades the *Tuyi*. On the other hand, when there is darkness, the *Tuyi* shines more brightly and the fish see the net and run away. This results in a very poor catch of the skates and rays. The Mappilas also expect good catches of the skates and rays when the sky is auburn just before sunrise and the sunset. The *Tuyi* fades away with the glow of the sun at these times. Since the fish cannot see the nets in the absence of the *Tuyi*, they move into them and are thus caught. These fishermen believe that the *Tuyi* does not shine up to 7.30 p.m. even on full-moon nights. In the months of *Kumbham, Minam, Medam* and *Idavam*, the skates and rays are not caught during the night but are hauled only during the day. The

Mappilas say the *Tuyi* shines brightly during these months on account of the appearance of the *Ponal* (plankton) and *Kedu Nir,* which are produced by the south-west monsoon. The fishermen maintain that when there is *Kedu Nir,* the *Tuyi* shines like a 1,000 W bulb. The *Tirandis* see the net when the *Tuyi* shines brightly and they run away from the fishing grounds. But in the months of *Mithunam, Karkkitakam* and *Chingam* they do not employ the skates-and-rays net because the sea is choppy. *Makaram* has been regarded the most favourable among the remaining months for getting bumper catches of skates and rays. As has already been stated, the *Kara Kattu* (eastern wind) blows during this month from east and west and the skates and rays move to the shore waters to enjoy the breeze. The *Tuyi* shines in a subdued way during this month because the *Kara Kattu* causes the mud to come up from the bottoms. This diminishes the luminescence of the *Tuyi* so that the fish easily move into the nets. In other months the fishermen get poor catches of the skates and rays. In the month of *Kumbham* they are caught from a depth of 10 fathoms.

The gill-net is employed to catch larger fishes like *Ayikkora* (seer), *Palamin, Sravu* (shark), *Etta* (cat fish), *Kanamin* (tunny) and *Neymin*. The use of this net and the quantum of the catch are determined by the appearance of the *Tuyi* and favourable wind. Unlike the skates and rays, the larger fishes mentioned above are caught in the darkness when the *Tuyi* shines more brightly. The brilliance of the *Tuyi* in dark nights induce the smaller fish to appear in shoals. Their appearance in shoals send the bigger fish running after them as lions after small animals in the forest. In the months of *Chingam, Kanni, Tulam* and *Vrischikam* small fishes like *Nethal* (white bait), *Calamathi* (sardine), *Chemmin* (prawn) and *Mantal* (sole) appear. When the larger fish chase their prey they do not see the nets and get entangled in them. Thus, the traditional gill-net is used on dark nights and when there is brighter *Tuyi*. Nowadays nylon gill-nets are used for catching the bigger fishes mentioned above whether there is *Tuyi* or not. Small fish escape through the big meshes of the traditional gill-net when they are chased by their predators. Each mesh is visible to them in the light of the *Tuyi* as if it is magnified fourfold. These days the nylon gill-net is used indiscriminately for catching larger fish irrespective of the presence of moonlight or *Tuyi*. Even if it is employed in the light of the *Tuyi* the fish cannot see the meshes of the nylon net, which is inferior in temper. Above all, when the smaller fish appear in *Polappu*, the bigger fish are more anxious to catch their prey and hence do not care for the net.

The gill-net is generally floated according to the flow of the *Nir* but against the wind. The boat is rowed from west to east. If the *Puram Kattu* does not blow very violently and the *Kara Nir* flows roughly, the gill-net is launched according to the direction of the current. On the other hand, if the *Puram Kattu* is very strong and the *Kara Nir* flows slowly, the boat is floated towards the east (i.e., not against the wind) and the net just after it, but according to the flow of the current, i.e., towards the east. The net is generally shot for an hour or two and sometimes it is lowered up to a depth of 10 fathoms. The operation of the gill-net mainly depends upon the flow of the current and the nature of the wind. Earlier the traditional cotton *Ayila Chalavala* (mackerel net) was employed during dark nights for catching the mackerel with the help of the light of the *Tuyi* but this net was not used when there was moonlight. Today the traditional cotton net has been replaced by the nylon mackerel net and it is used for catching mackerel even when there is moonlight. When the cotton net was used, on dark nights the mackerel could not detect the cotton net with the help of the *Tuyi* and would get into it. But, the fish cannot see the nylon net even in the moonlight and so moonlight is favoured for employing the modern nylon mackerel net, because the light of the *Tuyi* fades away in such a way that the fish cannot see the net. Even if there is a strong *Tuyi* the fish are unable to recognise the net since its meshes are light. For catching sardine however, moonlit nights are not preferred. When there is moonlight, the *Tuyi* fades away and sardines, prawns and other fishes are unable to see the net. The striking behaviour of the sardines and prawns is that once they see the net, they run into it. The *Tuyi* and dark nights are thus most favoured for catching sardines. Modern sardine-nets are not employed when there is moonlight because the light of the *Tuyi* is counterbalanced by it. The fish cannot see the net and consequently do not move into it. The fishermen generally do not go out for fishing *Mantal* (soles) during nights whether there is moonlight or not. The soles are unable to see at night and their movement is restricted. The fishermen are also unable to locate the shoal of soles even when there is moonlight. Therefore they are usually caught in the daytime.

Fishing expeditions at night with the help of the *Tuyi* are known as the *Tuyikku Pokal.* These expeditions by the traditional fishermen generally take place from *Vrischikam* to *Idavam.*

The Mappilas say that the subdued light of the *Tuyi* is favourable for getting better catches. If the *Tuyi* is brilliant, it affects the catch. They maintain that the occurrence of various currents make the *Tuyi* shine either brightly or dimly. The Hindu and the Mappila fishermen alike believe that when there is *Tekkan Nir* (southerly current), the *Tuyi* shines moderately but the current turns the sea violent, particularly in the months of *Mithunam* and *Karkkitakam,* preventing them from going out for fishing. The flow of *Tengarakara Nir* accompanied with bright *Tuyi* gives the fishermen little hope of a good catch. The *Tuyi* which appears during the occurrence of the *Tengarakara Nir* is called *Mani Tuyi.* The *Mani Tuyi* sparkles when there is unison of *Kara Kattu* (easterly wind) and *Tengara Nir* (south-easterly current). Above all, when there is the *Tengarakara Nir* and the *Karakattu, Coris* also appear which help to increase the light of the *Tuyi.* All these factors contribute to a poor catch or sometimes to no catch at all. The Mappilas of Tanur say that when there is *Tembrayil Nir* (south-westerly current) and the *Tuyi* is gleaming they get a good catch. The Mukkuvas (Hindu fishermen of Blangad), on the other hand, maintained that the simultaneous appearance of the *Tembrayil Nir* (south-westerly current) and *Mani Tuyi* results in very poor catches. Both the Hindu and the Mappila fishermen hold that the occurrence of the *Vadavartha Nir* increases the brightness of the *Tuyi* more than required. This sometimes results in a poor catch because the fish swim to the *Puramkadal* when the *Tuyi* shines brilliantly.

The presence of the *Vadakkan Nir* (northerly current) makes the *Tuyi* shine with great luminescence. This *Nir* flows throughout the year, but more frequently from *Vrischikam* to *Makaram.* If the *Vadakkan Kattu* (northerly wind) blows along with the *Vadakkan Nir* the light of the *Tuyi* does not increase and the fishermen have to labour hard to get any catch. The presence of the *Vattakkara Nir* (north-easterly current) with moderate *Tuyi,* gives hopes of a good catch. Similarly, during the occurrence of the *Puram Nir* (westerly current), the *Tuyi* becomes moderate and the fishermen have little difficulty in getting a good catch. On the other hand, whenever the *Kara Nir* (easterly current) appears, the *Coris* go down to the bottoms. This diminishes the light of the *Tuyi* and it cannot be seen at all. Thus, the *Coris* are also partly responsible for the increase of the light of the *Tuyi.* These *Coris* are mainly zooplanktons. The Mappilas say that these *Coris* consist of *Anukkal* (germs), which are responsible for the production of the *Tuyi.*

It is evident from the above that the light of the *Tuyi* should not be brilliant for getting a good catch. But, the Mappilas say that the modern nylon nets are very useful since the sardines cannot see these even when the *Tuyi* shines with extreme brilliance. The *Tuyi* shines brilliantly if any current flows violently from any side and if the fish catch sight of the net in its light, they run away. However, the occurrence of the *Vadakkan Nir* (northerly current) makes the *Tuyi* shine with so much brightness that the fish cannot see the net. Thus, the fishermen get good catches under the concerted action of the *Vadakkan Nir* (northerly current) and the *Tuyi.*

The fishermen carry torches when they go out for fishing, especially during the months of *Chingam* and *Kanni,* in spite of the appearance of the *Tuyi.* They say that the *Kedu Nir* (dirty water) becomes a very solid substance in these months and appears on the surface. They call this solid substance *Kara*. The fishermen use torches in order to find out the *Kara* because the fish hide under it. The fish do not migrate to the net if the *Kara* is present despite the moderate light of the *Tuyi.* The fishermen beat the crossbar of their boats in order to frighten the fish and thus drive them into the net.

The fishermen compare the sea with the human body. They say a person suffers from influenza and fever because some polluted water enters his body. Similarly, when the *Kedu Nir* appears in the sea during the monsoon it becomes inhospitable to the fishermen. Sometimes the *Kedu Nir* takes the form of the *Kara* and moves from one place to another causing hardship to the fishermen, particularly when the *Puram Nir* (westerly current) flows at a faster pace (*Varakal*). The *Kedu Nir* in that case is called *Charipuli*. When the *Puram Nir* flows swiftly, the *Kedu Nir* travels to the sea coast and appears as a smoke screen. Thus, when the *Kedu Nir* moves fast the fish go down to the sea bottom. The sea then becomes inhospitable for fishing. The fishermen sometimes compare the sea with a woman in her period when the *Kedu Nir* appears. Similarly, the Mappila and Hindu fishermen say that the indiscriminate occurrence of the currents makes the sea sick. They say that during the months of *Mithunam* and *Karkkitakam*, as many as 48 currents occur in half an hour's time (*Nalikakku Nalppattettu Nir*). The current in the sea is compared to the circulation of the blood in a human body. Just as infection of the blood causes various kinds of illness, similarly, the occurrence of various *Nir* causes illness to the sea. This adds to the hardships of the Mappilas and the Hindu fisherfolk.

The Lunar and Tidal Influences on the Catch

The Mappilas of Malabar are quite conversant with the occurrence of various tides and their influence on the catches. They adjust their fishing techniques not only according to the nature and character of the tides but also in accordance with the different phases of the moon.

Both the Hindu and the Mappila fishermen of Malabar do not generally go out for fishing during the period of *Veliyettam* or *Vellam Velli* (high tides), particularly on the days of the *Karutha Vavu* (New Moon) and the *Velutha Vavu* (Full Moon). They say that higher the tide, greater is the depth of the sea and more is the likelihood of capsizing of their boats. That is the main reason why they do not venture to go out for fishing on the days of the *Karutha Vavu* and *Velutha Vavu*. These fishermen hold that on these days when the moon is vertically overhead, the highest tides occur, and when the moon sets, the lowest tides occur. These high and low tides are called *Ucca Veli* (spring tide). But when the tides do not rise so high or fall so low, they are known as *Necca Veli* (neap tide). The fishermen's knowledge about the cause of the tides and their effects on the catches corresponds to that of astronomers and physical geographers. As the moon travels westwards the locality of the high water also moves westward keeping pace with the moon.

The Mappila and the Hindu fishermen believe that when the *Nilavu* is vertically overhead there is commotion in the sea and water from the *Karakadal* sweeps the shores. In other words, when the moon is vertically overhead the *Vellam Veli* or *Veliyettam* occurs and consequently the level of the sea increases and the water rises to the shore. When the moon sets the tides fall and therefore the level of the sea also falls and thus the *Veliyirakkam* or *Vellam Irakkam* occurs. Like physical geographers and astronomers the fisherfolk of Kerala, particularly the Mappilas, maintain that the level of the sea rises and the *Veliyettam* occurs twice a day. Similarly, the level of the sea falls twice a day and consequently the *Veliyirakkam* also takes place twice a day. The fishermen, like astronomers and physical geographers, hold the view that at a particular area the interval between one *Veliyettam* and the next one is a little over 12 hours. They say if the *Veliyettam* takes place at 9.30 a.m. the sea rises very high at 12 noon and continues to do so upto 3 p.m. Just after 3 p.m. the level of the sea starts receding and falls rapidly till 6 p.m. Thus the *Veliyirakkam* reaches the maximum at 6 p.m. and continues to fall up to 9 p.m. At about 9.30 p.m. again the *Veliyettam* begins and it rises very high at 12 midnight. This level is maintained up to 3 a.m. After 3 a.m. the level of the sea falls and the *Veliyirakkam* reaches its lowest level at 6 a.m. This continues past 9 a.m. on the next day. On this day again the tides rise up and reach the maximum level at 12 noon and goes on rising up to 3 p.m. After 3 p.m. the *Veliyirakkam* begins and it falls to the lowest level at 6 p.m. and the fall continues up to 9 p.m. Again just after 9 p.m. the *Veliyettam* begins. The above analysis proves that the Hindu and the Mappila fishermen do not lag behind in their knowledge about the reason of the occurrence of the high tides and the low tides every day. On the day of *Karutta Vavu* (New Moon) and *Velutha Vavu* (Full Moon) the *Veliyettam* rises very high and the fishermen do not venture to go out for fishing, particularly in the months of *Idavam, Mithunam* and *Karkkitakam*. When the *Veliyettam* occurs during the days of the New Moon and Full Moon, the Mappilas say that the *Chemmin* (prawn) go to the bottom sea and they appear on the surface water again when the moon sets. During the time of the *Veliyettam* it is difficult to catch some of the important fishes. The Mappila and the Hindu fishermen hold that prawn, sole, jew fish, silver belly, etc., migrate to the bottom of the sea and live safely there during the occurrence of *Vellamveli* or *Velliyettam*. But sometimes, particularly in the months of *Mithunam* and *Karkkitakam*, the *Chemmin* and *Mullan* appear on the surface when *Vellamveli* occurs and so they are caught abundantly. Although it is believed that during the occurrence of the *Veliyettam*, particularly on the days of Full Moon and New Moon, the fish migrate to distant places, the Mappilas expect good catches when the *Uccaveli* and *Necca Veli* occur. They say that when the *Veliyettam* occurs on these days coupled with winds and currents, the whole environment of the sea changes and it gets agitated. All the bottom fish come up to the surface water during this agitation and sometimes they are caught in abundance. The Mappilas expect good catches in shallow waters and at times they are caught in abundance. They hold that when the *Veliyerakkam* occurs the fish from the *Puramkadal* migrate to the shore. In other words, when the moon sets the low tides occur and the fish, particularly, prawn, sole, jew fish and horse mackerel migrate from the *Puramkadal* to the *Kara Kadal*. As a matter of fact, the fishermen harvest their bumper crops from the nearest fishing grounds to the shore when the *Veliyerakkam* occurs in the sea. Our informant Enikutty Maulavi held that the *Chemmin* are caught in abundance during the nights of Full Moon compared to the New Moon when the low tides occur. Before the introduction of the nylon net, the Maulavi once went out for fishing in the daytime on the *Velutha Vavu* with the traditional cotton net and hauled a good catch of prawn as a *Veliyettam* occurred unexpectedly. Then he undertook a similar fishing expedition in the same night and prawn were caught abundantly during the time of *Veliyerakkam*. For the following 13 days he caught plenty of prawns during the times of *Veliyerakkam*.

The Mappilas and the Mukkuvas subscribe to the view that *Ayikkoras, Nemmin* and *Palamin* can be caught in large quantities during the nights of the *Karutha Vavu* compared to the *Velutha Vavu*. They also believe that the Seer fish can be caught to their hearts' content by employing the nylon gill-nets more favourably during the *Veliyerakkam*. This cosmological knowledge of the fishermen corresponds to that of marine biologists. Mathai and others (Mathai et al., 1971) observe: "It is a common belief that the seer are caught in abundance during the new moon-nights compared to full moon-nights. It is also believed that operations during the low tides are more favourable than those during high tide".

The catch of the mackerel and sardine is also greatly influenced by the phase of the moon. The Mappilas say that during *Veliyerakkam* as well as *Necca Veli*, sardine and cat fish appear on the surface water in shoals and it is easy to catch them.

The Mappila and the Hindu fishermen also correlate the catch of the mackerel to the phases of the moon. They believe that the mackerel shoal (*Ayila Polappu*) moves towards east. One of the expert fishermen told us that the mackerel shoal and seers generally appear on the surface water diametrically opposite to the moon and they can be caught before the moon is vertically over the head or when it sets. Since the introduction of the nylon mackerel net, the launching is done only when the moon shines without obstruction because in the moonlight the *Tuyi* fades away and the fish migrate to the net, which they cannot see.

It is evident from the above analysis that the use of nets and the resultant catches of fish have a bearing on the phases of the moon and tides. The question may arise now as to how the Mappila and the Hindu fishermen of Malabar use their nets and adjust the technology on different lunar days. They employ the sardine boat seine net on the lunar days following the *Karutha Vavu* but only after the moonset. They consider that the first ten lunar days, i.e., from Pradama to Dasami, is very conducive to fishing with the boat seine-nets particularly, with the *Mathikolli Vala*. They expect good catches during this period for two reasons. Firstly, the *Veliyerakkam* occurs when the moon sets in and during the low tides they usually get good catches. Secondly, they have sufficient time at their disposal to carry out their fishing operations with the help of darkness, which is absolutely necessary for using sardine nets. On the contrary, the fishermen cannot use the mackerel and gill-nets on the first three lunar days immediately after the New Moon. The main reason put forward by the Mappilas is that the duration of the moonlight during these days is quite insufficient for conducting the fishing operations. Hence they go out for fishing with the mackerel net from the fourth lunar to the tenth lunar day. They start at 6 p.m. and return to the shore as soon as the moon sets in. On the other hand, the Mappilas go out for fishing with the mackerel net during the first three lunar days following the Full Moon only after the rising of the moon. The Mappila and the Hindu fishermen maintain that the period between the first and tenth lunar days after the *Velutha Vavu* is good for catching fish with gill-nets. After Dasami (10th lunar day) they do not venture to go out for fishing with gill-nets as they do not have sufficient moonlight for conducting their operations. Contrary to this, they undertake fishing expeditions from the fourth lunar to fourteenth day after the New Moon. During this period they go out for fishing as soon as the sun sets and return to the shore with the catches immediately after the setting of the moon.

Thus, we note from the Chapters 1 and 2 that most fishing communities of Kerala follow a life pattern which is moulded by the cultural ecology. The life pattern, belief system and economic organisation of these fishing communities can be seen to be the direct result of the marine environment which they inhabit, and their reaction to it. The repercussion of the marine environment on them is an outstanding example of the intricate relationship that exists between man and environment. A fisherman's ecological knowledge is his main asset and provides the means for his livelihood. Besides, this knowledge also helps him in times of dangerous situations at sea.

The ecological knowledge of these fishing communities with regard to the marine ecology irrespective of their religious affiliations is vast and deep. They are exposed to these environments from their childhood and learn about them both directly and indirectly. They make attempts to rationalise much of this ecological knowledge with religious and supernatural beliefs. Most of these fisherfolk began their occupation in their early teenage. Zachariah (1996) observes: "The fisherfolk are very knowledgeable of about the topography of the sea. They are fully aware about the niches rich in marine life, sports that are dangerous due to strong currents, rocky reefs, unstable and changing seabed, depths, etc. These knowledge have been acquired as mentioned, partly from older and more experienced fishermen and through their own experiences. Most of them, especially the older members, are aware of the areas that are strewn with jutting and sharp edged rocks, and take precautions to lay their nets in the direction of the ocean currents flowing away from such areas. They are aware that these areas have abundant fish and other marine life, but according to them the risks and costs of damaging their expensive nets are not worth the efforts. In earlier times they used to create water turbulence and noise in such areas after setting their nets in order to scare fish out of the rock crevices and push them in the direction of the nets. But this method is resorted to less now. The reasons given for this were vague, but the general one was that the noise of the outboard motor engines of boats frequently scared away fish even before nets could be set. The older generation especially had keen perceptions about the nature and signs of the sea. They were able to tell from the shore the presence of fish shoal at almost the depth of five to six metres even at a distance of half to one kilometre, (according to them upto 3 *Maaru* – 1 *Maaru* = approximately 1.8 metre or 6 feet)".

Zachariah also agrees with our observations regarding different currents and winds which are related to fishing technology. She had said that "when they are no longer able to see the shore from the sea, they are guided by the currents, the sun during the day, and in the night by certain stars, for direction and also time. This is especially so at night when also most of the fishing is done. They knew of eight kinds of currents (*Nir Ozhukku*), based on the directions from which they flow, and corresponding to the eight points of the earth (*ashtadhikka paalakar*). They call the currents *Vadakkan Nir* (north), *Tekkan Nir* (south), *Kara Nir* (east), *Puram Nir* (west), *Vadakara Nir* (north-east), *Vadavartha Nir* (north-west), *Tengara Nir* (south-east) and *Themmara Nir* (south-west). Bovis-Mogeyar fishermen in the different villages have various opinions about these different currents. But they all agreed that if the currents are strong or warm, most fish 'sank' to the bottom, and that it is not always advisable or profitable to launch an expedition then."

According to the Hindu fisherfolk in general and Bovis-Mogeyar in particular, the seabed at Kozhikode and south is deeper, while it is shallow north of Kozhikode to Cannanore and beyond. Apparently the depth of the sea 2 kms offshore at Kozhikode and 3 kms offshore at Cannanore are the same. The seabed around Kasargod is apparently more rocky while the seabed to the south near Quilon and Trivandrum are full of rocks that pose deadly risks to fishing boats. The problem of a rocky seabed around Kasargod sea coast is also aggravated by the presence of foreign trawlers.

3

Ecological Context of Sociocultural Heritage of Hindu Fisherfolk

Fishing communities are found all along the coast and the backwaters that criss-cross Kerala. The coastline of the state is 590 kms long and the state, besides being a major exporter of fish, lands 30 per cent of the country's total catch. The Hindu fishing communities in the state is segmented into several sub-castes, viz., Arayan, Arayavathi, Bovis-Mogeyar, Mogavirar, Mokayan, Mukkuva, Nullayan, Panniakal, Valan, Valinjeyar, Bharathar, Kanakkan-Padannan, Padannan and the Paravan. These sub-castes have regional variations in their names and their own versions about their origin. Some of these fishing castes are also found in Tamil Nadu, especially at Kollengode, Colachel, Manvalakurichi, Eravinthura, Neerodi and Valavila, as these areas were part of the erstwhile Travancore State. The fishing castes of Tamil Nadu have social contacts with the fishermen of the Trivandrum district and have even the same house-names. Likewise, in the north, the Mogavirars and Bovis-Mogeyars have contact with members of their communities in Karnataka, especially at Malpe, Udippi and Mangalore.

The traditional Hindu society is a typical example of a closed system of stratification wherein different components such as caste, class and power combine in the same way (Beteille, 1966). This is the result of a combination of several factors, chief amongst them being the Hindu religious philosophy, technological backwardness, agrarian economy, and the resulting slow rate of social change. The configuration of all these factors has produced direct repercussions on the quantum and rate of social mobility and consequently, on the nature and form of the stratification system. In such a traditional society, where rank was determined on the strength of the caste system, purity and pollution, it was impossible by mere economic activity to promote or improve one's place in the hierarchy. There was, therefore, a barrier against change of rank or status. The Hindu fishing communities follow an unclean and despised occupation (Hutton, 1961). However, they claim superiority over the Brahmins as the twice-born were ordained from amongst the Arayans, whereas they were present right from the beginning, being the descendants of Vyasa, the great Hindu sage. Despite their claim of superiority, caste Hindus consider the fishermen polluting. In the traditional caste hierarchy, they rank below the Kammala (Asari, Thattan, Kollan, Thachan and Musari), Kaniyan and the Thandan. In some places, they are treated on par with the Ezhava or Thiyyas.

The first step necessary to remove the stigma of untouchability requires the discarding of their unclean profession. This is unthinkable on the part of the Hindu fishing communities. Although the younger generation resents the hard toil and stigma attached to fishing and prefers other occupations, the elders think that their traditional calling is honourable. In the caste-ridden hierarchical system of the Indian society, which is known to be one of the most stable in the world, a man is positioned for life through a cluster of ascribed statuses. Caste determines the membership of the community and the social status of each caste in regard to others. Combined with this is the ability of every Indian to accept this ascribed status and thus to resign to his fate. This is the most distinguishing feature of the Indian way of life over the centuries, new groups have arisen and been placed in the caste hierarchy; some groups have improved their rank, others have remained static, and a few others have changed their professions within the ambit of the traditional system. This group mobility has been far more important than individual advancement, which has not been so important until recently.

Srinivas terms this caste mobility as Sanskritisation and Westernisation. Sanskritisation occurs when mobility occurs within the framework of a caste, whereas Westernisation implies mobility outside the framework of a caste (Srinivas, 1968: 8). Both these processes are taking place in most Indian communities today. If a low-caste man wants to improve his socio-economic position, he has to do so by breaking off with his home and kindred. He has to get a new occupation that is not represented in the traditional caste system. If he succeeds in his attempt, he joins a milieu in which representatives of many castes live side by side in a new employment situation and where caste plays a subordinate role. This requires high education and investment of capital for travelling and facing competition. It is however impossible for the Hindu fishing community to do so since the economic condition is so poor that few have succeeded in individual advancement.

Social stratification is global; its specific feature depends on a combination of factors such as the nature of the political system, economic development, social organisation, occupational structure, technological advancement, value system and the demographic composition. All these factors undergo different degrees of change, bringing in their wake changes in the nature and type of combinations, which finally, in their turn, reflect changes in the social stratification. Thus, social stratification can no longer be said to have a purely ritualistic base.

Financial stability combined with political power wipes off the germs of pollution from a low caste. Pollution is as much a ritual barrier as an economic barrier, if not more. It is because of this that all those who have been historically and ritually branded as impure have also been economically backward. But the moment a ritually impure man becomes economically well off or politically powerful, he crosses the barrier of pollution. Thus, one remains ritually impure only as long as he is poor. Nowhere in the world are there people who are economically well off, and politically strong, but are considered ritually impure. Thus, it is pollution that always keeps company with poverty.

Eating together, accepting cooked food, drinking water or marrying somebody's daughter – these types of indices are used to determine the ranking of castes on the basis of purity. However, the pollution scale need not always yield the correct readings (Marriot, 1959: 92; 1968: 133). Social distance in the caste system is not merely ritualistic, rather it is a euphemistic characterisation of poverty in religious terms. The unequal distribution of wealth and power is what gives rise to social stratification. Caste is an ascriptive category, determining the ritual status of an individual or a group; status, in contrast to caste, is used as the socio-economic position of an individual or a group, as measured by the variables of land, income and education. In contrast to power, status is used as the position an individual or a group occupies in the hierarchy of prestige, as determined by his general social standing as well as by the specific style of life in his community. It is worth mentioning that India's great strides in the industrial progress in the 20th century have been made in complete disregard of the caste system, not as a result of it. Unfortunately, the Hindu fishing communities have been, because of the very nature of their occupation, bound to a specific kind of lifestyle. The primitive technology used by them is a major reason for their socio-economic backwardness since technology cannot be divorced from economy and economy from social relation. Technology, economy and society coalesce in a triumvirate reality of human societies.

The Hindu Marine Fisherfolk

Nullayans

The Nullayans, one of the Hindu fishing communities, are found exclusively in the southern part of the Trivandrum district. Although they are mainly marine fishermen concentrated in the villages of Karimam, Poovar and Kaladi, they often fish in the *Kayal* (lake) as well. Their castemen are also found in the Kanyakumari district of Tamil Nadu. For the past 35 years, they have been styling themselves as Arayan and many have also got this caste name entered in documents such as school certificates and service books. At Manavalakurichi, Kollengode, Colachel and Eravinchura, they are called Nullayans. At Pozhiyoor, Karimam and Poonthura both the names, Arayan and Nullayan are used indiscriminately but they are called Nullayan by the other communities despite their claim that they are Arayans. This is further substantiated from verification of their title deeds, in which this community is designated Nullayan. Anthropological investigations have also corroborated the documents.

Origin of Community and Name

Members of the Nullayan fishermen inhabiting the Trivandrum district originally came from villages further south, especially Colachel, Pozhiyoor, Kollengode, Pachallur, and Tengapatanam. Their mother-tongue was originally Tamil though they speak Malayalam fluently. A small trace of Tamil is still present in the language spoken by them. They have however mostly retained their traditional Tamil kinship terms. The origin of the name Nullayan has not been explained satisfactorily, but a couple of elderly informants confessed that they were once untouchables. Despite their claim to superiority now over high-caste Sudras, they were earlier considered a low caste, at par with the Pulayas. The same informants also recounted some instances, which showed a lower status assigned to them long ago. They seem to have adopted the name Arayan to elevate themselves or simply to change their degraded status.

Descent and Inheritance

The Nullayans have four exogamous *illams* or *gotrams* (clans), viz., (1) Panchatharayar or Panchatharasar, (2) Kanyoorarayar or Kanjee-puratharasan, (3) Koodathukiyarayar or Kochandarayan, and (4) Urandarayar. They are matrilineal and follow Marumakkathayam. Most of their rituals relating to the onset of puberty, marriage, birth and death have a good deal of Tamilian influence. They have common names with the Tamils, but this is due to their close proximity to the Tamil-speaking areas. Vannans are their washermen on ritual occasions. They had no *Mattu* system. The headman or his wife had to sprinkle water mixed with cowdung on individuals who suffered from ritual pollution.

Political Organisation

The headman of the Nullayans is called Amavanmar and it is he who decides their caste disputes. The Amavanmar is assisted by the elders of the caste in the village. The office of the headman is not hereditary and he is selected from among the elders. The selection is based on seniority and knowledge of the caste customs, besides his capacity in dealing with caste matters.

Relationship with Other Fishing Communities

The Nullayans do not have much social contact with the Christians and Muslims, which are the other fishing communities found here. Some Ezhavas are also engaged in fishing at Pachallur but the Nullayans consider them inferior in social standing. These Ezhavas are popularly called Pachalli Ezhavas and subsist on fishing. Some Ezhava families were excommunicated for committing adultery with the fishing communities. The Pachalli Ezhavas are the descendants of these excommunicated families. Though Nullayans claim to be Arayans, investigations have revealed that their claims cannot be substantiated and they are below the Arayans in social standing. However, there is free social intercourse with the Arayans, with the exception of marriage. This, in itself, is conclusive proof that the Arayans enjoy an elevated status compared to the Nullayans.

Ezhavas are treated as untouchables by the higher castes and they also do not take food from the Nullayans. The Nullayans address Ezhava as Panicker and the Ezhavas call the Nullayans 'Arayan'. However, both Ezhava and Nullaya women work side by side while making coir-ropes and visit each others, houses as well. This may indicate a status equal to that of the Ezhavas but they claim a higher status. Earlier Vannan, Thandan, and Pulaya castemen were considered lower than the Nullayans and were not allowed to enter their houses and compounds. They confessed that higher castes like the Nairs, Kshatriyas and Namboodiris practised untouchability with them in common with the other fishing communities. They claimed that it was one of their caste woman who helped king Marthanda Varma to cross a river while being pursued by his enemies and that the land near Vellachipara (6 kms south of Trivandrum) was given to her as a reward for this service. Most of the Nullayans of Poonthura had no land whatsoever and their *Karayogam* purchased land, which was distributed among the landless. Though they live so close to the capital city of the state, the lot of the Nullayan is no better than that of the other Hindu fishing communities.

Religious Beliefs

Unlike most of the Hindu fishing communities, the Nullayans do not attach much importance to goddess Bhagavathy or Bhadrakali. They worship most of the deities commonly adored by the Tamilians, like Arumugan, Ganapathi, Vishnu, etc., and celebrate all Tamil festivals. Some of the Nullayans at Poonthura and Veli celebrate Onam and Vishu like other Malayali Hindus. They have their own temples, managed by a committee and celebrate different festivals by collecting subscriptions. This committee is elected annually by the members of the community.

Arayans

Arayans are a sub-division of the Hindu fisherfolk found all along the coastal areas, mainly south of the Nattika Panchayat, Thrissur district, Kerala. Towards the northern coastal districts, the Mukkuvas style themselves as Arayans making it appear that these two names are interchangeable. They are mainly marine fishermen, though some of their members are engaged in fishing in backwaters in Thiruvalla and Kuttanad taluks of the Alleppey district. They are also called *Kadalarayans* in order to make a distinction between Mala Arayans, a hill tribe found in the Kottayam and Idukki districts. Besides fishing, they were engaged in ferrying boats for passengers and commodities. In some places the Arayans were royal ferrymen, as they rowed the boats of the local kings. These rulers conferred *Arayasthanams* or the tittle of Araya headmen on their leaders e.g., Purakatt Arayan, Kayamkulam Arayan, Chembil Arayan, etc. The Arayans were engaged for guarding the coast-line by the kings of Travancore and Cochin in former times.

Origin of Community and Name

Anantha Krishna Iyer (1909: 261) has stated that the Arayans have four distinct appellations – Sankhan, Bharathan, Amukkuva and Mukkuva. The Arayans belong to Sankhan and the Valans to Bharathan, while Amukkuva is a sub-caste of the Arayans performing priestly functions. The Arayans have no idea as to the origin of their name but a few informants hold that Arayan means 'king of the sea'.

There are several stories about the origin of the Arayans and their relationship with the other sub-groups. One story says that Arayans were present from time immemorial. Vyasa, son of Parasu Rama by a Kali Arayati, is claimed to be their progenitor. This is pointed out to substantiate the antiquity of the Arayan community. They say that Brahmins were ordained from amongst themselves to perform priestly functions and study the holy books. They also claim that four Brahmin youths ('Balans' youth) who ate meat were excommunicated from their community and they were the first Valans. For this reason, some Arayans claim that they are superior to the Valans.

Another story about their origin has it that the Mukkuvas were the original sub-caste present at the creation and had three *illams*. A new group was ordained to perform priestly functions, giving rise to a new *illam*. The members of the new group called themselves Arayan (leader), and the name stuck to them. However, some informants said that leaders of the Mukkuvas were called Aravan/Arayan/Arasuraja and they enjoyed a superior status and position. This group was called the Arayans later. The Mukuvas were present in the north and those who migrated to the southern districts, adopted this title. This is the reason why in the south only the Arayans are found, whereas in the north, both Arayans and Mukkuvas, are living side by side. Because of their different social status there is no intermarriage between the Arayans and the Mukkuvas.

One informant had an interesting story to relate as to wherefrom the Arayans came and how they migrated to Kerala. According to him, Minervars and Pandavas were living side by side during the Harappan times along the Indus River (*Sindhunadi*). The Minervars were a fishing community (*min* – fish). When the Harappan civilisation declined with the advent of the Aryans, the Minervars fled southwards, traversing along three routes, one by the sea along the western coast, and settled themselves along the coastal areas. Another group which went along the mountain tracks and settled in these terrains are now called Mala Arayans. The third group travelled through the jungles to the East Coast (Tamil Nadu) where, it is claimed, that they are still known as Minervars. The two groups of Minervars took the name of their conquerors, Aryans.

From these migrants from the north came the first seafarers and marine fishermen who are at present called Arayans. Till their arrival fish was caught from rivers and estuaries by local fishermen. These local fishermen maintained a social distance with the Arayans. It appears that they were probably the Mogeyars in the north and the Paravars/Bharathars in the south. The Arayans are generally fairer in complexion and handsome in features than the indigenous fishing communities. It is said that this is due to their migration from the north. The Arayans and the local fishermen lived with little social intercourse in the past and it is only during recent times that some intermarriages have taken place.

Throughout Kerala the fishing communities are generally called Marakkar although in Trivandrum district the term 'Marakkar' is used to designate Muslim fishermen and Mukkuva for the Christian fishermen. This term is taken to be honorific in the northern districts, but it is maintained by some that it is a general and loose designation for all fishermen. In the southern districts, however, fishermen in general, take strong exception to being called Marakkar. Their contention is that it is a derogatory word used by the upper castes to indicate their lowly profession. Some even demanded a ban on the usage of this designation. In the district of Palakkad all untouchable castes, particularly Ezhavas, call the Tamil Muslims Marakkar, which is in fact, an honorific title.

Sub-division and Inheritance

The Arayans are said to belong to *Naalillakar*, i.e., four *illams*, which are exogamous. They are: (1) *Ponillam/Ponotillam*, (2) *Chembillam/ Chembotillam*, (3) *Kachillam/Kattotillam*, and (4) *Karillam/ Karottillam* (Iyer, 1909: 261). Marriage within the same *illam* was prohibited in the past but no such taboo is enforced today. Younger generations are not even aware of the name of their *illam* and what it means. Some of the elderly people too did not know their *illams*. Succession among the Arayans is from father to son. Previously, daughters had no claim to a share in their father's property but with the enactment of new laws they are able to claim an equal share with their brothers.

Political Organisation

The headman of the Arayan community is called Arajan/Aravan/ Arasuraja. These titles are similar to those among the Valans and the Mukkuvas. They enjoyed special privileges and performed certain duties at their religious festivals. They settled caste disputes and commanded a very influential position.

Relationship with Other Fishing Communities

The Arayans consider themselves superior to the Mukkuvas, the former being *Nalillakar* (four *illams*) and the later *Moonillakar* (three *illams*). They claim superiority over the Valans as they say the Valans suffered social ostracism in the past. On the other hand, the Valans claim a superior social standing over the Arayans, as they claimed that they were closely associated with certain Namboodiri families. This was stated by our informants in the district of Ernakulam. The general consensus was that they were royal ferrymen and both the Arayans and the Valans were at different times employed as such. Taking into consideration the roles played by the Arayans and the Valans in historical events and their relationship with other caste groups, it may be asserted they enjoyed an equal status. Moreover, many of the customs and rites of both these sub-groups are similar in nature, indicating, in fact, that they belonged to the same group in the past. But the closer contact of the Valans with royal and upper caste families has given them a superior status compared to the Arayans.

The Arayans are found mainly in the southern districts and they do not have much contact with the other Hindu fishing communities in the north, like the Mogeyars and Mogavirars. The Nullayans in the south rank below the Arayans as they are migrants from Tamil Nadu. Migrants are always relegated to a lower status everywhere. The Arayans consider the Muslim and Christian fishermen inferior to themselves purely on account of their religion, but they work cohesively.

The Arayans, especially the more well-off ones receive the services of washermen and barbers called by different names in various areas such as Valinjiar and Panniakkal (barbers) in the Thrissur district, Paravar and Cherupanar in the Alleppey district, Velan, Mannan and Pipanmar (washermen) in the Quilon and Ernakulam districts. The washerwomen perform certain ritual duties during birth, menstruation and death ceremonies, for which they enjoy certain rights and privileges during these periods and at festival times. They also used to supply *Mattu* during the periods of pollution resulting from menstruation, birth and death. It is interesting to point out that the Ezhavas and the Hindu fishing communities have a common washerman community. The pollution from birth and menses vary from place to place. Obsequies for the most part are performed on the eleventh and sixteenth days. The dead are mostly buried though the richer people prefer cremation. A grand feast is given on the eleventh or sixteenth day after the death, when dishes relished by deceased are served. Fishing is suspended when a member of the caste dies.

Relationship with Other Castes

The social status of the Arayans vary from place to place. In places like Punnapra, and Ellamkunnapuzha they are treated at par with the local Ezhavas by the high-caste Hindus. Both these communities are treated in the same way by the upper castes in these places. During the sociocultural solidarity movements among the Ezhavas under the leadership of Sree Narayana Guru and Dr. Palpu, the Arayans lent them their moral support, especially at Punnapra.

The Ezhavas, though treated and considered at par with the Hindu fishing communities at Ellamkunnapuzha, consider themselves superior, as they seldom admit the Arayans, Valans and Arayavathis to their houses or draw water from their wells.

Both the Arayans and the Ezhavas (Chovans) in the Thrissur district contest for superiority on the basis of their occupation. The Arayans claim that it was they who brought the tappers from Ceylon (Izham), to cultivate toddy and coconut palms and draw the juice, the favourite brew of the fishermen. The Ezhavas, on the other hand, claim that a king of Ceylon sent them to Kerala to protect the interests of the fishermen at the latter's request. As far as could be ascertained the Kammalas, Kaniyans and Thandans are higher than the Arayans. The Arayans consider the Pulayan/Cherumakkal below them and eschew any social relationship with them. The Pulayans are not allowed to enter their houses, temples and the fishing community never take food and water from them or from any of the Scheduled Castes with the exception of the Thandans. Denial of services by washermen and barbers to a caste is the major criterion for the inclusion in the list of Scheduled Castes. The Thandans, Ezhavas and the Hindu fishing communities receive the services of a common washerman caste but they are served by their own barber castes.

Despite the important role the Arayans played in the history of the Cranganore (Kodungallur) Bhagavathy temple, they are generally not allowed to enter its precincts as they are considered untouchables by the higher castes. If they have any offerings to make, they have to place them on a banana leaf at a specific place far away from the temple, move back and invite the attention of the temple servants, who, after ascertaining that they were far off enough, fetch them. The *Prasadam* is rolled into leaves and thrown to such devotees. However, the Hindu fishing communities are allowed to enter the temple during the Bharani festival and an Arayan headman has certain privileges during this time. At the same time, the Ezhavas were never allowed to enter this temple. The Kodungallur Bhagavathy temple was the only one place and Bharani, the only festival, at which the Hindu fishing communities are allowed into its precincts.

Iyer (1909, I: 261) wrote that the Arayans "were, in former times, considered as an inferior race, and, as such, precluded from travelling along public roads, and consequently obliged to keep to the sea-coast". When they travelled on public roads, they had to shout aloud to acknowledge their presence and maintain a distance from the higher castes. Their movements were restricted to the coastal area, where members of other castes had no occasion to go. Besides, the stink and filth associated with fishing kept them away. It was also due to this factor that they were considered inferior and kept far away.

Many of these caste pollution and segregation have died out greatly as well as the practice of untouchability, excepting in certain pockets of fishing villages. For instance, at Trikkunnapuzha and Pannathura the Hindu fishing communities were not allowed to draw water from the wells belonging to the upper castes.

Religious Beliefs

The Arayans have special regard for the goddess Bhagavathy/ Bhadrakali but they do not neglect to worship all other Hindu deities. It is said that the Bhagavathy temple at Kodugallur owes much of its present-day importance, rituals and legends to the Hindu fishing communities. This temple, perhaps, originally belonged to the Buddhists and Jains who advocated the principles of non-violence.

There are stories to the effect that the local rulers employed Hindu fishermen to drive away the Buddhists and Jains from

Kodungallur. The fishermen used their oars as weapons, and conducted sacrifice of goats and fowls, making the temple premises bloody. Intoxicated fishermen hurled abusive language and flung dirt on the temple walls and thus desecrated it. These could not be tolerated by the Buddhists and Jains and they fled away. The practice of singing obscene songs and offering bloody sacrifices which started as a means to drive away the Buddhists and Jains, were continued till recent times in the same way as they were held acceptable to the deity installed there.

The ritual defilement of the temple is conducted on the festival of Bharani, which falls in the month of *Meenam* in the Malayalam calendar. This festival is considered very important and people from all over the state, including the Hindu fishing communities celebrate it with much pomp and fanfare. The sacrifice of goats and fowls in the temple is prohibited now.

There are two idols at the Bhagavathy temple of Kodugallur, which are said to have been caught in a fisherman's net from the sea. The local ruler installed these idols in the temple, one facing the north and the other to the west, the north being the direction from where the fishermen originated and the west being from where they earn their livelihood.

The rope that is used to hoist the temple banner for the Bharani festival at Kodungallur is provided by an Arayan headman of a particular *Taravad* living there. He has a privilege over the other Arayans to enter the temple. Despite the fishing communities' important role in bringing the temple to its present status, it was only during this festival that they were allowed to enter it.

The Arayans and the Valans claim that the *Aravana Payasam,* which is distributed as *prasadam* at the shrine of Ayyapa at Sabarimala, was originally made for them. They used to convey the Raja of Pandalam in a palanquin to the temple over the mountain terrain trekking all the way, for which they were given the reward of this special *Payasam* in which a lot of jaggery was added. The other devotees were given just *Palpayasam* and *Panchamritham*. Now, every one is given the *Aravanapayasam*, they say. Besides this temple they offer worship at all other important shrines of the Hindus.

Mukkuva

Mukkuvas are found all along the coastline north of the Nattika Panchayat in the Thrissur district. They are generally marine (*Kadalkar*) fisherfolk, but at Thoyakav and Pavarti (in Marthayoor and Peringad wards respectively) they take to backwater *(Kayal)* fishing along with the Valans there. The Mukkuvas have fishing for their main occupation, but they also ply ferry boats. The Mukkuva women, unlike their Arayan counterparts, participate quite actively in the fish trade. They salt fish and sell it in the interior areas from door to door. In the northern parts like Kasargod, they work together with the Mogeyar fisher-women.

Origin of the Community and Name

Iyer (1909, I: 266) has stated that the word Mukkuva is connected with the Canarese Moger, which means 'to dive'. Many of our informants confirmed that Mukkuva means 'one who dives'. Some others maintained that the word comes from *Muthu* (pearl); 'Muthuvan' was corrupted into Mukkuva. In this context *Muthu* (pearl) means fish. Tradition and literature seem to point out that the Mukkuvas and the Thiyyas migrated from Ceylon but this contradicts their Canarese origin.

Buchanan (1807) states that "the Mucua, or in the plural Mucuar, are a tribe who live near the sea coast of Malayalam (to the inland parts of which they seldom go and beyond its limits any way they rarely venture)". The *Gazetteer of Malabar* (1908) describes the Mukkuvas as a caste of fishermen following *Marumakkathayam* (inheritance through the maternal line) in the north and *Makkathayam* (inheritance through the paternal line) in the south. This is confirmed by the present study. In the Thrissur district they are patrilineal, whereas in the Kozhikode and Cannanore districts they are matrilineal.

The Mukkuvas are said to be *Moonillakkar* or belonging to three *illams*, but they have all the four *illams*, which the Arayans also possess. This is especially so in the northern parts. In the southern parts they have three *illams*, namely, *Ponodillam, Chembodillam* and *Marodillam*. The northern Mukkuvas have *Kachodillam* as well. There seems to be a close relationship between the number of *illams* and the law of inheritance. Where the inheritance is matrilineal there are four *illams* and where it is patrilineal, there are only three *illams*. How this kinship relationship has come about would make an interesting study. The four *illam* people are considered to be higher in status than those belonging to the three *illams* and in the past there was no intermarriage between them. Iyer (1909: 267) had made a reference to this aspect. Marriage within the *illam* was not allowed, though this restriction is breaking down now. Children belong to their mother's *illam*.

The Mukkuvas had the custom of brideprice in which the groom paid the bride's father a fixed amount of 26–3/4 *Panam (kanapanam* or *kanamkodukal*). Brideprice is no longer given under extraneous influences and the table has been turned. Now, the bride's father has to pay a dowry, provide ornaments and meet the marriage expenses. An average wedding expenditure comes to above Rs 8,000. When a marriage contract is broken a compensation between Rs 90 and Rs 100 is given to the aggrieved party and this is called *penninte ajaram neekal.*

Traditionally, The Mukkuva women observed seven days of pollution during their monthly periods. During the first confinement, pollution was observed from 41 to 90 days, and for subsequent deliveries, 28 days. A Thiyya woman or a Velathi acted as the midwife during the delivery. Pollutions at these times were removed only with the supply of a *Mattu* and sprinkling of water by the women of the Mannan caste. The Velan community performed this service in the southern parts, especially in the Thrissur district. Aduthavan (female-Aduthaval) was their priest and barber and they had specific roles during the birth and death. Earlier, the dead were buried while the better off were cremated. The place for burial is called *thambram.* Death pollution is observed for 11 days. During special occasions the members of the servicing castes were given gifts of clothes and money. A measure of rice or one *Panam* used to be given for their services. Pollution is no longer observed and the social barriers are being broken down. Earlier, there was no intermarriage between the Mukkuva on the one hand and the barber and washing communities on the other, but nowadays many such marriages are taking place.

Political Organisation

The headman of the Mukkuva is called Aravan, Arasu or Arajan, the same being the case with the Arayans and the Valans. In some places (Cannanore district) they are called *Karanavar*. At times the headman performed priestly duties as well. Earlier their powers were tremendous and their word final. They presided over the caste council and pronounced the final judgement. Their prior permission had to be obtained for holding any social ceremony. The caste council had various names like Kadalkodi, Karayogam, etc. Those who violated the rules of the caste were excommunicated and many of those who suffered this penalty embraced Islam or Christianity. At times they were taken back into the community on paying a fine. The headmen were entitled to special privileges and payments during different functions and festivals. The position of the headman was hereditary but now he is elected on the basis of seniority and consensus of the community members.

Relationship with Other Fishing Castes

The Mukkuvas occupy a rank lower than that of the Valans and the Arayans, and even below the Mogeyars and the Mogavirars. They were treated at par with the barber community, Aduthavans, Arayavathis, etc. Nevertheless, the Mukkuvas consider themselves higher than the latter and there were no intermarriages with them.

Though there was no touch pollution between the Thiyyas and the Mukkuvas, the former were considered superior. The most common reason given was that they had eight *illams*, the criteria being that more the *illam*, the more superior the community. Though the Thiyyas and the Mukkuvas interdined with one another, accepted food and water from each other, the Thiyyas never drew water from the wells of the Mukkuvas. The Pulayas and the Parayans had to keep a distance of 10 feet from them and were not allowed to draw water from their wells. The members of these communities called the Mukkuvas *Thamburan* and their women *Thamburati/Amma*. Mannans, Velans and Valinjiars are the washermen castes who supplied the *Mattu* to the Mukkuvas. Their barbers and priests belong to the Aduthavans, Kavuthiyans, Khadeer and Panniakkals.

Iyer (1909: 275) wrote that the Mukkuvans "occupied a low state in the estimation of the high caste men and were precluded from passing along the public roads. Being obliged to keep to the coast, and unable to bear the social disabilities, many became Christians and converts to Islam. They had to stand 32 feet away from Brahmins, Kshatriyas and high caste Sudras" (Ezhavas too had to stand 32 feet away from Brahmins).

During the British regime, the Mukkuva children were not allowed to attend schools where high caste boys read. Moreover, these schools were located too far away. To overcome these social and physical restrictions, the British government built numerous fishery schools all along the Malabar Coast. People were especially employed to recruit children to these schools and, therefore, a number of them were able to get some basic education in those days. Till Independence, these schools provided free food, books and clothes; they were well managed and well-attended too. The post-independence days saw a drastic drop in their attendance, when free aids were withdrawn. Those who had managed to get education and other employments were able to break the caste barriers and educated their children also. There are, therefore, among this community some better off individuals. This is an important pointer to the urgent need of imparting a sound education and providing an alternate occupation to help this community come out from its present-day poverty.

Religious Beliefs

The religious beliefs of the Mukkuvas are similar to those of the Arayans and the Valans. Bhagavathy is their patron goddess. The Bharani festival at the Kodungallur temple is very important to the Mukkuvas and devotees throng to this shrine from every fishing village carrying their votive offerings, with the chant of the goddess's eulogies. During this festival, like the Arayans and the Valans, traditionally they too had a preference over the Ezhavas to participate in the ceremonies. This was probably due to the historic role played by the Hindu fishing communities in making the temple what it is today. In the past, the Mukkuvas, were not allowed to enter temples and had to stand at a distance. Instances have been narrated of cruel punishments meted out to those who violated this social distance.

The Mukkuva fisherfolk also offer worship at the temples of Sabarimala, Guruvayoor and Ponnani. They observe almost all the festivals popularly celebrated by the caste Hindus. They worship most of the Hindu deities, besides goddesses like Gouriamma, Kiriyaniamma, Bhadrakali, Durgadevi and Gopaliamma. Their ancestors are also worshipped with feasts on certain days of the year. They have their own oracle (*Velicapad* or *Komaram* plural: *Komarangal*). The first days of all Malayalam months are special days, some more than the others. The Mukkuvas do not eat beef and pork. On rare occasions they eat buffalo meat. Fish is a must for them and is invariably eaten on all important occasions.

Bovis-Mogeyar

The Bovis-Mogeyars inhabit mainly the northern extremes of Kerala along the coastal fishing villages of Trikaripur in Kasargod district. They occupy the highest social status among the Hindu fisherfolk, followed by the Mogavirar. The Bovis-Mogeyar have ten *illams*, while the Mogavirar have only eight. The Mogavirar accept the superoior status of the Bovis-Mogeyar. They speak a dialect of Malayalam in which Kanada and Konkani words predominate. They have a temple at Kadangodu, dedicated to Bhagavathy. Perhaps this has led the fisherfolk at Madappally and Edakad to identify them with the Kadangkodu Bovis-Mogeyars. The fisherfolk of Kadangodu on the other hand, do not seem to be aware about the Hindu fishing communities of Madappally and Edakkad and say that these places do not come under any of their eleven *sthanams* (original settlements).

The Mogeyars are marine fishermen everywhere except at Kadangodu and Trikaripur where a few of them fish in the rivers. Their womenfolk are actively involved in the fish trade, much more so than their counterparts among any other community. They buy fish from the boat owners, take the catch from their menfolk and sell them in the interior villages. They are a real economic asset to their families and greatly augment the meagre family income.

Origin of the Community and Name

This community is called Bovis-Moyer or Mogavar in Kasargod and places north of it. They are also called Malayala Moyer/Malamoyar to differentiate them from the Tulu Moyer. They are believed to have migrated from Mangalore and its surrounding areas. The word "Mogeyar" is said to mean a diver; Bovis, on the other hand, had its origin from the word "Boyi" which means bearer. It is said that they were ferrymen and palanquin-bearers in the past.

The Mogeyars seem to be a sub-division of the fishing community of Karnataka called Bestha, and were called Boyas when Mysore was part of the Madras Presidency. The Mysore Census Report of 1901 mentions a caste called Besthas, who are "fishermen, boatmen and palanquin-bearers, and who are known by different names according to the localities they live. The Telugu-speaking population call themselves Boyi – their chief occupation being fishing, palanquin-bearing and lime-burning". Buchanan (1808) mentioned that palanquin-bearers were of Telinga descent and in their own dialect called themselves Bai, and that they did fishing besides carrying palanquins. He also stated that Mogers were fishermen, were a caste of Tuluva origin. According to Thurston, they resembled the "Mucuas" (Mukkuvas) of Malayalam but had no communion with each other. The Mogeyars are boatmen, fishermen, porters and palanquin-bearers. Some Mogers abandoned their hereditary profession of fishing and took to agriculture, oil-pressing and playing musical instruments. Some were employed as palanquin-bearers. The oil-pressers called themselves Ganigas, the musicians Sappaligas and the palanquin-bearers Bovis. These are all occupational names. Some Bestha immigrants from Mysore had settled in Pattur Taluk and were also known as Bovis. The word Bovi is a form of Telugu Boyi (bearer) (Thurston, 1909: Vol. V, p. 67).

The Bovis-Mogeyars are mostly dark complexioned, tall and angular, whereas a number of Mokayans are lighter-skinned, light-eyed and copperish-haired. The latter's blondism is attributed to the admixture of blood in the colonial days. The Bovis-Mogeyars were strictly endogamous and until recently were not very communicative even with other fishing castes. Those who married from outside their community were excommunicated along with their family members. Of late, such extreme punishments are seldom inflicted.

The Bovis-Mogeyars speak Malayalam generally; at Kasargod, Uppala, Kunjathur, etc., in the extreme north, their language has an admixture of Kannada as well. Their womenfolk pierce their nose, and wear nose-rings, and also claim that their attire of *Sari* is more decent than Mukkuva women's *mundu*. Their Saris, especially those of the middle-aged and elderly, are often bluish in colour with silver borders. The men wear a special head-dress made of areca-palm flower spathe.

Descent and Inheritance

The Bovis-Mogeyars claim that they have 10 *illams* and 11 *sthanams* and are, therefore, also called *Pathillakar* and *Pathinonnusthanakar*. Each *illam* is an exogamous clan. According to the Bovis-Mogeyar fishermen of Kadangod, there are 44 *kulams* and these *kulams* are probably their lineages. This has to be verified from further in-depth studies. Members of the same or related *illams* cannot intermarry. Children belong to their mothers' *illam*. Cross-cousin marriages, either patrilinear or matrilinear, are preferred. The Bovis-Mogeyars, to a great extent, even today look up for the *illams* before marriage. They, more than any other sub-groups of the fishing communities, still maintain and adhere to their traditional customs and rituals. They are also very particular about pollution resulting from birth, death and menstruation, and observe the connected rites with great care. Birth pollution lasts for 28 days and the *Mattu* is supplied for five days on the 4th, 9th, 16th, 19th and the 29th day. The menstrual pollution lasts for three days. The pollution resulting from death is removed by purification ceremonies on the third and the seventh days. The *Mattu* is supplied by the washer-women who also purify the house and compound by sprinkling water. The Vannathis were supposed to wash for them. Valinjair and Korakadavamar were their traditional barbers.

Political Organisation

The headman of a Bovis-Mogeyar village is called Karanavar or Achan. At Kunjathur and Kasargod they are called Gurikallar or Gurikammar. They have their own *Velichapads* (oracles) who help the Karanavar to run their caste affairs smoothly. The life in their village revolves around its temple and its activities. All their activities in connection with their livelihood and caste are controlled by the temple-keepers, namely the *Karanavar* and the *Velichapad*. Like the other fishing communities, the Bovis-Mogeyars are deeply religious and steeped in superstitions. This is due to the precarious life and the risks involved in their occupation. The office-bearers of the temple are elected or appointed on the basis of their wisdom, knowledge and efficiency. The Bovis-Mogeyar Association provides help to educate the bright and promising children in good schools and meeting their expenses for accommodation in places like Bombay, Mangalore, etc. This is because of the paucity of such facilities in the coastal areas. The Association also advances loans for those who have secured jobs abroad. Because of the self-help, this caste is definitely better off than the other Hindu fishing communities.

Relationship with Other Fishing Castes

The Bovis-Mogeyars kept aloof, but despite this relative isolation, they among all the Hindu fishing communities, studied and occupied the highest social position. This is due to their having the maximum number of *illams* i.e. ten, while the rest of the communities have less than this number. They never took food from the Thiyyas, Mukkuvas and the Mogavirars but their food was acceptable to these people. At the same time, it has also been reported from Kunjathur and Uppala that their members carried the palanquins of the Nairs till as late as 1950. They were not allowed to wear any upper garments or foot-wear, to roll their *Mundu*, tie a head cloth, enter temples and go near higher castes. They had also to bow low and cover their faces whenever they encountered high caste members. According to some informants they were not allowed to walk along public roads nor cross the fields of a Nair or a Namboodiri. In Kodangodu, Nilewswaram and Chervathur they were not allowed to enter the compounds of the houses of Nairs and Namboodiris. Despite such restrictions to which other fishing communities were also subjected, the Bovis-Mogeyars had a higher status than the rest of their brethren. They were served by the washing communities of Vannan and Valanjiar but had their own barbers (Aduthavan). They neither allowed the Cherumans and Kanakkans to enter their houses nor draw water from their wells. The Kaniyans (astrologers) were consulted for finding out the causes of the occurrences of calamities, for selection of sites for construction of temples, houses, wells, and for fixing auspicious days. They do not seem to have suffered much indignities at the hands of the upper castes as did the other fishing communities because of their sturdy independence and solidarity.

Religious Beliefs

The Bovis-Mogeyars have 11 *sthanams,* in consequence of which they are also called *Pathinonnusthanakar*. Of these 11 *sthanams,* where the shrines dedicated to goddess Bhagavathy are located, seven are north of Kasargod and four south of it. The seven temples in the north are located at Adhur, Kumbla (Mora), Omanjur (Uppala), Kerala, Udayavaram (Kunjathur), Unhillam, and Vittila. The temples in the south are located at Kadangodu (Nellikal), Trikkaripur (Koyangara), Nileswaram and Manikoth (Kayangad). Uchillam is in the state of Karnataka. These temples or *sthanams* have their own groups of villages affiliated to them.

The Bovis-Mogeyars are deeply religious and the slightest indignation is sufficient to start collection of funds to construct a shrine to appease the displeased gods or goddesses. Among these temples the most important ones are located at Kadangodu, Nileswaram, Adhur and Uchillam. The Bovis-Mogeyars of Kunjathur believe that the goddess Bhadrakali came from the south to Manjeswaram with the intention of settling there, but noticing the strange language spoken (Tulu) there and different customs observed (those of Mogavirar), she returned home. On her way back she built a temple at Kotikulam, followed by another at Kalandad and finally settled down at Kasargod. It is true that Kasargod has the highest concentration of Hindu fishermen in Kerala, estimated to be over 5000 individuals. The goddess is supposed to have come from Nileswaram where her largest temple is situated. In Adhur (28 kms from Kasargod), there is a temple belonging to the Brahmins, where the *Ulsavam* (festival) cannot be held prior to the one at the Bovis temple at the same place. It is believed that if the *Ulsavam* at the temple of the Brahmins is held before that of the Bovis without obtaining the concurrence of their headman (Gurikkalar/Gurikara), that temple would catch fire and misfortune would befall on the Brahmin community. Once the *Ulsavam* at the temple of the Brahmins was conducted contrary to this injunction and it caught fire. Many Brahmins died in this accident. This year (1982) the priest of the Brahmin temple came to Kasargod to request the Bovis' headman to hold his *Ulsavam* early so that the Brahmins could hold theirs. This is the only place where the Brahmins are obliged to a lower community to hold their religious festival. This seems to be a very important point.

The Bovis-Mogeyars have their own caste association with its headquarters at Kadangodu. This association has its sister organisations in Karnataka and Bombay. They are highly organised and every year each male member has to contribute to their respective temples Rs 50 and ladies Rs 25. An additional amount Rs of 100 to 150 per family is levied for the maintenance of the temples. The defaulters are not allowed to go to the sea, women are not allowed to buy fish from the fishermen and traders not allowed to sell their goods. The defaulters are excommunicated by the other members and therefore it is very rare that such lapses occur. The temple authorities inform everyone concerned about defaulters. The members do not, however, need any compulsion to pay their subscriptions in time as they know that their temples cannot be maintained without their contributions. Besides these payments, they also contribute generously for the temple festivals.

The main festival at the temple is *Purolsavam* celebrated in the month of Meenam according to the Malayalam calendar. It is also called the *Nadavalli Ulsavam.* The Bovis-Mogeyars also celebrate Vishu, Deepavali and *Thullupathu* on a grand scale. They do not celebrate Onam. The first day of every Malayalam month (*Shankranthi dinam*) is also celebrated, but on a smaller scale. The festivals are celebrated in common by the Bovis-Mogeyar, the Mogavirars and the Hindu communities of Karnataka.

Mogavirar

The Mogavirar are found only in the northern extreme of the Cannanore district. Their total population is very small in Kerala. There are twenty five houses at Uppala, eighteen in Bangara, Manjeswaram, and seven in Hosbetto, the total coming to about eight hundred in all. Many of their members (estimated to be around five hundred), are living across the border in Karnataka with their relatives. Uppala is the southern extremity where this fishing community is to be found. Their castemen are also found at Ullalam, Mangalore, Malpe and Uchillam and the community in both states is a single cohesive social group. They speak Tulu and are, therefore, known as Tulu Moyer, to differentiate them from Malayalam Moyer or Malamoyer, whom we have designated Bovis-Mogeyars.

The Mogavirar settled in these places in the remote past. They are marine fishermen and their members in Kerala employ traditional fishing techniques and crafts. In Karnataka most of them own motor boats and purse-seine boats; some of the caste members living in Kerala go to Karnataka to work on these units. A couple of Kerala Mogavirars are partners of purse-seine boats which are stationed and registered at Malpe in Karnataka. The Fisheries Department of Karnataka has extended financial help to the members of this caste and they have undertaken numerous projects to improve their fishing technology. The vigorous and steady implementation of these programmes have benefited a large number of the Mogavirars here. Stringent measures were taken to give preference to genuine and traditional fishermen. Whatever the reasons, it is nothing short of a wonder that vast improvements have been brought about in this field and consequently in the lives of the Mogavirar fishing community here.

Origin of the Community and Name

Usha Rao (1981: 75), in her book, *Deprived Castes of India*, mentions that the Mogers of Karnataka call themselves Mogaveeraru in Kannada and Marakkalar in Tulu. The Mogavirars in Kerala have no idea whatever about a fishing community called Moger in Karnataka. It is probable that the Mogavirars in Karnataka have styled themselves Moger, which is a Scheduled Caste there, in order to grab benefits earmarked for them. Marakkalar is a general term applied to all fishermen in Karnataka just as Marakkar is the term used in some parts of southern Kerala to designate certain Hindu fishing communities. The Mogavirars are also called Marakadu, another general term applied to fishermen in Karnataka.

Descent and Inheritance

The Mogavirars have eight *illams* or *bhari* and are, therefore, also called *ettillakkar* or eight-*illamed* people. These *illams* are exogamous and children belong to their mother's *illam*. The descent is reckoned through the female line. The eight *illams* have a number of lineages but this aspect could not be studied. Cross-cousin marriages of both patrilinear and matrilinear kinds are popular. Instances of a few polygamous marriages have been noticed.

Menstrual pollution is observed for four days, birth pollution for 41 days and death pollution for 12 days. On these days and also on the fourth and thirteenth day after birth, their washer-women perform certain rites to free them of their pollution. They have their own washermen community called Madelar/Madiwala. At some places, as at Bangara-Manjeswaram, Tulu Banathis, who are the washer-women of the Billavas and the Thiyyas, perform this purification ceremony for the Mogavirars.

Political Organisation

The Mogavirars call their headman Gurikaramar. They are responsible for the upkeep of their temples and act as their custodians. Wherever the Mogavirars are settled they have a temple there and a temple committee. Caste disputes are settled within the village by the Gurikaramar. If disputes are not settled at the primary level, they are taken to their regional council of elders, who represent fourteen villages. This council is called the *Pathinaluvashpam,* and it meets once in every year at one of these fourteen villages.

Relationship with Other Fishing Castes

Traditionally the Mogavirars of Kerala had no contact with other Hindu fishing communities except the Bovis-Mogeyars. They acknowledge the superior status of the Mogeyar/Bovis (Malamoyer) as the latter have ten *illams.* The Mogavirars claim that in Karnataka they are numerically strong and that they have a superior social status among all other fishing communities with the exception of the Bovis-Mogeyars.

The Mogavirars and Billavas, who are equivalent to the Thiyyas, stand on the same social level. Both accept food and water from each other and were allowed to enter each other's houses even in former days. In Kerala, both of them have a common washerman community, namely, Bannathi. Besides, they both have eight *illams.* There is apparently no difference in the social intercourse between these two communities. On the other hand, though Bants and Brahmins everywhere never allowed Billavas to enter their houses and compounds, the Mogavirars were in some places allowed free access to their houses. In some other places, the Mogavirars were not allowed to wear shirts, footwear, headcloth and wear their *dhoti* below the knees just like other low castes. This was, however, strange and was reported only from the village of Hosbetto. This could have been the result of the attitude of the higher castes in general in this part of the state to their occupation. Otherwise, our general conclusion would substantiate Usha Rao's (1981: 75) findings that the Mogavirars were never regarded as untouchables nor were they subjected to such maltreatment. Besides they also deny that they are low castes though they do not claim to be equal to the Bants.

The majority of the Mogavirars in Kerala are very poor and also suffer from educational backwardness. There are some who have taken degrees but they were all educated in Karnataka. A few of them are employed there, but none in Kerala. There are ten Mogavirar graduates who have followed their traditional calling of fishing for want of better employment. Though they never suffered any indignities socially at the hands of higher castes, they are poor on account of their uncertain occupation and isolation from other communities. The Fisheries Department in Karnataka has improved the lot of this community. The affluent members, in turn, have been able to help their poor relatives to a certain extent across the border. This has helped to improve the socio-economic level of the Mogavirars in Kerala, in comparison to the lot of the other Hindu fishing castes. Nevertheless, they have yet to achieve a reasonable level of education.

Religious Beliefs

Unlike most other Hindu fishing castes, the patron goddess of the Mogavirars is called Malaraya Devi. This may be yet another name for Bhagavathy or Bhadrakali since their most important temple is the one at Uchillam (between Mangalore and Udipi in Karnataka State) where the presiding goddess is Bhagavathy.

All Mogavirar villages have a temple and a *Sabha* (committee). At Uppala and Bangera-Manjeswaram these are called the Mogavira Sabha Bajana Mandiram. The patron goddess is Sharada Devi in both the places. Besides collecting subscriptions from their members for the upkeep of their temples, the Mogavirars have a workshop at Bangera-Manjeswaram where they make and repair nets for other fishermen, especially Muslims. This is a very active centre during the lean months and the earnings are set apart for their temple. Their *Sabha* gives financial help to the members in times of need.

The most important festival of the Mogavirars is the *Malaraya Ulsaga Jatra,* held in March coinciding with the Bharani festival at Kodugallur. Deepavali, Vishu, Ashtami, and Ganesh Ulsaga Chowthi are their other important festivals. The *Shankaranthi dinam* (the first day of every Malayalam month) is celebrated to a limited extent. They do not celebrate Onam just as the Bovis-Mogeyars.

The Hindu Backwater Fisherfolk

Valans

The Valans are Hindu backwater fishermen found in the Cannanore, Thrissur and Ernakulam districts. A few of them are also found in the Kottayam and Alleppey districts. The Valans inhabiting the Cannanore district are all immigrants from the erstwhile Cochin state. They inhabit the taluks of Cannanore and Taliparamba. The Valans say that they were first taken by Hannabi Kotta Mohamed to Cannanore to fish for him. Another version is that the Valans, being expert boatmen, were brought to Cannanore to defend the coastline against foreign invaders and finally settled down there. They still keep contacts with their kinsmen at Vypin, Ernakulam and Kodugallur. They have similar house-names in the north. Marriages are still contracted between members from both the areas. The Valans are settled on the river-banks, estuaries, and backwaters, within easy reach of the water bodies. Recently, some of them have ventured to go out to sea on mechanised vessels. They have specialised in the use of stake nets and are adept in Chinese nets. Their women-folk in Cannanore do not participate in the fish trade, but at Njarakkal, Ellamkunnapuzha and Methala their pinching poverty forces them to sell their menfolks' catch by themselves.

Origin of the Community and Name

There are different stories as to how the name Valan originated. One is that it is derived from the word Vala (net) or Valuvala (tail net – a kind of net). Another version is that the Valans were originally Brahmins. Once four Brahmin youths were playing by the side of a river and they caught a fish with the help of their cloth. The catch was put on the banks. Some Arayans, who were passing by the banks, asked them for the fish. The youths gave them the fish but they were curious to know what the Arayans did with the fish. So, they followed them and saw them frying the fish and eating it. The youths were surprised, but the Arayans told them that it was delicious. The Brahmin youths were persuaded to eat the fish. Their Brahmin priest, on hearing that four youths had thus transgressed their religion, excommunicated them and settled the boys (Balans) in four different places and hence their four *illams*. These Balans, in due course, became Valans. Yet another version has it that the Cochin rulers of yore wanted some strong Arayan youths (Balans) to ferry their boats and they were, in consequence called (Valans). They also claim to be descendants of Vyasa just like the Arayans and in all probability belong to the same original stock.

Descent and Inheritance

The Valans have four exogamous *illams* (clans) and their names vary from one place to another. The Valans of Njarakkal say that their *illams* are Puthenpelikara, Nanappalli, Kattikunnu and Chenthi. The Valans of Fort Cochin, on the other hand, say that these are the *sthanaperu* of their progenitors, who were settled by the Brahmins and that the names of their *illams* are Ennalu, Allayinkadukiriam, Vaishyakakiriam and Vashapallikiriam. Some other informants hold that their *illams* or *moolakudumbaperu* are Mannarkad, Pambukad, Ameda and Vellarvattam. Iyer (1909) has mentioned the names of the four *illams* of Valans as Alayad, Ennalu, Vaishyakiriyam, and Vazhapally. The Valans at Pazhayangadi could only name two of their *illams*, i.e., Vaizhyakiriam and Nullikutti. All these people are Valans. But how these different names of the *illams* originated could not be ascertained. They follow the patrilineal order of descent.

Political Organisation

The social and political organisation of the Valans is similar to that of the Arayans. Their headman is called Arajan, Aravan, etc. The headmen had special privileges and duties to perform at different occasions and festivals. They wielded considerable power and their word was final in all caste disputes. The office of the headman was hereditary but now it is conferred on individuals on the basis of seniority and efficiency.

Relationship with Other Fishing Castes

The Valans everywhere claimed to enjoy a higher status than the Arayans. They were royal ferrymen and were closely associated with the royal and Namboodiri families. However their association with royal families was reported only by the Valans of the Cochin area. The Arayans and the Valans did not intermarry in the past but nowadays no such restrictions exist.

In the district of Cannanore the Valans are a minority community. They rank below the Mogeyars found here, which is evident from certain caste practices and rituals. As the Mogeyars have ten *illams* and the Valans only four, the former claim a higher social status in the caste hierarchy of the Hindu fishing communities. On this basis the Mogavirars who have eight *illams* would also rank above the Valans, but these two groups have no social contact. The Valans in the south had the services of the washermen who supplied them with *Mattu*. The *Mattu* is very important for the Mogeyars and the Mogavirars but this is absent among the immigrant Valans for whom none of the local washermen would render this service. Hence, there is a comparative difference in the status among the Valans living in the north and the Mogeyars, on the one hand, and the Valans in the southern districts, on the other. This aspect must be kept in mind when considering their social status. The Valans were higher than the Mukkuvas in the districts of Thrissur and Ernakulam and they also had the services of the washing community. The Valans at Fort Cochin and Vypin had both Velans (washerman caste) and Mannans (Theenda Mannans) as their washermen. The washermen also supplied the *Mattu* to the Valans to remove pollution from menstruation on the fourth day, on the twenty eighth day after birth and on the sixteenth day after death.

The Valans had their own barbers and priests who were called by several names in different places. At Vypin and Alleppey they are called Arayavathi. In some places here they are also called Aduthavan (one who is near). In the district of Thrissur they are called, Panniakkal, and at Pazhayangadi, Khadeer. In this region the Valans were higher than the barber and washermen communities and there were no intermarriages, acceptance of food and water, nor interdining between them.

Relation with other Castes

The status of the Valans in the caste hierarchy was similar to that of the Arayans, both being treated at par by the upper castes. Unlike the Arayans in the same places where they were considered at par with the Ezhavas, nowhere, besides in the Cochin area, has it been stated that the Valans have enjoyed a higher status than the Ezhavas. This fact is corroborated by Iyer (1909: 259): "The members of the fishing castes come in the order of social precedence below the Ezahavans and have to stand at a distance of thirty six feet from Brahmins, Kshatriyas and high caste Sudras, so as not to pollute them. The Valans were not allowed to enter the temples of caste Hindus and had to stand outside the compound walls. They accepted food and water from the higher castes but not from the Pulayas/Cherumans and Parayans. A number of atrocities were committed on the Valans at Vypin by caste Hindus on the score of untouchability."

At Fort Cochin where the Valans ferried the royal families and caste Hindus, members obviously enjoyed a higher status than the local Ezhavas (Chovans), but the situation was different on the nearby island of Vypin. The caste Hindus treated the Ezhavas and this fishing community at par. They were neither allowed to enter their houses, nor draw water from their wells. They had to stand at the same distance from the temple premises. Despite this, the Ezhavas never allowed a Valan to enter their houses nor their temples. The Arayans and Arayavathis were treated in a similar way as the Valans by the Ezhavas.

The Valans at Methala say that inferiority to the Ezhavas has been further aggravated by the latter's enormous economic uplift. This has been made possible by the organisation of trade unions among the Ezhavas and their monopoly in tapping toddy. The economic advancement of the Ezhavas and their social elevation were mentioned in many other fishing villages.

Presently, there is an increasing trend in the number of intermarriages between the different fishing communities and other castes. This is keeping in tune with the changing times and the society at large.

Religious Beliefs

The role played by the Valans in the history of the Kodugallur temple, like that of the Arayans, was very significant. They too were actively employed in chasing away the Buddhists and Jains, and in initiating the obscene rituals practised at this shrine. The Valans had special rights in the celebration of the Bharani festival at the temple. Like the Arayans, They were not allowed to enter the temple except on the occasion of this festival. However they claimed that the *Aravanapayasam* distributed at Sabarimala was meant exclusively for them. The legends of both the Valan and the Arayan communities are similar. Their patron goddess is the Bhagavathy and they worship other Hindu deities as well. They also practice ancestor-worship. Like the Arayans again, a major part of their income is spent on religious activities. One reason may be that their livelihood being so hazardous and uncertain, they take refuge in religious rituals.

The Hindu Marine-cum-Backwater Fisherfolk

Mokayan/Mogeyan

The Mokayan fishermen are found in the Cannanore district, and in some northern parts of the Kozhikode district. They identify themselves with the Bovis-Mogeyars found further north, though this study reveals that they are two distinct ethnic groups. This community is called Mogeyan, Mokeyan and Mokaya (as verified from school certificates), in different places. At Mambram, Peringathur, Mozhapinangadi, and Edakad, the members of this caste have styled themselves as Mogers, which is included in the list of Scheduled Castes. The members of the fishing community who are found at Nileswaram, Kadangodu, Kasargod, and Kunjathur are the Bovis-Mogeyar, but they call themselves Mokayans. The Mokayans are found in the villages of Puthiyappa, Korapuzha and Kappad (Kozhikode district), Perlasseri, Mellur, Mambram, Peringathur, Mahe, Mozhapinangadi, Pookkad, Chombal, Thuyapara and Edakadavu (Cannanore district). Among none of the other subcastes discussed is there such a diversity in occupational mobility as is found among the Mokayans. Among them there are both marine and river fishermen; and among both, *Chundakkar* (hook-and-line) and *Valakkar* (nets). The Mokayans at Korapuzha use stake-nets in the rivers like the Valans. They learned this technique from the Valans who have settled across the river and are called 'Cochin Moyer'. At Puthippa and some villages near the coast the Mokayans collect cockles and shell-fish as well.

Mathur (1978) mentions that among the Mappila fisherfolk, the *Valakkar* and *Chundakkar* (Beppukar) do not intermarry and that the *Chundakkar* are of a higher status than the *Valakkar*. He inferred that these two groups are converts from two distinct castes, probably fishing communities, who were originally *Chundakkar* (Beppukkar) and the other, *Valakkar*. It is probable that the Mappila *Chundakkar* are converts from the Mokayan community and the Mappila *Valakkar* from the Mukkuvas. After their conversion in the past, they have maintained this distinction and remain endogamic.

Our informants at Puthiyappa say that they are immigrants from Mangalore, whose original profession is medicine. Some took to river fishing to make both ends meet and later ventured out to the sea, having specialised in hook-and-line fishing. The Mokayans work as building-labour, head-load coolies, coir-makers, sand-scoopers, prawn-peelers and domestic servants. They do all sorts of odd jobs. A somewhat similar situation obtains at Korapuzha, Kappad, Peringathur, and Mozhapinangadi. Thurston has noted that the Moger, also known as Mogeyan, spoke Canarese and hailed from South Canara. They were found along the Malabar coast. Whether this reference was to the community under discussion or to the Bovis-Mogeyar is not clear.

Descent and Inheritance

The Mokayans are divided into four clans, namely, (1) Chathanathukulam, (2) Kurumbanathukulam, (3) Padithanathukulam, and (4) Cheruvandiam. In some places the last mentioned *Kulam* is known as *Thayanathukulam*. They are matrilineal and practise clan exogamy. Pollution resulting from birth, menstruation and death vary from place to place. Menstrual pollution lasts for 7 to 14 days. During the first delivery pollution lasts from 14 to 90 days, and for subsequent births, 14 to 28 days. Death pollution lasts for 7 to 11 days. The *Mattu* was received by the Mokayans everywhere though this custom has now died out to some extent. Their washermen are called Marakurup, Parathi, Peithichi, Parathiyama, or

Velathi at Puthiyappa. This community only supplied the *Mattu*, whereas the Vannathis washed for them. The *Mattu* was supplied by Vannathis or Peithichis at Korapuzha. The community that supplied the *Mattu* were given either one to four *Annas* or a small measure (*nazhi*) of paddy.

The Aduthavan or Valanjiar performed the funeral and related purificatory rites at Madapally and Puthiyappa. They are different from those that serve the Mukkuvas who are their barbers.

Political Organisation

The Mokayans have a village headman whose title is either Gurikaramar (Puthiyappa) or Karanavar (Korapuzha). They are the custodians of their temples. Every village has a *Kshetram* (temple) dedicated to the goddess Bhagavathy and each *Kshetram* has a council. This council adjudicates all their caste disputes. A move is on foot to amalgamate the smaller factions with the caste council at the village level and affiliate such councils to the Bovis-Mogeyar Association. The headman of the village is selected on the basis of his ability to deal with their caste affairs and his knowledge of religious matters.

Relationship with Other Fishing Castes

The Mokayans are in contact with the Mukkuvas but whom they consider inferior. That is why they never accepted food and water from the Mukkuvas. The Mukkuvas were allowed to enter their compounds but not their houses. The Mukkuvas however accepted food and water from the Mokayans. The Mukkuvas and Mokayans lived on equal footing at Kappad but the Mokayans considered themselves superior. Both are treated alike by the caste Hindus. The Mokayans hold a higher status at Korapuzha, across which live the Valans. There is little social contact between the two communities.

The Mokayans' relationship with other castes vary from place to place. At Puthiyappa they were considered higher than the Thiyyas. The Thiyyas were not allowed to enter their compounds and their food was not acceptable to Mokayans. The Thiyyas, like the Mukkuvas, were not allowed to draw water from their wells. The Thiyyas and the Mukkuvas were not allowed to enter the compounds and houses of the Namboodiris and the Nairs, but the Mokayans had access to them, which definitely indicated a superior status.

In some other places the Mokayans were considered superior to the Thiyyas though they were not allowed to enter freely the houses of the Namboodiris and the Nairs. The Kanakkans, Cherumans, and the Parayans were not allowed to enter the compounds of the Mokayans and they avoided social contact with them. The Mokayans stood socially high among all the fishing communities. Despite their financial backwardness, however, a few Mokayans became tahsildars, station masters, collectors, doctors and lawyers.

Religious Beliefs

Like most of the other fishing communities, Mokayans are devout worshippers of Bhagavathy. All their temples so far visited, are dedicated to this goddess. However they also offer worship to other Hindu deities and celebrate all Hindu festivals. A major portion of their income is spent on religious festivals, which are celebrated with great pomp and show. Membership to their *Kshetram* (temple) council is compulsory. This council comes to the aid of its members in times of need like weddings, deaths, etc. As mentioned earlier, these *Kshetram* councils are being affiliated to the Bovis-Mogeyar Association and the process is gaining momentum.

Servicing Castes of the Hindu Fishing Communities

The main servicing castes of the Hindu fishing communities are the washermen and barbers. The names of these castes vary from place to place and all of them are separate endogamous groups. The barbers are called Arayavathi, Aduthavan, Kavuthiyan, Khadeer and Panniakal while the washermen are called Valinjiar, Mannan, Velan, Paravan and Vannan. Some of these servicing castes render service to a particular fishing community only. For example, the Arayavathis are the barber caste of the Arayans and the Valans. They do not serve the other fishermen although this situation has changed a great deal now. Some of them like the Kavuthiyans also serve the communities like the Thiyyas.

The servicing castes had to perform special rites and duties at birth, marriage, puberty and death of the members of the Hindu fishing communities. These functions were very important and in return they received remuneration during the life crises and important festivals like Onam and Vishu. In the past these communities were paid uncooked rice, clothes and sometimes cash in varying quantities. Today the relationship between the servicing castes and the Hindu fishing communities has slackened considerably.

The barbers, besides cropping the hair and shaving, also performed the ritual of cutting the hair of the one who lighted the funeral pyre. The barber-women were, in many places, the midwives to the Hindu fishing communities. The washermen, besides washing the clothes, also often supplied the *Mattu*, the supply of which was necessary to remove ritual pollution brought about by birth, menstruation and death. The pollution-period varied from one Hindu fishing community to another and from village to village. The *Mattu* (washed cloth) was supplied to the members of the Hindu fishing communities suffering from pollution, for their ritual purification. The wearing of the *Mattu* was essential to remove the pollution. The washerman or his wife sprinkled Ganges water, in the absence of which water mixed with cow-dung, over the individuals concerned and the whole house, as a purificatory rite. The services of these two servicing castes were very important in the past. Denial of their services to a particular caste was one of the reasons cited by the Government of India for inclusion into the list of Scheduled Castes.

Of all the servicing castes studied, none except the Arayavathis went out for fishing. All the others were devoted to their particular professions. When questioned about this, the Arayavathis said that unlike the other servicing castes, they served only a particular community and it was not an everyday job to attend to a birth or death. They took to fishing in order to supplement their income and also to engage themselves. Another group, the Panniakals, could not be included in this study. The Mannans and the Vannans are the servicing castes of the upper caste Hindus like Nairs. They are presently included in the list of the Scheduled Castes.

The Arayavathis claimed that they were the *Vathiar* (teachers) of the Arayans and that they were lured into this profession by crafty Brahmins. They were known as Aduthavans (one who is near) in the past. They were also called Khadeer in the district of Cannanore. They are a very small community, whose membership is limited to about 2,500, spread all over the coastal belt. One story goes that a Namboodiri lady had illicit relations with a Nair. The penalty for violation of this caste norm was death for the household. She was whisked out of the *illam* (Mana) with the Nair without the knowledge of the castemen. The children of their union were ordained priests of the fishing communities. One day one of them was pressed to perform some death rites. When the time came to shave the person who set fire to the funeral pyre, no barber was available and the priest himself cut the hair. Since then this job has fallen to their lot and thus originated the name Arayavathi. The story seems to be imaginary as they claim to have three *illams*.

Although the Arayvathis claimed to have three *illams,* they were only able to identify two, viz., (1) Mangaimana and (2) Azhanjarimana. These *illams* are exogamous groups. Formerly marriage within the *illam* was punishable with excommunication. There was no intermarriage between the Arayavathis and the Arayans on the one hand and the Arayavathis and the Valans on the other.

The Arayavathis are found mainly at Kallikadu where there are about 50 families, and on the Vypin Island, where they have about 300 families. No such information could be collected about their distribution elsewhere. Arayavathi households are reported to be scattered in the districts of Cannanore, Kozhikode and Thrissur.

The Arayavathis at Kallikadu are much more advanced in comparison to the rest of the Hindu fishing communities. Eighteen of the Arayavathis at Kallikadu subsist on subsidiary occupations besides their traditional one, and 28 of them have studied upto Matriculation (High School). These figures are higher in comparison to the rest of the Hindu fishing communities. Their living standard is comparable to that of the general middle class. A significant fact worth mentioning here is that the parents of those who are employed in other occupations never went out for fishing. They had a regular source of income, however, small it may be, to meet their basic needs, which fishing does not and cannot give.

Some of the Arayavathis still perform certain rites during weddings and deaths. Some go out for fishing but do not own any craft or gear; instead they accompany the Arayans to sea as their labourers.

The Arayavathis of the Vypin Island, on the other hand, are comparable to the rest of the fishing communities covered by the present study. At Ellamkunnapuzha and Njarakkal, they are mostly engaged in fishing, but a few are professional barbers as well.

Social Position

The different servicing castes of the Hindu fishing communities ranked very low in the general caste hierarchy. The Arayavathis ranked below the Arayans and Valans, both because of the lesser number of their *illams* and by virtue of their profession. Their *illams* have a common suffix *Mana* which means the household of a Namboodiri. In terms of *illams*, they are comparable to the Mukkuvas. They too suffered from indignities arising out of untouchability at the hands of the high caste Hindus.

Communities Engaged in Fishing along with other Occupations

There are a number of communities engaged in the backwater fishing along with other occupations like lime-making, ferrying, coir-making, etc. Some communities engaged in such fishing are: Bharathar/Paravar, Kanakkan, Padanna/Padannan, Thiyyas and Pachalli Ezhavas. The Bharathar, Paravar, Kanakkan, Padanna and Padannan are included in the list of the Scheduled Castes. Fishing among these communities may be either a major or a subsidiary occupation. The techniques employed by them are not comparable with those used by the fishing communities discussed earlier, but nevertheless, the degeneration of caste, based upon occupation has made them experts in their own spheres. Several other communities have taken up fishing, while some of those whose hereditary occupation was fishing have changed their jobs. Though the scope of the present study did not allow a detailed investigation on all these communities, some information have been collected regarding them. In the course of the investigation, some important aspects of their life have been brought to light.

It has been found that the Bharathars and Paravars, two castes presently included in the list of Scheduled Castes, are in fact one and the same. Most of them are Christians, though there are a few Hindus as well in their fold. The Bharathars and Paravars intermarry, whether they are Hindus or Christians. They are also known as Chozhavar in Tamil Nadu and in the southern parts of the Trivandrum district. They have community temples and also a common association, the headquarter of which is at Vaikom. They are found in notable concentrations from Ernakulam to the south. It is claimed that their members are also found in Cannanore. There is an account of this community given in the Census of India 1971; Series 19 – Tamil Nadu, Part VB; Ethnographic Notes, Scheduled Castes of Tamil Nadu (Vol. I, pp. 19–64). The legend about their origin is identical with the Arayans and the Valans, Vyasa being their progenitor (p. 24). They had fishing and ferrying boats as their traditional occupation (pp. 20, 32), besides making lime and rattan works. Agricultural work and labour is of a comparatively recent origin. "Brahmins are invited to officiate at their marriages. The barbers and washermen serving the Caste Hindus also serve them. They are allowed to take water from public wells and to enter all the temples found in the area". "The Paravars of Manavalakurichi are not treated as untouchables locally. They do not appear to be suffering from social disabilities. They live in caste Hindu localities" (p. 39). Anthropologists like Iyer (1909: 42–46) have stated that the Paravar and Bharathar are the same. They were palanquin–bearers. The Paravars/ Bharathars did not suffer any social disability like untouchability, and they should not have been included in the list of the Scheduled Castes. Their disabilities are purely economical and mainly due to this they are educationally backward.

Many Christians are getting documents made in the name of Hindu Paravar/Bharathar by nefarious means in order to avail of all the benefits earmarked for Scheduled Castes. Latin Catholic Mukkuvas too are believed to have infiltrated into their fold by similarly producing false certificates. The Paravar/Bharathar have four exogamous *illams* or *Kulams* (clans), namely Azhathurparavar, Zshurparavar, Kollaparavar and Kandaliparavar. Until 60–70 years ago they followed the *Marumakkathayam* law of inheritance, but now they follow the *Makkathayam* system. They observe birth, death and menstrual pollutions for 28, 5 and 4 days respectively. The *Mattu* is supplied by the Mannan community. Puberty rites are elaborate among them.

The Paravar, like the other Hindu fishing communities, offer worship to the deities of the lower Hindu pantheon like Bhagavathy, Bhadrakalli, Chamundi, Murthi, etc. They have their own temples and the festivals are celebrated on grand scales. Large-scale infiltration of Latin Catholics into their fold is depriving them of their rights and increasing competition among themselves. They complain that those of them who had rattan work as their principal occupation are neither supplied with this raw material nor are they allowed to cut any from the jungles as they used to do in the past.

Socially, this community was not considered inferior, nor were they considered untouchable. In fact, they had a relatively high position. But economic depression and the resulting educational backwardness have made them lag behind other communities.

Both the Kanakkan Padanna and Padannan are included in the list of Scheduled Castes. The study of these two communities in the present survey was confined to the Thrissur district. These two communities are in fact one and the same. Their members call themselves the Padana Vettuvan in the Ponnani taluk with whom their cousins in the Thrissur district intermarry. The Vettuvans (or Kanakka Vettuvans) are coconut-palm climbers and have a lower status than the Padanans. They do not intermarry.

"Though formerly a cultivating caste, many are now engaged in boating, fishing, and other kinds of labour, and are, perhaps for this reason, superior to the rest of the agricultural classes. They belong to the slave castes, and are even now attached to some landlords in the State. In the taluks of Thrissur, Mukkundapuram, and Cranganore, where information was obtained about them, it is reported that they are *atiyars* (slaves) of Chittur Manakkal Namboodiripad at Perumannom near Thrissur" (Iyer 1909: I, p. 138). Padannans were originally salt-makers at Mallipuram (Ernakulam district), Koyambaram and Erikammana (Malappuram district). The place where salt is crystallised is called *Padana* and this is how they got their name *Padannan*. As to how the name Kanakkan got attached to them is not known. The story goes that the Padannans were high class Sudras, employed by the Namboodiris to ferry their boats. Once there was a quarrel over the fare and the boatman broke the Namboodiri's *Poonool* (sacred thread). The Namboodiri cursed him thus; "*Uppali Kurup Uppollam Varumbol Vellamayi Potte*". Since then, they have been degraded and deprived of their title Kurup.

Among the Cherumans there is a sub-division called the Kanakkan. These Kanakkans in the Palghat district are segmented into five sub-castes, viz., Pathinezhu Kanakkan, Ombathu Kanakkan, Pala Kanakkan, Patta Kanakkan and Padanna Kanakkan. According to our field study, the Padanna Kanakkan of Palghat and those styling themselves Kanakka Padannan in Thrissur are not one and the same community. The Padanna Kanakkan of Palghat never received the services of barbers and washermen who served the Thiyyas or Ezhavas. It is said that there are, four divisions among the community of Kanakkans in Thrissur, namely, Padannan, Vettuvan, Era (also called Molayamar) and Chavala. These groups are strictly endogamous and among them Padannans claim to be the highest. According to one informant the local landlords had imported these Padannans from Chingelput where they were salt-makers and fishermen, to keep temple accounts. They were given the title Kanakkapillai, which they attached to their name as Padannan. (Some of their kinship terms are of Tamil origin, e.g., *Appa* is used for father). The title of Kanakkan must have been adopted in order to get the caste included in the list of Scheduled Castes.

Along with salt-making, Padannans used to ferry boats for goods and passengers along the backwaters and along the river route connecting Tirur and Cochin. This was a very busy route and an important trade link till the advent of the railways and the highways. Their womenfolk in some areas busied themselves with coir-making. They used to go occasionally for agricultural work and are active in the *Punja* cultivation. A number of the Padannans have become agriculturists in the area. It is claimed the Padannans of Ponnani never had fishing as their occupation. They plied ferry boats and their womenfolk were mainly engaged in salt-making. Now, most of them are idle for want of any occupation despite the fact that the Padannans are included in the list of Scheduled Castes. Some Padannan Vettuvans go out to sea in boats at Padinyarakara and Porathur near Ponnani. They claim that they have 28 *kiriams* but only the names of four could be obtained, viz., Painikiriam, Cherakiriam, Palokiriam, and Vengatkiriam. These clans are all exogamous. They observed birth pollution for 28 to 90 days and the *Mattu* was supplied on the 7th and 15th, 28th, 56th and the 90th day. Menstrual pollution lasts for three to four days and death pollution for 11 days. Earlier, death pollution lasted for 15 days. The Mannathis removed pollution by sprinkling water on the individuals concerned and the house. Nowadays, their Karanavan performs this rite. The Panniakals were their barbers.

The Ezhavas were not allowed to enter the houses of the Padannans whereas Nairs admitted them to their courtyard. However, Padannans were not allowed to touch the main pillars of the Nair houses. Both the Ezhavas and the Padannans were debarred from entering the temples of the Nairs and the Brahmins.

The headman of the Padannans is called Karanavan and the meeting of their council of elders is termed Panchayat. These elders decided caste disputes in the past. Political affiliation and infiltrations have rendered the Karanavan effeminate.

There are about 150 families of the Thiyyas at Peringathur who have been fishing from the rivers for the past three generations. They have not given up their traditional occupation of toddy-tapping, but it is only secondary to them. Their womenfolk make coir-ropes. The fish is sold to Muslims who hawk them on bicycles in the interior villages. It may be mentioned here that about 25 Thiya families of Azhikkal fishing village of Azhikode (Kannur district) have adopted fishing during 2006–07.

There are a number of Ezhavas at Pachallur (Trivandrum district) who have been fishing from time immemorial. They are said to have been excommunicated from the Ezhavas for their illicit relations with the fishing communities. The Pachallur Ezhavas are their descendants (Mathur, 1975).

A number of individuals belonging to the non-fishing communities have taken to fishing in recent times with the advent of mechanised boats. They are also involved in canning, prawn-peeling, trading, etc. This occupational mobility is found lacking among the traditional fishing communities, notably among the Hindu fishing communities. This is mainly due to the fact that the establishments are rarely owned by their members. Private owners of factories hire or employ their own members; they do not recruit hands from the Hindu fishing communities.

At all phases of history human beings have adapted to their ecological environment to earn a living, adjusting himself and inventing at the same time. Many communities, especially those that lived on the coastal areas, rivers and lakes, took up fishing, either to supplement their income or for their own consumption. Villages in the past were self-sufficient and unlike today, they merely had to cater to their own needs. Nothing was produced on a large-scale for sale. Likewise, fishing was not undertaken for sale, but to satisfy their own consumption needs. Gradually, as the demand for fish increased, more people took to fishing and became experts in it. Along with this some retained other occupations like making coir-ropes and lime, salting of fish, plying ferry boats, etc. Today, high caste Namboodiris and Nairs too have taken to fishing. There are about 150 Thiyya families at Peringathur who have been following the occupation of fishing from the river for at least three generations. They also continue to climb coconut trees in the off-season and their womenfolk make coir-ropes. This kind of adjustment to new occupations is generally encountered everywhere. It is regretted that while the fishing industry is progressing rapidly in every direction, the Hindu fishing communities are left out of its scope, giving room for non-fishing communities, whose interests are getting vested day-by-day.

Intercaste Relationship

One of the characteristics of Hinduism is that castes are graded according to their ritual status in relation to each other. Every caste had a traditional occupation and the various castes of a village or region were, in the usual course, mutually dependent. A man inherited an occupation, and its skills and secrets were transmitted to him by his father, uncle or elder brothers. There was a feeling that traditional occupation was the proper or natural one for members of the caste and they took pride in it with a sense of its importance. While occupational specialisation resulted in the interdependence of castes, hierarchical ideas, especially as expressed in endogamy, which was the rule, and restrictions in interdining, emphasised their separation from each other. A caste was also a commensal group in the sense that food cooked by one of its adult members was acceptable to all. This is distinguished from cases where there was only a one-way movement in cooked food and drinking water. Here, the acceptors of the food were ritually inferior to the givers.

Caste hierarchy is rooted in the notion of superiority, which in itself is a consequence of the claims of purity and holiness. High castes stand aloof from castes which are ritually impure. This idea of purity and pollution led to the maintenance of the identity of castes and of the social distance between them. A caste was pure in relation to a lower caste and impure in relation to a higher one. The higher caste was prohibited from accepting cooked food or drinking water from the lower, let alone have connubial relations with it. In extreme cases a minimal physical distance had to be maintained. For example, the Hindu fishing communities had to keep a distance of 36 feet from the high caste members. Violation of the rules governing the acceptance of cooked food and water, and other forms of contact, resulted in the member of the higher caste suffering pollution.

The idea of hierarchy is present everywhere. Each man belonged to a caste, which formed part of a whole. Particular elements of culture such as diet, occupation, customs and rituals were distinguished as higher and lower: thus there was a higher as well as a lower diet, and superior as well as inferior occupations. However, despite all this, the notion of superiority and inferiority is not a rigid or a clear-cut one.

The caste system gives to the Hindu society a segmentary character. One level of segmentation is relevant for purposes of marriage, another for religious and sectarian affiliations and a third one for political purposes. Again here, neither is this pattern of segmentation a rigid one nor does it apply uniformly to all categories. Being segmentary in nature, casteism evinces the two characteristic processes associated it: fission and fusion. Historically, the process of fission seems to have been a fairly generally widespread feature of the Hindu society (Beteille, 1966). Hutton (1961) refers to the fissiparous tendency as an inherent feature of the caste system. In the past the number of orders of segmentation, as well as the number of segments, tended to increase progressively. Today, greater importance seems to attach to the opposite process of fusion. The relationship between traditional and actual occupation has always been of a somewhat uncertain nature, and this is particularly so today.

Principles of caste ranking rests essentially on conceptions of social honour or esteem. Social esteem is attached to particular life styles and groups are ranked as high or low according to how or whether they pursue such disciplines. What is highly esteemed differs from one society to another and depends ultimately on its value system as a whole. In India, ritual elements have historically played an important role in the styles of life, which have enjoyed high social esteem. The whole caste system rests on this aspect. Status honour in a caste hierarchy was based upon wealth to the extent that the possession of a certain amount of it was a necessary condition to the pursuit of a certain style of life. Status honour, in addition to being associated with specific lifestyles has in every society, a strong traditional bias. Hence, it is not merely sufficient for a caste to adopt a particular style of life in order to achieve a higher social rank.

Although hierarchy is an important feature of the caste system, it should not be assumed that wherever there is segmentation we can rank the segments as higher or lower. There are often conflicting claims to superiority and it is often impossible to speak of a consensus. It so happens that two castes put forward rival claims to superiority with regard to which members of other castes affect indifference or consider themselves as incompetent to decide either way. This feature was frequently encountered in the course of the present study.

An important point to be borne in mind about the hierarchy of the caste system is its ambiguities beyond a certain level. These ambiguities are essential in a system, which seems always to have permitted a certain degree of resilience. They are further increased by the fact that the basis of social honour or esteem, the entire value system on which the ranking of castes depends, has in recent times been undergoing important changes (Beteille, 1966). The caste system is not only hierarchical but it is also the basis of an elaborate division of labour. Every caste, however broadly or narrowly one may define it, evinces two characteristics: (1) autonomy in certain spheres such as marriage and kinship; and (2) dependence upon other castes in other spheres, such as economic and ritual service.

Pride in one's caste was common and along with it went a devaluation of the other castes and their customs. Fishermen are commended by other castes for their sense of unity. Along with the Hindu fishing communities inhabiting the coastal areas of Kerala, there are present Christian and Muslim fishermen. In their social standing they are above the Hindu fishing communities, as they are not considered untouchables by the caste Hindus. As a matter of fact, these Muslims and Christians were members of the Hindu fishing communities who underwent conversion centuries ago to escape social opprobrium and untouchability. The Hindu fishing communities hold the Muslim and Christian fishermen inferior on account of their converted status. Despite such claims and counter-claims, in the sphere of their common occupation, they are highly cooperative with one another and form a very close-knit unit.

Despite their religious differences, all these fishing communities have certain belief systems, rituals and supernaturalism, and are very deep in their religious beliefs and faith. Alongside, Despite their conversion, they have retained a lot of their Hindu heritage (Mathur, 1978).

Among the Hindu fishing communities, there are different social levels. Some like the Mokayans, Bovis-Mogeyar and Mogavirar had a higher social status than the rest, though in some places they were discriminated against. This was the case with the Mukkuvas, Arayavathi, Aduthavan and the Panniakal. These communities were not allowed to walk along public roads, enter temples, etc., the caste Hindus observed touch pollution against them. The Arayans and Valans exhibit a strange intermixture of both these extremes within the Hindu fishing communities. That is, they claimed a higher status than the Mukkuvas and Arayavathi, by virtue of their having an extra *illam*. But, on the same basis, they ranked much below the Bovis-Mogeyars and the Mogavirars. The Arayans and the Valans of the erstwhile Cochin state however claimed superiority over the Bovis-Mogeyars and Mogavirars because the former were royal ferrymen and, at some periods of history, rather close to the royal families. This feature was however not regular nor consistent everywhere. The Mokayans also have four *illams/kirims* like the Arayans and the Valans, but while the former were never labelled untouchables by the high caste communities, the latter were considered to be so. Earlier writers like L.A.K. Iyer and Thurston have noted that the Valans and Arayans were taken to be of low caste and untouchable.

The Bovis-Mogeyars, Mogavirars and Mokayans did not accept food from the Mukkuvas. The Valans never accepted cooked food from the Mukkuvas. The Bovis-Mogeyars and Mogavirars accepted cooked food from one another but never ate together. The Arayans did not accept food from the Mukkuvas and Valans, and the Valans eschewed food from both the Arayans and the Mukkuvas. It was taboo on the part of a Mukkuva to take food from the washermen and barber castes like Arayavathi, Panniakkal, etc. The Arayans could not take food from the Nullayans.

The present study shows a certain social order among the Hindu fishing communities from the anthropological point of view. The Bovis-Mogeyars, Mogavirar and Mokayans occupy a higher position among the Hindu fishing communities. Neither did these communities consider themselves to be untouchables nor did they, barring in a few instances, suffer from much social disabilities on the basis of the number of *illams* or *Kulams* each of these communities had. The Bovis-Mogeyar ranked first having 10 *illams*, followed by the Mogavirar (8 *illams*) and the Mokayans (4 Kulams). The Valans and the Arayans constituted the next category. Both have four *illams*. Iyer had noted that they were considered an inferior and polluting caste. (Iyer, 1909: 259–60, 265). The Valans and Arayans stand on an equal footing, though their members contest superiority over each other. This claim is based on their past service to some royal and Namboodiri families. But such claims could not be advanced everywhere nor was it a common feature. Members of these communities were chosen as royal ferrymen because of their expertise in that art.

Next in precedence come the Nullayans and the Mukkuvas. The Nullayans have four *illams*. Among the Mukkuvas there are both 3-*illam* and 4-*illam* groups. The *moonillakar* (3-*illam* people – Mukkuva) are considered to be inferior to *naalilakar* (4-*illam* people – Arayans). The Arayans and the Valans stand on a higher plane than the Mukkuvas and the Nullayans. The Nullayans and the Hindu Mukkuvas have little social contact and their relative positions could not be ascertained. But, if we take their social relationships with the Arayans into consideration, both these communities stand on an equal footing. On the basis of the criterion of *illams*, the Nullayans are equal to the Arayans but, in fact, it is not so and the Nullayans are socially below the Arayans. They cannot claim a higher position than the 3-*illam* Mukkuvas.

The lower category among the Hindu fishing communities is composed of the Arayavathi, Panniakkal, Valinjiar, Khadeer and Aduthavan who are barbers and washermen. They were the exclusive servicing castes of the fishing communities. In addition to them, other servicing castes attached to the Ezhavas and Thiyyas like

Mannan, Velan, Vannan and Kavuthiyans also served some of them. Among the servicing castes, only the Arayavathis are actively engaged in fishing in some areas as marine fishermen. The Panniakkals are said to be at times engaged in fishing but they confine themselves to the backwaters of the Thrissur district.

Social Stratification

The following configuration of the Hindu fishing communities' social standing has been obtained:

1. (i) Bovis-Mogeyar
   (ii) Mogavirar
   (iii) Mokayan
2. (iv) Valan
   (v) Arayan
3. (vi) Mukkuva
   (vii) Nullayan
4. (viii) Arayavathi
   (ix) Panniakkal
   (x) Valinjiar
   (xi) Khadeer
   (xii) Aduthavan

The Hindu caste system may be categorised into three groups, namely, Brahmins, non-Brahmins and untouchables, for the purpose of determining their social hierarchy. The Namboodiris are the genuine Brahmins of Kerala; others are immigrants. Among the non-Brahmins are included a large number of communities like the Kshatriyas, Nairs, Ambalavasis, etc. These were all clean castes. They had their servicing castes of washermen and barbers.

The so-called untouchables consisted of a number of castes who were all considered unclean and polluting. This stigma of uncleanliness and pollution was in relation to their occupations. They had their own social hierarchy on the basis of their occupation and purity-pollution. The Ezhavas (Thiyyas or Chovans) were supposed to be the upper polluting caste. Pollution began from the Ezhavas and all those below them were also polluting castes. The caste Hindus observed pollution and untouchability against them. Their attitude towards the fishing communities was one of contempt since fishing was considered an unclean and despicable occupation.

The Bovis-Mogeyars and Mogavirars have closer affinities with communities in Karnataka where they do not suffer any social backwardness, which is the reverse in case of fishing castes in Kerala. The situation in Kerala varies from place to place and from community to community.

In the erstwhile Travancore state are found the Nullayan and Arayan fishermen. The Nullayans, despite their claim to equality with the Arayans and Ezhavas, were in fact slightly lower in esteem. The position of the Arayans differed widely from equality with the Ezhavas (Alleppey and Quilon districts) to further relegation in Ernakulam, Thrissur and the rest of the districts. The Valans have nowhere been considered nor accepted as equal to the Ezhavas/Thiyyas but in the former Cochin state they claimed a superior status by virtue of their being royal ferrymen. Iyer (1909, I: 259) had noted the Valans came in social precedence below the Ezhavas. The Mukkuvas and the Thiyyas in the district of Kozhikode had almost an equal status, though the Thiyyas never allowed the Mukkuva to draw water from their wells. The Mokayans, on the other hand, enjoyed a higher status than the Ezhavas and were even admitted to Nair houses. However, the Nairs neither allowed an Ezhava into their house, nor accepted food from them. In Peringathur the Mokayans who have styled themselves as Moger had an equal status with the local Thiyyas. The Mogavirars and the Thiyyas had an equal status since both have eight *illams*, whereas the Bovis-Mogeyars were considered higher as they had 10 *illams*. Though the fishing communities were regarded inferior in Kerala, they were not labelled untouchable in Karnataka. The washermen and barber castes attached to the fishing communities stood below them. The Kammallans, Kaniyans and Thandans were superior to the Arayans.

All the Hindu fishing communities stood low in the Hindu caste hierarchy and most of them suffered from untouchability; they themselves subjected the Cheruman/Pulayan, Paravan and other castes to such stigmas. These communities seldom had contact with the fishing communities, and were not allowed to enter their houses, nor was their food acceptable to them as their touch polluted.

Evidently, from the Valan downwards, these fishing communities suffered social disabilities, though not to the same extent as did the Parayans or Pulayans. They suffered from disabilities such as preclusion from use of public thoroughfares, entry into temples and admission to schools where children of caste Hindus read. That is why the colonial rulers established schools for them on the coastal areas, especially in the Malabar area.

Fishing is an occupation in which risk and uncertainty is writ at large and the men engaged in it have to subserve the forces of Nature. They insure against uncertainty and risk by appeasing the supernatural powers on a grand scale, spending a major portion of their meagre income. Families are, therefore, left with little to eat, to wear and to house themselves. Their grinding poverty has landed them in the abyss of backwardness not only in economy but also in the fields of education and health. In the prevailing conditions, they cannot improve their lot; in fact they are going down. Unless they have an alternative source of income and are not provided with a better technology to pursue their traditional calling, it is next to impossible on their part to earn sufficiently to meet their basic needs. Whatever may have been the position in the past, today the Hindu fishing communities are at a very low socio-economic level vis-à-vis the general population. Exceptions can be seen among them too but they are few and far between. On the whole, the Hindu fishing communities are backward in all spheres – educationally, economically, and socially.

Infiltration into Scheduled Castes

Investigations have revealed that a number of fishing castes have intruded into the fold of the Scheduled Castes and Scheduled Tribes in order to get the benefits earmarked for them. Some of the infiltrations are of recent origin, but the process has been going on for nearly 51 years. The exact extent of infiltration is not known but it is believed to be considerable.

In the Trivandrum and Quilon districts there are many Christian members among the Bharathar/Paravar community. Numerous intermarriages have taken place among them. Besides these Christian members, Latin Catholics in these areas are also making false claims as Hindu Bharathar and Paravars. A number of Latin Christians have managed to become Hindu Paravar and Bharathar by getting false certificates, it is believed.

In the taluks of Thiruvalla and Kuttanad (Alleppey district), there are a number of Hindu backwater fishermen who are locally called the Arayans and several of their members have obtained certificates and identifications as the Mala Arayan. The Mala Arayans are a hill tribe inhabiting the hilly terrains of the Kottayam and Idukki districts. They, however, do not have fishing as their occupation and it is surprising how the backwater fishing community of Arayans of the low lands have succeeded in getting false certificates in the name of a non-fishing tribe. A number of them have got admission into professional colleges and obtained high government jobs on the basis of their false certificates. When some elderly members of Veeyapuram village of the Kuttanad taluk were asked if they were called Valans also, they replied that the Valans and the Arayans are one and the same. As the Valans are the traditional backwater fishermen in Kerala, it is possible that this community belong to the Valans. It is also probable that the coastal Arayans had migrated inwards and had taken up backwater-fishing.

Panniakkals serve as barbers to the Hindu fishing communities, especially the Mukkuvas, and are found mostly in the district of Thrissur and northwards. Several members of this community in Ponnani have got certificates in the name of the Scheduled Tribe, Paniyan, who are found mainly in the Palghat and Wynad districts. The name Panniakkal is often misspelt or misinterpreted as Paniyan and such infiltration has been reportedly going on for sometime past. Those authorised to issue community certificates must be especially warned against such manipulations and falsifications by simulation.

The caste name Moger has been used indiscriminately, especially by the Hindu fishing communities in the district of Cannanore, to get benefits set apart for the Scheduled Castes. The castes who have been found to have infiltrated into the Moger fold are the Mokayans, Bovis-Mogeyars, Valans and the Kaipudakars. The caste name Moger has been used by these forward and backward communities to misrepresent themselves and this nefarious activity began almost fifty years ago.

The Mokayans of Madapally, Peringathur, Mambram, Edakad, and Perlasseri have, since the 1950s, styled themselves as Mogers, claiming that Moger is another name of their caste. In the early 1950s the Mokayans of Perlasseri and Edakad, formed a Moger cooperative society and enrolled their members as Moger. These members here have the same customs and traits as the Mokayans. The Mokayan members in other places like Mambram and Peringathur followed suit and several of them got false certificates as Moger, reportedly with the connivance of Marxist leaders. The Mokayans of Madapally have also started such infiltrations after contracting matrimonial alliances with the Mogers of Perlasseri. The children of these marriages are registered in schools as Moger on the strength of either one of the parents being a Moger.

No less than 350 families belonging to the Mokayan have grabbed benefits earmarked for Scheduled Castes by producing false certificates as Moger. A large number of them have got service in the Central and state government establishments, nationalised banks and public sector undertakings on the strength of their bogus certificates. The Bovis-Mogeyars of some villages like Kadangkode, Cheruvathur and Cannanore made a futile attempt to claim themselves as the Mogers, *en masse* like the Mokayans of Perlasseri and Edakad. It is believed that some individuals among the Bovis-Mogeyar have succeeded in getting false certificates as Moger in Cannanore.

The Valans of the Cannanore and Taliparamba taluks have also infiltrated into the fold of the Moger. This was started before Independence when this area was under the Madras Presidency. The Valans, being immigrants from Cochin, were not included in the list of the backward classes, but they identified themselves with the Moger. Authorities then included the Valans along with the Moger in the list of backward classes. The Valans get their children entered as Moger in school records. An initial impression one gets at their villages is that the Valans and Mogers intermarry a great deal. Many Valans, on the strength of their false documents, have got jobs in the armed forces, central and state government establishments, etc. It is estimated that 250 Valan households in the Cannanore district have infiltrated into the fold of the Moger.

The Kaipudukar, living in the Kasargod and Manjeswaram taluks, are farm labourers. They too have styled themselves as

Mogers for the past 20–25 years and are grabbing benefits earmarked for the Scheduled Castes. They are also called Mera, which is a sub-caste of the Holeyas of Karnataka.

Certain queries and investigations have led to the following revelations regarding the inclusion of the Moger as a Scheduled Caste. Buchanan (1808), Thurston (1909), Madras Fisheries Bureau, Bulletin 9, and Madras Presidency (1916) have cited the Mogers as a traditional marine Hindu fishing community of Karnataka (South Kanara district). This could not, however, be verified minutely during the present investigation.

Usha Rao (1981: 75) writes that in South Kanara, the Mogers are also called the Mogaveeraru in Kannada and Marakkalar in Thulu, but the Mogavirars of Kerala do not seem to know who the Mogers are. Marakkalar is said to be the term used loosely for all fishermen. She has noted that the Mogers "were never regarded as untouchables. So, it is interesting to know how this group came to be listed as Scheduled Castes". As stated elsewhere, the Kaipudukar a sub-group of the Holeya, are also called Mera, and Mugayaru (Thurston, 1909 II: 345). Rao says that instead of the Mera/Mogayaru being included as a Scheduled Caste, Moger was listed as such by mistake (ibid., p. 75). "Here, we find 'Mogare' being mentioned as a sub-group of the Holeys. It is not clear from the Census whether the Moger mentioned therein, are a sub-division of the Holeyas of South Kanara or the traditional fishermen caste itself. However, the Ethnographic Notes on Scheduled Castes and Scheduled Tribes (Census, 1961), published recently clarifies this doubt. It states the local people both laymen and officials are aware of the fact that the fishermen community of Mogers, has never been subjected to such treatment as would relegate it to the status of a Scheduled Caste … . There is another indigenous caste group locally known as *Mugera, Muggera* or its synonym Mera which has been suffering social disabilities (such as denial of services of barbers, washermen, of access to public places and temples, etc.) for generations. Stuart (1893, XIII) writes about Mera as being one of the major sub-divisions among the Holeyas of South Kanara (cf. Thurston, 1909, V: 62). But later publications do not contain any information on the Meras … . It is this community of Mugers that has been deemed as a Scheduled Caste ever since 1936, though perhaps due to a spelling mistake, the term 'Moger' has been adopted in various orders. It is quite likely that the 'Meras' preferred to be known as 'Muggera' as they were likely to pass off as 'Mogers' since the latter enjoyed a higher social status. However, with the prevailing confusion regarding the caste 'Moger' as well as its spelling (as admitted by the Census itself) it is quite likely that the benefits disbursed so far may not have reached the 'Muggeras' at all to whom it was actually meant and may have been availed of by the 'Moger' caste since their name has been listed in the Schedule. It is interesting to note that the population of 'Moger' which was 7.800 in 1961 has more than doubled, and is 16,994 according to the 1971 Census, while 'Mera or Muggera' is not mentioned in the list of Scheduled Castes even in 1971 .... "It is surprising how such an error has been allowed to continue in the successive Censuses" (ibid., pp. 145, 57).

This error was committed in 1936 and possibly the Mogeyar and Mogavira (and perhaps other similarly pronounced castes) have infiltrated into the fold of the Moger. The question now is, is there actually a caste called Moger? If there is, what is their traditional occupation and are they known by any other name as well and are they eligible for inclusion in the list of Scheduled Castes? Although Thurston has stated that the Mogers are fishermen (1909 V: 65), during the present survey no fishing community has been identified who are genuinely Moger. Of the two fishing communities claiming the Moger status, one was found to be Mokayans and the other, the Valan. Another community identifying itself as Moger in Kerala are actually Kaipudukar or Mera, a group of agricultural labourers found in the extreme north. On the basis of Rao's findings, the inclusion of the Moger was a mistake (pp. 11, 75–76); consequently, the Mogers must be struck off from the list of the Scheduled Castes and the Mera substituted. It is strange that for the past 50 years this mistake has remained unrectified. There are many anomalies in the present structuring of the list of Scheduled Castes and Scheduled Tribes. Lapses and loopholes in the general administration had led to a number of mistakes and deliberate manipulations. It is evident from the cases cited in this study that a good deal of infiltration is going on. This must be stopped without further delay if the purpose of really helping the Scheduled Castes and Scheduled Tribes is to be achieved.

4

Technology of Fishing in the Ecological Context

The kinds of crafts, gear and techniques employed for fishing in Kerala are bewildering in their variety. The number and methods of their employment depend upon factors such as marine environment, climatic conditions, the species of fish, capital at the command of the fishermen and the cultural ecology of the region. This chapter examines whether the technique of fishing is related to the traditional knowledge of fishing communities, behaviour of the ocean, varieties of the fish, ecological systems, and cosmological knowledge.

We shall make an attempt in this chapter first of all to give the parameters of fishing technology in selected fishing villages of Kerala in order to show the correlation between production and distribution which characterise the Mappila, Hindu and Latin Christian fishermen. We shall also focus our attention to delineate the differences in methods of fishing on the Kerala coast as there is a tendency for different types to dominate in given ecological situations. Certain generalisations have been attempted on the regional and district level taking into account the whole state of Kerala, on the basis of the field work conducted during 2006–07.

Tanur Fishing Village, Malappuram District

Hook-and-Line Fishing

During the 1970s the fishing lines used in Tanur were of three types, viz., (1) *Kaiccunda* or *Otta Cunda* (hand lines), (2) *Beppu* or *Ayiram Cunda* (long lines), and (3) *Cangala Cunda* (chain lines with big shank hooks) (Mathur, 1978). Mathur (1978) has described the following traditional hook-and-lines employed by the Mappila fisherfolk of Tanur:

Kaiccunda

The simplest method employed by the Mappila for catching fish was the *Kaiccunda* (hand line) in Tanur. Since hand lines are shot from a single dugout and generally remain unwatched, either in shallow waters or in the bottom of the sea for a long time, they are known as a kind of set lines. The hand lines were used in different coastal regions by the Mappila fisherfolk, Latin Christians of Travancore and Hindu fishermen, particularly Bovis-Mogeyar of Malabar.

Beppu or Ayiram Cunda (Long Lines)

The *Beppu* (long lines) in Tanur in the past constituted a special form of set lines consisting of about 1,500 feet length of fishing twine made of cotton yarn usually fixed on planks made from arecanut barks. The long lines were provided with equidistant thinner branch lines called *Chattu* (snoods) to which were attached the fishing hooks.

Cangala Cunda (Chain Line)

The *Cangala Cunda* or chain line with a different kind of strong hook was used for catching sharks. The line was three times heavier than the long lines. Unlike the long lines the snoods of the *Cangala Cunda* were made of chain or thicker yarn. The *Cangala Cunda* was used in Tanur whenever larger sharks were sighted. It consisted of 6,000 feet length of fishing line with a maximum of 100 hooks.

These traditional hook-and-lines are now replaced by drift lines. In other words, the *Kaiccunda, Beppu* or *Ayiram Cunda* and *Cangala Cunda* have been replaced by the *Otta Cunda* (drift line) which is employed by the Mappila fisherfolk and Latin Christian fishermen in different regions of Kerala, particularly in the southern districts of the state. One end of the drift line is fastened to a fishing boat while the other is held by the fishermen. In the line fishing technique fish are baited in such a manner that they are neither able to take it away nor can they escape once it is taken, but in the case of *Otta Cunda* drifting plastic fish is the bait.

Beppukar, the hook-and-line fishermen of Tanur, have discarded the traditional *Ayiram Cunda, Kaiccunda* and *Cangala Cunda*. About 100 *Beppukar* families adapted the new technology of line fishing 20 years ago. In the past they could go with hook-and-lines and a dugout (*Vallam*) upto 20 fathoms. Nowadays, not only the *Beppukar* but also *Valakkar* have adapted modern hook-and-lines and fibre *Vallam* (boats) fitted with outboard engines of 9.9 HP. In fact, with line fishing, they can go upto 35 to 40 *Maaru*.

Our informant, Mohammad Kutty, Moosante Purakkal of Tanur, told us that he has preserved the traditional *Ayiram Cunda* for presenting it to the Indira Gandhi National Museum of Mankind, Bhopal. The traditional *Beppu Toni* or *Vallam* (dugout) and hook-and-lines have given room for fibre boats costing between Rs 65,000 and 80,000 and *Otta Cunda*. A fibre boat is made of *Ayini* planks and is covered with fibre.

In the technique of line fishing fishes are lured by offering artificial fish. The bait may be either with hook or without it. Hooks help to capture the fish by ripping. We have found that hand lines, set lines, troll lines and drift lines are used by different fishing communities in Kerala. The hand-line is operated with a few hooks by a single man. In the set-line fishing several hooks are set in a fixed position at regular intervals with or without baits. In the southern districts of Kerala and also in Tanur drift lines are mainly employed over larger areas. These lines are fastened to a fishing boat. They have considerable lengths and the snoods are generally longer.

Hand line is the simplest form of fishing. However, the number of lines, which can be operated by a person is limited. With the development of set lines it is possible to use more lines. From simple fishing lines, capable of catching a single fish, have emerged the long lines, which have several hooks and extend to a considerable length. These are a kind of unwatched set lines. The need to operate over a larger area using a movable fishing line led to the development of trawling.

Otta Cunda (Drift Lines)

The Mappila fisherfolk, both *Beppukar* and *Valakkar*, have adapted the modern technology of hook-and-line fishing, popularly called *Otta Cunda*. Apart from Mappilas, the Latin Christians of Quilon and Trivandrum have also started employing *Otta Cunda* for catching larger fish.

The *Otta Cunda* is made up of an artificial plastic fish and a plastic rope measuring 25 to 30 *Maaru* and steel hooks. Three plastic lines are tied to a small artificial fish whereas six plastic lines are attached to a big artificial fish like sardine. Artificial fish made of steel measuring three inches is also employed for hook-and-line fishing. The lines are tied to the head of the artificial fish and the hooks are pierced into its tail. Then the fishermen employ the *Otta Cunda* as hand lines which are cast from a single boat with an outboard engine. Sometimes, a number of *Otta Cunda* are launched into the sea from a boat for catching larger fish like *Ayikkora, Sravu* and *Tirandi*. The *Otta Cunda* is shot from a single fibre *Toni* fitted with an outboard engine of 9.9 HP with the fisherman holding the rope in a standing position. The *Otta Cunda* is operated from west to east, either in shallow waters or in the bottom of the sea.

The Mappilas of Tanur launch the *Otta Cunda* (drift line) by holding the lines in a standing position. The artificial fish with three to six lines is dragged with the help of the 9.9 HP motor. Two lines are operated from a mechanised fibre boat on both sides and the third line is employed from the centre of the boat. They go for fishing upto 30 kms in this manner usually at 6 a.m. and return at 6 p.m. GPS (Global Positioning System) are employed particularly in the months of *Chingam* and *Kanni* to find out the fish shoal in rocky places. They get catches worth between Rs 5,000 and Rs 10,000 on an average per expedition. *Ayikkora* is the major catch from this technique.

The Mappilas catch bottom fish with the help of GPS. The *Beppukar* say that the rocks of the sea bottom can be located by it. This knowledge is used later on when the fishing unit reaches over the same rocky places. They also believe that *Ayikkora* and larger fish generally inhabit the rocky places. They hold the view that from Tanur to Blangad the sea bottom is filled with *Chali* (mud) which can go upto 35 *Maaru* in the muddy bottom. After 35 *Maaru* the bottom is sandy. According to our *Beppukar* informants, GPS helps them to identify different currents, winds, tides and the distance covered during the expedition and their home. *Ottabeppu* is employed for catching *Ayikkora* and *Chuda* in the months of *Chingam* and *Kanni.* They go upto Ponnani, a distance of 20 miles from Tanur. The fishing boat is fitted with a 9.9 HP motor for hook-and-line fishing.

Fishing Crafts

The sea is generally very rough on the east coast and there are few landing stations. The main crafts employed in the East Coast for fishing are the *Catamarans* (floating crafts) and the non-rigid country crafts. The sea is usually calm on the West Coast compared to the East Coast except during the south-west monsoon when fishing is almost suspended for all practical purposes. Hence, small canoes with an outboard engine below 10 HP are used, particularly during the period of trawling ban.

Mathur (1978) has classified the traditional fishing crafts into two categories, namely, (i) the indigenous or non-mechanised, and (ii) mechanised. The non-mechanised crafts are again divided into three types: (1) *Catamarans*, (2) the dugout canoes (*Beppu Toni* or *Odam*) and (3) plank-built *Tonis*. Thus, the fishing boats may be divided into five categories, viz., (1) *Catamaran*, (2) indigenous small non-mechanised boats, (3) outboard small fibre boats, (4) *Valiya Vallam* with Leyland engine, and (5) trawlers with inboard engines.

Catamaran

Catamaran is one of the indigenous fishing crafts employed along the East Coast from Orissa to Kanyakumari in Tamil Nadu and Trivandrum and Quilon districts of Kerala. The *Catamarans* are 2′ and 1.4″ in length. Four to five logs of light-wood are tied together with coir. When the logs are tied together they become curved and shaped like a canoe. The craft is rowed with the help of a bamboo pole. When the *Catamarans* return after fishing expeditions, the logs are separated and kept in the sun for drying. A single sail is used. Such *Catamarans* are widely used in Virinjam of Trivandrum district. In Virinjam of Kerala and Kanyakumari, Latin Christians (Mukkuvar) are the principal users of this fishing craft. The Latin Christians mainly go out for fishing against the wind in their *Catamaran.* These *Catamarans* are employed even today for hook-and-line fishing and also for launching the gill-nets. This craft is widely used in Trivandrum district for shooting seine-nets called *Tattu Vala.* Several *Catamarans* have been mechanised in Virinjam as a part of the Tsunami Rehabilitation Programme.

Fishing Boats

The most common types of boats used in Tanur during the 1970s were *Vallam* or *Beppu Toni* (dugout canoes) and *Toni* (plank-built). The *Beppu Toni* was used mostly for traditional hook-and-line by the *Beppukar* (the Mappila hook-and-line fishermen). These two types of boats have been replaced by plank-built fibre boats in Tanur.

Dugout Canoes

Dugout canoes were made from large logs of *Ayini* (Ailanthus), Mango and *Chini* (Austiarie) trunks by scooping out the insides. All over Kerala the *Odam* or *Vanji* (dugout canoes) had a length varying from 32 to 42 feet. The *Ottathadi Vallam* or dugout canoe of Alleppey was generally of 42 to 52 feet. The *Odam* of Malabar and *Ottathadi Vallam* of Alleppey were the most important fishing crafts used in Kerala. The *Ottathadi Vallam* was also found in Quilon district, particularly in Azheekkal, Alappat Panchayat. *Beppu Toni* of Tanur had a length of 27′6″ and a depth of 1′11″. The *Odam* of Cannanore, Kasargod and Kozhikode had a length of 32 to 42 feet. A crew of 12 to 18 men was employed for the operation of these boats in these districts. But for the *Beppu Toni* (dugout) only four persons were employed for fishing in the sea. Three sails (masts) were used for rowing the Beppu Toni whereas a single sail was employed for bigger canoes like *Odam, Vanchi* and *Ottathadi.* Sails were used for rowing the boats in the same direction of the wind. But in the case of unfavourable wind the sails were reduced from three to one in the case of the *Beppu Toni*. The sails are seldom used against the wind. A crew of four to five men used to go out for fishing in the *Beppu Toni*, while the *Odam* of Cannanore, Kozhikode and *Vanchi* of Cochin, Trivandrum and Quilon, employed a crew from 9 to 12 men. The *Ottathadi* of Alleppey had a crew of 18 men. The *Ottathadi Vallam*, the largest dugout in Kerala, had a length ranging from 12 Kol (1 Kol = 2 ½ feet)–22 ½ Kol.

Plank-built Boats

In Tanur, *Cheriya* fibre *Toni* (non-mechanised boats) are made of *Ayini* wood covered with fibre. It has five *Kallis* (compartments) and six *Padis* (cross bars). Two men are employed for going out for fishing *Cheriya* Mathi (small sardine), *Kora* (Jew-fish), *Mullan* (Silver Belly), *Cheriya Chemmin* (small prawn), *Sura* (Cuda), *Karikadi* (one of the species of *Chemmin*) and other small species of fish. The Mappilas still use the traditional sails, oars and rudders for rowing the *Cheriya Toni* to the distant seas. Sometimes *Ayila Vala,* which is made of plastic known as *Tangis,* is also used from this fibre covered boat. There are about 500 such traditional fishing units in Tanur. The nets used in this boat are popularly known as zero point plastic *Vala* and nylon *Vala*. The Mappilas go out on fishing expeditions with *Cheriya fibre Toni* upto a distance of 10 kms.

Plank-built boats (*Toni*) were widely used by the Mappilas of Tanur, particularly by the *Valakkar* (those who used exclusively nets and boats). But these boats which were formerly known as *Tonis* in the districts of Malappuram, Ernakulam, Trivandrum, Alleppey and Quilon have now been replaced by *Kettuvallam*. In this connection Mathur (1978) observes: "In Alleppey and Quilon Districts they are called *Kettuvallam*. The planks are fixed to each other in Tanur with the help of coir ropes, copper nails and a gum called *Pantham*. This gum is obtained from the Pine tree (*Acanthus ilicifolius*). The gum mixed with coconut oil is boiled and the resultant solution is used for fixing the planks with the help of pieces of clothes. The wooden frames are fixed with thin coir ropes. Thus the *Tonis* of Tanur are carefully built, the planks being secured by copper nails. The copper nails are non-corrosive. The plank-built boats in Malabar are popularly called the Tanur Toni and they are made of *Ayini* and *Chini* trees, as stated earlier" (Mathur, 1978: 133).

The *Kettuvallam* was introduced in Blangad, a fishing village in Thrissur district, during the 1980s. The dugouts are greatly enlarged by fixing wooden planks to each other with the help of iron nails and coir ropes. Nine to eighteen planks are fixed together to manufacture a *Kettuvallam.* The big canoes like *Valiya* fibre-covered *Vallam* or Valiya glass *Vallam* are employed in Tanur and the neighbouring fishing villages with a crew of 40 to 50 fishermen. Two carrier dugout fibre boats are used for collecting the catches from the big canoes and for transporting them to the landing station. These transport vessels are fibre-covered boats with three to four men on them. There are about 1000 fishing units employing fibre-covered *Tonis* in Tanur. Today a single fibre *Toni* is used only for launching *Ozukkuvala* (gill-net), *Peyithan Vala* (sole net – Disco net) and also *Pothan Vala* in Tanur and Blangad.

Mechanisation of Boats

Fibre Toni

The traditional Tanur plank-built boats have been replaced by plank-built fibre *Tonis*. These boats are covered with fibre. Plywoods are fitted to each other and fibre is applied on them. In these boats four men can go for fishing. A fibre *Toni* consists of seven to eight bars or *Padis* and five to six compartments or *Kallis*. Each *Kalli* has a definite and prescribed function during the fishing operation. The *Kallis* from *Amaram* (stern) to *Aniyam* (prow) are appropriated for keeping the engine, nets, floats, sinkers, oars, rudders etc. by different members of the crew. "The *Chukkankaran* (steersman) directs the fishing operations from the *Chukkankalli* which is the nearest *Kalli* to the *Amaram*. The *Chukkankaran*, being the captain of the boat, detects the fish shoal, the blowing of the wind, the flowing of the current and tides and gives instructions to the other members of the crew to adopt and adjust their know-how according to the contingencies from the *Chukkan Kalli.* The floats and sinkers are generally kept in the *Chukkankalli.* It is from this *Kalli* that they are tied to the net just before the launching. They are generally dropped into the sea by the *Chukkankaran* with the help of *Tammankallikaran*. The *Tammankallikaran*, one of the important oarsmen who occupies the *Tammankalli*, hauls up the net when the fish is entangled and puts the net back into his *Kalli*. The third *Kalli* is called *Edakalli*. This is the place for the third member of the crew. The cost of repairing a dugout is higher than that of a plank-built boat. A damaged dugout is repaired by fixing a wooden plank, but this weakens the boat. The more the number of planks fixed to a dugout, less the number of years it survives. However, it has been estimated by expert fishermen that the life-span of a dugout canoe is 20 to 25 years whereas a plank-built boat lasts only for 10 to 15 years" (Mathur, 1978: 134).

The outboard engine varies from 8 HP to 9.9 HP. Nylon *Mathi Vala* and nylon *Ayilachala Vala* as well as plastic *Vala* or *Tangis* are used for catching different species of fish. *Pothan Vala* and Disco *Vala* are also used for catching fish from the fibre *Tonis*. For shooting the *Pothan Vala* two fibre *Tonis* are employed. During the 1970s a number of cotton nets were employed for catching different species of fish by engaging two plank-built boats. Nowadays all the cotton nets have been replaced by nylon and plastic nets. Traditional plank-built *Tonis* have been discarded by the Mappilas and they have adapted modern plank-built fibre covered *Tonis,* popularly called fibre *Tonis*. In each of the two fibre covered *Tonis* a crew of four men are employed as fishing labourers.

Mohammad Kutty owns a fishing unit which consists of a small fibre *Toni*, a *Pothan Vala* and a *Cheriya Peyithan Vala*. For launching the *Pothan Vala,* two fibre *Tonis* are required whereas for a *Cheriya Peyithan Vala* only one boat will do. The *Valiya Pothan Vala* has a weight of 15 kgs whereas the small *Peyithan Vala* weighs only 5 kgs. The *Pothan Vala* helps in hauling up demersal fish like *Mantal* (sole), and Karikkadi that are caught. This is no less a trawl net according to some traditional fishermen. The cost of a *Pothan Vala* is Rs 350 per kg weight. Normally it weighs 10 to 15 kgs. It will have a 5 kg rope and 10 sets of floats and four sinkers. This unit has a five-member crew including the owner (Muhammad Kutty). The other members are Moideen Bava, Anwar, and Rafi (Muhammad Kutty's sister's son). This *Cheriya* fibre *Toni* including its motor cost Rs 25,000 and it can operate upto a distance of 30 *Maaru*. The crew go for fishing at 5 a.m.

and return by 10.30 a.m. This unit went for fishing on 28 February 2007 and they got a catch worth Rs 450. At the same time they incurred an expenditure of Rs 500 towards fuel and food, incurring a loss of Rs 250. Muhammad Kutty told us that his unit could not go for fishing on 26th and 27th February, 2007 due to bad weather that led to a poor fish harvest. His boat is fitted with a Yamaha 8 HP engine that runs on Kerosene oil. The boat was purchased 20 years ago.

Valiya Vallam

The Mappilas of Tanur have adapted *Valiya* glass fibre *Vallam* (big fibre glass canoe) which costs from Rs 35 to 40 lakhs. The crew number from 30 to 35 men and they go out for fishing using *Tangu Vala (Valachilakkunu Vala* – purse seine nets) for catching *Ayila* and *Mathi*. The *Valiya* fibre glass *Vallam* is fitted with a Leyland engine. Purse-seine nets are encircling gears provided with a purse line at the bottom, which are pulled through the purse rings close the bottom of the gear preventing the escape of the fish. They can be operated from one or two boats.

During the period of ban of trawling from June 15th to August 15th, 2006, particularly after the Supreme Court verdict, the Mappilas could not employ big glass fibre canoes because the boats are fitted with Leyland engines which are more than 10 HP. But after the lifting of the ban, the Mappilas have been engaged in fishing with *Valiya* Glass fibre *Vallam* and *Tangu Vala.* According to our informants, the Mappilas owned 85 *Valiya* glass *Vallam* in Tanur. It is interesting to point out here that each *Valiya* glass *Vallam* is supported by two carrier fibre *Tonis*. The main function of the carrier fibre *Tonis* is to collect the catches from the *Valiya Vallam* and transport them to the auction centre of the landing station. The Mappilas normally operate the *Valiya* fibre glass Vallan in the day time. They have to incur an expenditure of Rs 5,000 to 10,000 towards fuel and food charges. The Mappilas of Tanur have not adapted the trawler boats and trawl nets, unlike the Christian and Muslim fisherfolk of Beypore, Puthiyappa, Koyilandi and Mappila Bay of Cannanore district.

Today, the Mappila fisherfolk of Tanur own no less than 100 fibre glass *Vanchi or Vallam* and *Tangu Vala (Valachilakkuna Vala* –

encircle purse net). As many as 35 to 40 men go out to catch *Mathi, Ayila, Chemmin, Para* and *Avoli* in a big *Vanchi*. The traditional *Vallam* of *Beppukar* has been replaced by fibre *Vallam* which costs between Rs 40,000 and 65,000 each. Both *Valakkar* and *Beppukar* have adapted the *Ayila Vala* (the yarn count is 56 mm) costing about Rs 50,000 against plastic nylon nets costing Rs 10,000 each.

The traditional *Beppu Toni* made of *Ayini* wood has been replaced by fibre *Tonis* for employing *Ozukkuvala* (gill-nets). The *Beppukar* and *Valakkar* use *Ozukkuvala, Ayilakkoli* and *Pothan Vala*. In the past *Vakkuvala* (hemp net) was used for catching *Etta,* and *Peyithan Vala* for Sole. The *Etta* and *Sravu* are generally found on the rocks at the bottom of the sea. In consequence of overfishing by foreign trawlers, *Etta, Tirandi, Sravu, Mullan, Avoli (Karutha Avoli* and *Velutha Avoli*) have almost disappeared, according to the Mappilas. Similarly, *Udumbu,* which was used for medicine (Jguana) has also disappeared. The Mappilas have voiced their demand for prohibition of foreign trawlers in deep sea fishing. Larger boats including trawlers from the southern district, of Kerala appear in Tanur and the neighbouring villages, which affect the catch of the Mappila fishermen. Thus, there is a shortage of fish in the sea. Mappilas welcome ban of trawling during the spawning period since a bumper catch is generally expected during that period.

Fibre Tonis

A fibre *Toni* in Tanur costs between Rs 65,000 and 80,000. The planks are made of *Ayini* wood (plywood) and are covered with fibre. *Ayila Vala* (mackerel net, in which the nylon yarn count is two) is used in fibre *Tonis* for catching mackerel.

The Mappila fishermen of Tanur demanded that apart from banning trawling, commercial fishing by trawlers with foreign ships should be banned. The poaching by the fishermen from Tamil Nadu in Tanur and the neighbouring villages should also be stopped.

Impact of Modern Technology in Tanur

The Mappila fishermen of Tanur have discarded their traditional wooden dugouts and plank-built boats and have adapted small fibre boats and nylon *Pothan Vala*. The advantages of the fibre boats are:

1. Less weight,
2. Coating on the outer part is not needed,
3. With outboard engine it can be operated to any distance,
4. Sea water will not enter into the boat, and
5. The fibre boat may not capsize whenever violent winds occur in the *Karakadal, Puram Kadal* and *Yama Kadal.*

Adaptation of Bigger Vallam

Over 85 Mappila fishermen in Tanur have adapted fibre glass boats in each of which 40 men can go out for fishing. But, at the same time, they have not adapted trawler boats in Tanur. Two small boats are used to tranship the catch to the landing stations. Since there are no harbour facilities in Tanur, fibre glass boats with inboard engines are operated only from Ponnani and Beypore harbours. The Mappilas have adapted plank-built fibre boat in which a crew of four men can go for fishing. There are more than 500 such fishing units at Tanur with outboard engines. *Peyithan Vala*, in which the yarn count is 18, is used for catching *Mantal* and *Chemmin*. Two plank-built boats were used in non mechanised fishing in olden times but today only one fibre boat is used for launching the *Peyithan Vala*.

Economics of the Bigger Fibre Glass Boats

Thirty to forty fishermen jointly own a *Valiya Vallam* (fibre glass boat) with Turbo engine and *Tangu Vala*. The cost of a *Valiya* fibre glass *Vallam* with inboard engine is around Rs 15 lakhs and the cost of a *Tangu Vala* is Rs 7 lakhs. The cost of two small transhipment fibre boats would be Rs 3 to 4 lakhs in case they are fitted with Yahama Turbo engine having 22 HP. Thus, the cost of a fishing unit consisting of a *Valiya* fibre glass *Vallam* with inboard engine, a *Tangu Vala* and two fibre outboard engines would be Rs 30 lakhs. We were in Tanur on 20th July, 2006 when six fishing units undertook their expeditions from the Elaram sea-coast. Each unit had 30 men, besides their two lighters with their three men crew. They went out for fishing at 5 a.m. and returned on the same day at 9 p.m. They went upto 30 *Maaru*. They got mackerel worth Rs 65,000.

The cost of fishing expedition:

Diesel	=	200 Litres	=	36 × 200	=	Rs. 7,200
Petrol	=	15 Litres	=	42 × 15	=	Rs. 630
Kerosene oil for lighters	=	125 Litres	=	125 × 27.50	=	Rs. 3437.50
Food			=		=	Rs. 3100.00
Total expenditure					=	Rs. 15,000
Total income					=	Rs. 65,000
Less expenditure					=	Rs. 15,000
Balance					=	Rs. 50,000

The owner of the unit had to pledge to a bank his property for the purchase of equipment and the interest charged was six per cent of the catch. The middlemen (*Hawalakkar*) who advanced loans to the owner of the equipment get 10 per cent of the catch. Then the commission agent got three per cent and money-lender who advanced Rs 1 lakh two per cent of the catch. After deducting these percentages of the catch, the remainder is shared by the owner of the boat and net (60%) and fishing labourers (40%).

A head-load worker gets the value of one basket of fish for carrying ten baskets of fish from the boat to the landing station. The expenditure incurred by the fishermen for votive offerings (*Vazhipadus*) to the local mosques and temples is not included in this division of catch.

Another fishing unit went out for fishing on 15 July 2006 and returned on 21 July 2006 after incurring a total expenditure of Rs 60,500.

Value of the total catch of mackerel	=	Rs. 1,20,000
Fuel and food charges	=	Rs. 60,500
Actual income	=	Rs. 59,500

The owner got 50 per cent and the remainder was divided between 35 labourers. Each got Rs 1000 for seven days.

Payment to the 28 labourers	=	28 × 1000	=	Rs. 28,000
Payment to additional seven men crew – for lighter boats	=	7 × 300	=	Rs. 2,100
Total			=	Rs. 30,100
Owner's share of the catch	=	Rs. 59,500 × 50/100	=	Rs. 29,750

The loss amounting to Rs 350 would be adjusted in the next fishing expedition.

Sardine, mackerel and prawn are mainly caught by the *Valiya* fibre glass *Vallam* and *Tangu Vala.*

Nylon Ayilachala Vala or Nittielakkal

The Nylon *Ayilachala Vala* is still used in Tanur. Writing on the technique of this net, Mathur (1978) observed: "This drifting technique is very commonly employed by the Mappila fishermen and has been introduced in Tanur in the year 1968. The employment of this technique depends upon the flow of the wind. The net cannot be released against the wind if the fishermen want to employ the *Nittielakkal* method. The net is shot according to the direction of the wind. For instance, if the wind is from east to west, the net is to be cast in the same direction anchoring the boat against the wind, i.e., on the western side. Again, if the fish migrate from north to south, the net is launched in that direction. It may be noted here that the *Nittielakkal* technique is only employed at night because the mackerel cannot see that net. As a matter of fact a single fishing unit can catch mackerel independently by this technique. As has already been mentioned, there are five *Murivalas* (larger piece nets) for a mackerel net. Thus, a fishing unit of *Ayilachala Vala* consists of a dugout, five *Murivalas* and a five man crew. As soon as the fishermen cover the predetermined depth of the sea, the fifth piece net (*Anchamathae Murivala*) is dropped first from east to west, if the wind is in that direction, followed by the other *Murivalas*, one after another serially. After shooting the entire net a 40-feet long *Kamba* (coir rope) is tied to the *Kadasu* (heavy head line) of the first *Murivala* and the other end of the *Kamba* is tied to the *Aniyapadi* (the cross-bar nearer to the stem of the boat). There is a hole in both *Amarapadi* and *Aniyapadi* and through this only the *Kamba* is tied to the boat. The lower end of the net is weighted with 100 granites or red brick sinkers, and the upper edge is raised by 100 sponge floats. Nowadays lead sinkers are also used. The floats and sinkers are attached to the net in equal distance. Each of the five *Murivala* has 20 sponge floats and 20 red brick sinkers. The distance between the net and the boat is about 40 ft. i.e., the length of the *Kamba*. The boat oscillates according to the direction of the wind. A bamboo pole is tied to the *Kumbidikalli padi* in order to prevent the oscillation of the boat. Once the bamboo pole supporter is tied and lowered to the sea, the fishermen wait for the migration of the fish into the net. They examine the net to find out whether the fish is caught after waiting for 45 minutes after anchoring the boat by pulling the end of the net. If the fish is caught they haul the net and remove the catch."

Valaccilakkal

Another technique used by the Mappilas of Tanur for fishing mackerel is called *Valaccilakkal* technique (encircling or entangling technique). In this connection Mathur (1978) has said that this technique is employed only in day-time without taking into consideration of the occurrence of the wind and current for catching *Ayila* (mackerel). As many as 16 fishing units jointly undertook an expedition in 1971 by employing this technique with the help of a corresponding number of boats. But they used only eight nets out of the total of 16 for entangling the *Ayila Polappu* (mackerel shoal). This technique was introduced in Tanur recently, though it was widely practised in the neighbouring villages as well as in the region of Kozhikode. Bumper catches were hauled employing the *Valaccilakkal* technique, particularly in the months of *Makaram, Kumbam, Minam* and *Medam*. The fishermen do not go out for fishing in *Minam* with the mackerel net because at night the sea is generally very rough and the net cannot be launched. All the eight nets were tied to the sixteen boats and the fish shoal was encircled in the *Valaccilakkal* technique. Even today the fishermen beat the cross-bars of their respective boats with oars and rudders and also make sounds to scare the fish and drive them into the nets. When the fish is entangled, the net is hauled up and the catch removed. The whole catch was divided into sixteen shares. Each unit divided its share among the members of the crew according to their custom. The maximum catch was hauled by employing this technique in the month of April. This technique has the advantage of preventing the fish from migrating to the bottom as well as to the distant areas of the sea. The *Ayilachala Vala* is not used for fishing in dark nights, but only in the day-time and on moon-lit nights.

Ozukkuvala (Gill-net)

Writing on the *Ozukkuvala,* Mathur (1978) has observed: "The *Ozukkuvala* is another important gill-net used for catching mainly *Ayikkora* (Seer) or *Nemmin* (Seer) in Tanur and the neighbouring villages like Blangad and Puthiyappa. This is a floating gill-net. The traditional *Ozukkuvala* was made of the hemp fibre. It has been replaced by nylon floating gill-nets. As the fish is caught by floating the net according to the current and wind or sometimes aimlessly, it is called *Ozukkuvala* (literally a floating net). The nylon *Ozukkuvala* consists of 10 to 12 pieces or *Murivalas*. For the first 10 *Murivalas,* the *Onpatamnumber* (the count of the yarn is 9) is used but in the remaining *Murivalas* the count of the yarn five is used. Each *Murivala* has a length of 72 to 84 feet and a width of 30 to 32 feet. The *Murivalas* are assembled together just before launching the net with the help of a joining string at a place called *Allu*. The size of the *Allu* is 7–8", and that of the net is 4–6". When a *Murivala* is attached to another a *Kavi* (a big wooden float) and 12 *Ovukals* (small wooden floats) are attached, in order to keep the net afloat. Each *Murivala* is weighted with five stone sinkers. The whole net is thus weighted with sixty stone sinkers and its upper end is raised above by 12 *Kavis* and 144 *Ovukal*."

There are two types of nylon *Ozukkuvala*s: (i) *Appiravala,* and (ii) *Aralavala*. The former has the mesh size of 6 inches for catching larger fish like *Ayikkora* or *Nemmin* while the latter with the mesh size of 3.5"–4" is used for catching *Avoli* (Pomfret), *Vala* (Sabre fish) and *Palamin,* besides small sharks like *Kalla Sravu, Pulli Sravu* and *Vala Sravu*. Thus the mesh size is a determining factor in hauling different species of fish according to their size.

We had also stated that the *Ozukkuvala* is generally employed from *Chingam* to *Edavam*. The net is launched at a depth of 7 to 20 fathoms from a single dugout or sometimes from a plank-built boat. Nowadays the *Ozukkuvala* is launched from a fibre covered boat. During the months of *Chingam* and *Kanni,* the *Ayikkora* come up to the surface water as *Polappu* (shoal). The float rope's length is then reduced and it is tied to the net very closely so that the net will be kept on the surface water. Again, 50 per cent of the stone sinkers are taken away for catching the fish which appear on the surface water.

In other months *Ayikkora, Vala* and *Avoli* appear in the *Idanir* (midwater) and bottom. In the month of *Tulam* small fish live in the *Idanir*. Bigger fish like *Ayikkora* and small shark prey on small fish. In such circumstances the float rope is extended from one *Maaru* to two so that the net is allowed to stay in mid-water. From *Vriscikam* to *Edavam* most of the fish including *Ayikkora* live in the bottom. Small fish like *Chemmin* and *Mantal* live in the bottom and the large fish like *Ayikkora* prey upon them. The float rope is extended and all the sinkers are tied to catch the bottom fish.

The *Ozukkuvala* is operated day and night but it is launched only when the night is dark. *Ayikkora,* which is caught by this net, moves towards *Tuyi* and gets entangled in it. During the moon light the *Tuyi* disappears and the fish see the net and run away.

Mathur (1978) observed that "The technique of launching and hauling the *Ozukkuvala* is similar to that of the *Ayilachala Vala*. As has been already stated that when the easterly wind blows the net is launched from east to west. In other words, the net is shot in the same direction of the wind. If the easterly wind is counter-balanced by westerly winds, the *Ozukkuvala* is not operated. When the easterly and westerly winds blow simultaneously, the fishermen haul up the nets, fall upon one another and the floats and sinkers get entwined. This is called the *Kettupenayal*. Once it happens the fishermen return to the shore and separates the net from *Kettupenayal* (entanglement)."

The *Ozukkuvalas* are launched even today from either a single fibre *Toni* or two fibre *Tonis* depending upon their length and width.

The technology of fishing in Tanur is still related to the traditional knowledge of currents (*Nir*) and winds (*Kattu*). The Mappilas hold the view that the following are the favourable and unfavourable *Nir* and *Kattu* for the operation of various techniques of fishing:

Favourable winds and currents	Nature of catch
Vadakkan Kattu and Vadakkan Nir	Good catch
Tekkan Nir and Thekkan Kattu	Poor catch
Vadakkan Nir and Vadakkan Katt	Poor or Nil
Tengara Kattu and Tengara Nir	Nil
Tembrayil Nir and Tembrayil Kattu	Bumper catch

Puthiyappa Fishing Village, Kozhikode District

Puthiyappa is one of the fishing harbours of Kerala. It was commissioned in 1985. The Mukkuva is the numerically dominant fishing community in this fishing village of Kozhikode district. There are no Arayans in this village. In 1971 when we did our fieldwork in Puthiyappa the population of the Mukkuvas was reported to be less than 5000. Now it has almost doubled.

Puthiyappa comes under the Elathur Panchayat and it is outside the Kozhikode Corporation whereas the Puthiyangadi, a neighbouring Mappila fishing village, comes under the said Corporation. According to our informants there are about 500 households of Mukkuvas in this village who subsist on fishing.

Puthiyappa fishing harbour is situated about 6 kms from the city of Kozhikode. Here traditional nets are still used for fishing various species of fish. More than 300 mechanised boats were under repair in June, 2006. These boats are fitted with outboard Yamaha engines of 9.9 HP motors and Leyland engines. An outboard Yamaha engine requires 300 litres of kerosene to go for fishing 20 kms away. On the other hand, a big *Valiya Vallam* with Leyland engine needs only 300 litres of diesel for covering a distance of 30 kms. There is a subsidy of Rs 1.50 for diesel, but limiting the amount to Rs 1,000 per month. Trawling was prohibited from 15th June to 15th August, 2006. However, Leyland boats were in operation with nylon nets for catching *Ayila, Mathi* and *Chemmin*. The sea was very rough in June. Five *Vanchis* or *Vallams* with a crew of 40 men in each craft went out for fishing on 26 June 2006. Five to six fishermen jointly owned a *Valiya Vanchi* and a *Tangu Vala* (Purse-seine net). The cost of a *Tangu Vala* was Rs 5 lakhs. The cost of a *Valiya Vallam* and a nylon *Tangu Vala* comes to Rs 25 lakhs. The trawl net could not be operated as it was the period of ban of trawling. A *Valiya Vanchi* alone would cost Rs 10 to 15 lakhs. They are manufactured at Munambu, Cochin, Beypore and Chalium yards. The cost of a small *Vanchi* made of *Ayini* wood used to be Rs 2 lakhs.

The traditional wooden *Vanchis* were made of *Ayini* wood by the carpenters belonging to the Arayan, Christian and Muslim communities. A *Valiya Vanchi* requires a crew of 30 to 40 fishermen. The price of diesel has gone up from Rs 7.50 to Rs 30 per litre today.

Though *Naran Chemmin* fetched Rs 400 per kg twenty years ago, today it has come down to Rs 300. The owner of the boats and nets get 40 per cent share while the labourers get 60 per cent of the catch. If a boat capsized in the sea, the next kin of the deceased used to get Rs 10,000 from the government as compensation. Now it has been increased to an amount ragging from Rs 50,000 to 1 lakh.

Our informant Baskaran (80) (Kuppante Akathu House, Puthiyappa) tells us that the *Valiya Vanchi* was introduced at Puthiyappa in 1990. Majority of the Mukkuvas of Puthiyappa are employed as fishing labourers. The owners of nets and small *Vanchis* get 10 per cent of the catch. The cost of the diesel is met by the owners of the boats and nets. At that time the price of a kilogram of *Naran Chemmin* was Rs 300 whereas that of *Poovalan* Rs 100. *Ayila* and *Mathi* are also caught during the monsoon seasons. Sometimes the fishermen at Puthiyappa cannot go out for fishing due to the roughness of the sea.

The Mukkuvas of Puthiyappa own only one fibre glass *Vanchi*, but about 500 *Valiya* fibre glass *Vanchis* belonging to the neighbouring villages of Quilandi, Elathur, Puthiyangadi and Beypore visit this harbour for fishing in different seasons. Four men called MP Brothers jointly own the fibre glass *Vanchi* here, which cost Rs 25 lakhs each.

There are about 200 fibre *Tonis* engaged in fishing at Puthiyappa. One such *Toni* costs between Rs 50,000 and 1 lakh. These small fibre glass lighters are used for transhipping the catch from the *Valiya Vanchi*. Two boats are used for launching trawl nets for catching *Karikadi, Poovalan Chemmin, Mullan* and *Mantal*. These double trawl nets cost from Rs 5,000 to Rs 7,500 and are used for catching demersal fish like *Naran Chemmin, Poovalan, Karikadi, Mullan* and *Mantal*. The length of the trawl net used here comes upto five *Maaru*. The Matsyafed or government gives loans from Rs 10 to 15 lakhs to groups of educated unemployed fishermen under the scheme of self-employment. Under this scheme, at Elathur, a *Vanchi* was purchased by a group of 10 fishermen for Rs 25 lakhs. The scheme has been implemented at Puthiyappa and several young fishermen who passed out from high schools have availed of it. Either the *Valakkuna Vala* (encircling net) or *Tangu Vala* is employed for catching *Ayila, Mathi, Chemmin, Chuda, Kadalmin, Avoli* and

Kadappa in June and July when they appear in midwater. Mechanised boats are used for launching the surrounding net (*Tangu Vala*). When the sea becomes cold during the monsoon the fish become pelagic and they migrate to the shore from June to October.

The catch is shared between the fishing labourers and the owner of the *Valiya Vanchi* on the basis of 50:50 in this harbour. Three per cent of the catch is set apart as votive offerings to the local Sree Bagavathi and Neelakanda temples and mosques by the owners of the fishing gear.

A large number of fishermen have migrated to Puthiyappa from the neighbouring fishing villages of Tanur, Quilandi, Beypore, Maradu, Puthiyangadi, Elathur, Kappad and Kozhikode. Even Latin Christians and Arayans from far off Quilon and Cochin are found at Puthiyappa.

Apart from the Mukkuvas there are 50 Bovis-Mogeyar families settled at Puthiyappa but there is no intermarriage between them. At the time of the study the Bovis-Mogeyars were hook-and-line fishermen but had adapted nets and small fibre boats. They had not yet adapted *Valiya Vanchi* (big canoe).

According to our Bovis-Mogeya informant the number of currents during June-July at Puthiyappa is about 40. Therefore it cannot be said that there is only one *Nir* in a day. On 26 June 2006 when we visited the village there were several *Nirs*. The *Vadakkan Nir* and *Vadara Nir* (north-western current) occurred on that day besides the *Vadakkan Kattu* though few people got catches. The traditional net weighed 70 to 80 kgs. Today, even if there are unfavourable winds and currents, the *Valakkuna Vala* is sometimes used.

The Mukkuvas of Puthiyappa do not go out for fishing during the night. Out of 500 trawlers found here, the Mukkuvas of Puthiyappa own 100.

The traditional small *Kettuvallams* at Puthiyappa are replaced by small fibre boats fitted with Yamaha outboard engines of 9.9 HP. Four men can go out with an *Ozukkuvala* and *Pothan Vala* in a small fibre boat for catching *Chemmin, Kora, Mantal* and *Ayila*. The launching of *Peyithan Vala* requires two *Vanchis*. Larger fishes like *Avoli, Sravu* and *Tirandi* are caught with the help of the *Pothan Vala*.

The *Valiya* fibre boats and *Valiya* fibre glass *Vanchis* are also employed in Puthiyappa with *Valakkuna Vala* (purse seine net) with a crew of 50 men. Both larger and smaller fishes are caught, which include *Ayila, Mathi* and *Chemmin. Chemmin* is preserved in the freezing plant at Puthiyappa which is owned by a Hindu fisherman. It is then sent to different parts of Kerala. Apart from this there are four ice factories at Puthiyappa of which three are owned by a Mukkuva and one by a Thiya. Approximately Rs 40 lakhs are required for establishing an ice factory.

The Mukkuvas of Puthiyappa have adapted the *Valiya* fibre glass *Vallam* in which from 30 to 40 men are employed for fishing. The cost of a fibre glass *Vallam* according to its size varies from Rs 25 to 30 lakhs. In this village two brothers own a fibre glass *Vallam* in which 40 men are employed for fishing. This unit went out for fishing on 15 July 2006, paying for 200 litres of diesel. The net employed was the *Tangu Vala.* Apart from this unit about 200 fibre glass *Vanchis* with *Tangu Valas* also went out for fishing from Puthiyappa on the same day. The Hindu fishermen did not go out for fishing on 15th July, 2006 because of *Sankramam,* whereas their Muslim counterparts from Puthiyangadi and Quilandi employed their fibre glass *Vallams* for catching *Ayila* and Mathi on that day. On 17th July 2006, as many as 15 units took to fishing with fibre glass *Vallams.* Each unit got a catch ranging from Rs 10,000 to Rs 40,000. Two units from Vellayil, near Kozhikode, got catches worth Rs 40,000 each. Fish traders from non-fishermen Muslim communities from Kozhikode, Quilandi and Talassery had advanced loans to the owners of these fishing boats and gears against 10 per cent commission of the catch.

At that time the price of one box of *Mathi* (40 kgs) was Rs 200 and *Chemmin* Rs 85 per kg. The price of *Ayila* varied from Rs 20 to 40 per kg. One of the fishing units got a catch worth Rs 30,000. Out of this, 60 per cent was given to the owner of the craft after deducting the expenditure for food and fuel and the balance of 40 per cent went to the labourers. If the owner of the fishing crafts gets Rs 30,000, the owners of the lighters get Rs 2,400.

Expenditure

Diesel for the Valiya glass Vanchi	=	Rs 5,000
Kerosene oil for two fibre covered boats	=	Rs 2,000
Vazhipadu or votive offerings	=	Rs 500
Food expense of 50 men	=	Rs 1,000
Commission to the agent @ 5 per cent	=	Rs 1,500
Total expenditure incurred	=	Rs 10,000
Balance	=	Rs 20,000
Owner' share 60 per cent	=	Rs 12,000
Labourers' share 40 per cent	=	Rs 8,000

Let us examine the working of the Elathur Gurusakthi unit at Puthiyappa. It has a fibre glass *Vanchi* and a *Tangu Vala.* The unit is owned by 35 shareholders. The total outlay of the unit is Rs 17 lakhs. The share-holders have taken loans from banks and private persons to cover this expenditure. 10 fishing labourers also contributed Rs 50,000 each towards the cost of the unit. Our informant Sahadevan advanced Rs 1.5 lakh to the fishermen for the purchase of the equipment. As the commission agent and auctioneer he gets 10 per cent of the catch.

The *Gurusakthi* unit could not go out for fishing on 17 July 2006 as their net was destroyed by the flowing of *Vadakkan Nir* and blowing of *Vadakkan Kattu*. In fact the Gurusakthi was disabled for 10 days. Several fishing units go out for fishing at 4 p.m. after the monsoon season for night fishing. But during the monsoon they start at 5.00 a.m. and return with their catch between 4 p.m. and 6 p.m. They generally go out for fishing upto Tanur and Quilandi if fish is available which depends upon the behaviour of the ocean. The *Vadakkan Nir* and *Vadakkan Kattu* occurred on 17 July 2006. Some units went out to a distance from 12 to 15 *Maarus* and they got catches worth between Rs 10,000 and Rs 40,000. Out of 50 units, which took part in the expedition, only 10 got catches.

The main catch of the *Valiya Vanchi* was *Mathi* which is caught by employing inboard *Valiya Vanchi* and *Tangu Vala.* The price of Mathi is unpredictable as in September 2006 it had become very low, fluctuating between Rs 50 and Rs 60 per box of 40 kgs. On 18

September 2006 there was a bumper catch of Mathi and the price rose from Rs 50 to Rs 150 per box. There are over 500 trawlers operating from this harbour for catching *Koonthal/Kanava* and other species of fish. The price of *Kanava* ranged between Rs 150 and Rs 170 per kg during September.

If a trawler spends two weeks in the sea, the fishermen sometimes get catches worth between Rs 10,000 and Rs 25,000. It is believed that whenever the *Karakattu* and *Vadakkan Nir* occur there will be a good catch. On the other hand, when the *Tengara Nir* occurs the fish will migrate to the bottom of the sea. When the *Tembrayil Nir* and *Tembrayil Kattu* occur simultaneously there will be a bumper catch. Similarly when *Vadavartha Nir* and *Vadakkan Nir* appear there will be good catches. The traditional knowledge of the fishermen is still preserved by the fishermen and they make use of it for catching different species of fish.

A fishing trawler owned by Biju of Puthiyappa went out for fishing during our field work there. For five days they consumed 500 litres of diesel and five litres of engine oil. The total expenditure for fuel was Rs 28,000. Two hundred blocks of ice were purchased which cost Rs 2,000. Another Rs 5,000 was spent on food. The crew had seven members. The fishing labourers working in this unit hailed from Karunagapally, Quilon district. They were among the 100 Arayans who had temporarily settled at Puthiyappa for working during the peak season. They usually received Rs 600 for working from five to seven days. The unit got Kanava worth Rs 1.5 lakhs for its five-day operation. It was equipped with echosounder, GPS and wireless. The other fishes they caught were *Kora,* Puthiyappa *Kora* and *Chemmin.* The *Kora* fetched between Rs 1500 and Rs 1800 per box of 40 kgs.

We have already mentioned that there are 150 fibre glass *Vallams* and 500 trawler boats working at Puthiyappa. Thus, it appears that the Hindu fishermen of Puthiyappa have adapted modern technological innovations unlike the Arayans and Nulayans of the southern districts and Mukkuvas of Kasargod. The fibre glass *Vanchi* can go only upto a distance of 25 to 30 *Maarus* whereas the trawlers can go upto 50 *Maaru.* A fibre *Vallam* employs from 30 to 40 men for fishing whereas a trawler requires a crew of seven only. Our informant Sahadevan disclosed to us that the Mukkuvas of Puthiyappa owned 150 fibre glass *Vanchis* (inboard *Vallam*) and 100 trawlers.

It was also observed that the *Odam* has been replaced by *Maravanchi* or *Kettuvallam* (wooden *Vanchi*). The *Kettuvallam* has a crew of six to nine men. Nylon nets are used for catching sardine, mackerel and other species of fish. The *Kettuvallam* was introduced in this village in 1984 by the migrant fishermen who came from the southern districts. This boat has been mechanised with outboard engines having 9.9 HP, 25 HP and 40 HP motors. Kerosene is the fuel used in these boats. The *Kettuvallam* is now being replaced by *Valiya* fibre *Vanchi* and *Valiya* fibre glass *Vanchi*. The *Valiya* fibre glass *Vallam* can give employment to 40 to 50 fishermen for using *Valakkuna Vala* or *Tangu Vala*. Two lighters accompany these boats for collecting the catches and transporting them to the auction centre. The *Valiya* fibre glass *Vallam* uses an inboard Leyland engine of 110 HP. The *Ozukkuvala* (nylon gill-nets) are used for catching larger fishes like *Ayikkora*. These are operated either from a *Valiya* fibre boat or from a *Valiya* fibre covered *Vanchi*.

The Mukkuva (Hindu) fishermen of Puthiyappa have adapted trawler boats and trawl nets. This community owns more than 100 trawlers fitted with 110 HP motors. Seven to eight fishermen are required for fishing in each trawl boat while a fibre glass *Vanchi* requires a crew of 40 to 50 men for fishing from August to January.

Rani Vala, an improved traditional *Peyithan Vala* is used for catching *Mantal* and other demersal fishes in a *Kettuvallam*. Fibre covered *Vallam* and *Valiya* fibre glass *Vallam* have now been introduced in this village. Medium size fibre covered *Vallams* have a crew of nine for employing *Peyithan Vala*.

We met the members of some of the crew of trawlers, who went out for fishing in March, 2007. They disclosed the following details of their income and expenditure:

The total catch of the Nakshatra boat	=	Rs. 60,000
Expenditure towards fuel and food	=	Rs. 30,000
Number of days spent in fishing	=	3 weeks
The number of men who went for fishing	=	6

The remaining Rs 30,000 was distributed among the six fishing labourers (40 per cent of the catch) and the owner of the boat got 60 per cent after deducting the expenditure.

The Tankam trawler of Puthiyappa went out with six men for fishing for four days from 25 February 2007 to 1 March 2007. The trawl net that was used was made of plastic. The inboard engine fitted in this boat was Ashok Leyland Turbo. The cost of the boat was Rs 25 lakhs. The fishing unit auctioned its catch for Rs 25,000 which consisted of *Kanava, Koonthal, Valiya Kili, Cheriya Kili, Unnimeri* and other small species of fish. They went upto 40 fathoms for fishing. During their four day expedition the unit consumed 1200 litres of diesel. The total cost of fuel was nearly Rs 45,000 besides food and ice, which came to Rs 10,000. The trawl *Vala* is locally called *Chuda Vala*. This net is dragged and it is used a little above the sea bed after removing ten floats. Thus the *Chuda Vala* is a kind of trawl net.

The Mahavishnu trawler is being operated from Puthiyappa and its owner belongs to the neighbouring fishing village of Puthiyangadi. It went out for fishing for four days with its crew of seven from 26 February 2007 to 1 March 2007. The fishes caught by this unit were *Valiya Kili, Kilimin, Unnimeri, Kanava, Koonthal*, etc. They spent Rs 40,000 towards fuel and food and their catch was worth Rs 39,000. The total loss incurred was of Rs 3,000. In the previous expedition lasting for five days they had incurred a loss of Rs 4,000. The accounts are usually settled once in a month. One of the major catches of this expedition was *Kilimin* which fetched a price of Rs 600 per box (1 box is 40 kgs). The price of the *Puthiyappala Kora* was Rs 650 per box. The fishing labourers told us that the Mahavishnu trawler incurred a loss of Rs 1 lakh in the month of January, 2007. It is very interesting to point out here that even when the unit incurs a loss, the fishing labourers are given their wage at the rate of Rs 100 per day.

The Aparna trawler went out for fishing for three days with a crew of six and spent an amount of Rs 2,000 for fuel and a sum of Rs 1,000 towards food. No fishing was done during nights. Their total catch was worth Rs 20,000 of which the major catch was Kilimin. However they incurred an expenditure of Rs 20,000.

The trawl net is popularly called *Lajavathy* in this region. It has 40 mm mesh size and it is made of plastic. The yarn count is ½. Five men go out for fishing in a small trawler with *Lajavathy*. The cost of this plastic net was Rs 235 per kg at the time of the study. A minimum of 8 kgs of plastic yarn is required for a small *Lajavathy* net.

At present, the *Valiya* Fibre Glass *Vallam* with Leyland engine of 110 HP is incurring loss due to the high cost of fuel and food charges. The commission agent, who is also the auctioneer, gets five per cent of the catch in this *Vallam*. In a fishing expedition carried out in February 2007 sardine was the major catch. Some units got catches worth from Rs 50,000 to Rs 1 lakh. The lighter costs Rs 250 per trip and it is paid to the owner of the *Vallam*. The commission agent gets a share of the wages of a fisherman for his investment. In case he advanced Rs 5 lakhs to the fishermen for the purchase of equipment he gets the wages of five fishing labourers from the sale proceeds of a fishing expedition. If his investment goes up to Rs 10 lakhs, he gets the wages of 10 fishing labourers from the total sales of the expedition.

There is a rumour that *benami* (illegal) investments are rampant in the marine fishery sector. Peethambaran, Thayil Parambu House, Puthiyappa, owns a small wooden trawler which cost him Rs 2 lakhs, and which is operated upto a distance of 8 to 10 kms. It is fitted with a Leyland engine of 90 HP. His unit went out for fishing on 27 February 2007 at 4 a.m. and returned at 10 a.m. His total catch was worth Rs 4,000 after incurring an expenditure of Rs 2,000 towards fuel and food. Peethambaran told us that the *benami* boat owners go for fishing expeditions both during the day and night. However, fishing at night is detrimental to the growth of the fish resources, causing decline in the fish harvest. His unit generally gets catches of *Karikkadi, Chemmin, Mantal* and *Poovalan*. The *Kanava* is also caught during this period. Soon after the lifting of the ban on trawling from September to October, year before, a number of trawlers got good catches worth between Rs 50,000 and Rs 1 lakh. According to Peethambaran, February again sees a decline in the fish harvest. The major catch by trawlers in the first week of March 2007 was small *Puthiyappala Kora*. GPS is used in almost all the trawlers.

Technique of Tangu Vala or Valachilakkuna Vala (Boat Seine Net)

The *Tangu Vala* is a boat seine net and it is made of square or rectangular pieces joined together. These nets are larger with heavy gears and a number of persons are required for their operation. These nets encircle a considerable area, narrowing the space for catching the fish. The unique characteristic of the seine net is that it is equipped with a strong central bag for preventing the fish from escaping. There are two long wings on both sides of equal length. The wings enable the net to encircle the fish in a larger area. Warps are attached to these wings which help in driving the fish together. In contrast to the trawl nets, the wings of seine nets are substantially larger.

The *Tangu Vala* is generally shot around the fish shoal and is pulled together and the catch hauled. A seine net can be hauled from the same place where the setting started, owing to its circular shape. The area to be fished is generally selected in advance, although the seine net is shot opposite to the target. It is significant to note that during the operation of the *Tangu Vala*, the fishing programme previously fixed cannot be altered. The purpose of the seine net is to fish the largest possible area, stretched over several hundred metres.

Every seine net has a head rope and a foot rope. Sinkers are attached to the foot rope and floats are tied to the head rope. The striking feature of the seine net is that from the top to the bottom (from the mouth to the cod end), the thickness of the yarn of the net increases. At the cod end the yarn is thicker. This helps to hold the entire weight of the fish caught. The meshes are smaller at the cod end in order to prevent the escape of the fish from the net.

The *Valachilakkuna Vala* or *Tangu Vala* is operated for catching *Ayila* (mackerel) and *Mathi* (sardine) only during the day time without taking into consideration the occurrence of the wind and current. Earlier as many as 16 fishing units jointly used to undertake expeditions by employing the *Tangu Vala* technique with the help of a corresponding number of boats. Nowadays, only eight nets are used with 16 boats for entangling the *Ayila Polappu* (mackerel shoal) and *Mathi*. This technique was widely practiced in the Kozhikode region, particularly at Puthiyappa. The bumper catches are hauled by employing the *Valachalikkal* technique and *Tangu Vala,* particularly in the months of *Makaram, Kumbam, Minam* and *Medam*. All the eight nets are tied to the sixteen boats and the fish shoal is thus cornered by the *Valachilakkal* technique.

The fishermen even today beat the cross-bars of their respective boats with the oars and rudders and also make loud sounds to scare the fish and drive them into the nets. When the fish is caught, the net is hauled up and the catch removed. Each unit divides its share among the members of the crew according to their custom. The maximum catch is hauled by employing this technique in the month of April. The *Valachalikkal* is also called the *Kettiyilakkal* in certain areas of Malabar. This technique has the advantage of preventing the fish from migrating to the bottom as well as to the distant areas of the sea.

Boat Seine Nets

The seine net operated from a boat is called the boat seine net. This is one of the commonest types of nets employed for catching prawns, sardines, mackerels, soles, etc., along the Kerala coast. Though there may be some minor regional variations, a boat seine net essentially is composed of a bag-like part or *Vaikettu*, (cod-end), *Mevala* (belly), two coir wings (*Kadankanis*), and two *Kambas* or dragline coir warps. This net is operated by 8 to 10 men from two boats. It is shot generally to a depth of 10 to 20 fathoms. At present one of the boat seine nets called *Tangu Vala* is operated by 40 to 50 fishermen from a *Valiya* Glass *Vanchi.*

Nethal Vala

Mathur (1978) classified the *Nethal Vala* (white bait net) into two types, viz., (1) *Oncuvala* or *Ilakkipidikkuna Nethal Vala,* and (2) *Korunna Nethal Vala* or *Peyunna Nethal Vala.* The former is used for catching the *Vella* (white) Nethal while the latter is meant for hauling the *Kari* (black) Nethal. *Nethal Valas* are launched at Puthiyappa from plank-built fibre covered boats.

Oncuvala

The *Nethal Vala* is used by the Mappilas of Puthiyappa and the neighbouring Puthiyangadi village. The *Oncuvala* is a nylon *Nethal Vala* and it consists of several *Murivalas* (piece nets). The net has two main parts, viz., (1) *Murivala* or *Tattu* (upper belly) and (2) *Vaikettu* (cod end). The Murivala consists of five pieces, each measuring 50 *Kanni* in length and 800 *Kanni* in breadth. Thus the piece nets of the

Oncuvala has a total measurement of 250 × 4,000 *Kanni*. The *Vaikettu* of the net also consists of five *Murivalas*. The *Murivala* at the end of the *Vaikettu* has a measurement of 50 *Kanni* in length and 50 *Kanni* in breadth. The size of the subsequent *Murivalas* gradually increases. The fifth *Murivala* is the biggest, having 100 × 600 *Kanni*. Thus, the entire *Vaikettu* of the *Oncuvala* is 100 *Kanni* in length and 3,000 *Kanni* in breadth. As a matter of fact, the *Vaikettu* of the *Oncuvala* is also called the *Muvayiram Kannivala* (net with 3000 meshes). The mesh size of the *Vaikettu* is ¼" whereas that of the Murivala is ½". This net's *Murivala* is tied to a *Kamba*.

Operation of the Oncuvala

We have already mentioned that the *Oncuvala* is used for catching the *Vella Nethal*. The net is used particularly when the fish migrate to the shore waters during the months of *Mithunam, Karkkidakam* and *Chingam*. The catching of the *Vella Nethal* by the *Oncuvala* is popularly known as the *Oncipidikkal* technique. No boat is used in launching the net but the crew itself operates it. First of all, the fishermen launch the *Oncuvala* in the fishing ground where the shoal is expected by swimming and diving upto a distance of two-and-a-half miles. They remain in water for 10 to 12 hours engaging themselves in diving and swimming freely. Their swimming is determined by the blow of the wind and the flowing of the current. For instance, if the wind is blowing from south to north, they dive towards north. In other words, both the Mappila and Hindu fishermen swim with the *Oncuvala* for launching it in the same direction as the wind. On the other hand, they always dive against the current with this net.

The net is also launched by the technique of drifting. The fishermen hold the *Kamba* (rope) in their hands and dive. When they dive the net behind them is allowed to drift along the muddy bottom or midwater. During the course of swimming and diving, the net is launched and the fish is caught. During the months of *Kanni* and *Tulam,* the *Vella Nethal* appear in the bottom as well as on the surface of the water. When the *Nethal* on the surface is caught the net is generally drifted towards the shore. Sometimes, when the net is drifted *Mantal, Karikkadi* and *Kora* are also caught.

Korunna Nethal Vala

The *Korunna Nethal Vala* is used for catching the *Kari Nethal*. The net has two main parts: (1) *Murivala*, and (2) *Vaikettu*. The *Vaikettu* consists of 10 *Murivalas*. Each of them has a different measurement. For instance, the *Murivala* at the extreme end of the *Vaikettu* has a measurement of 100 × 800 *Kannis*. But the tenth *Murivala* is 1,000 *Kannis* in length and 3,000 *Kanni* in breadth. On the other hand, the *Murivala* of the net has 15 *Murivalas*, and a measurement of 1,500 × 3,000 *Kannis*. But the *Murivala* where the floats are tied is usually of seven *Kannis* in width and 300 *Kannis* in length. This piece net is called *Arivala*. A *Kadamkani* (rope net) is tied to each end of the *Murivala*. Similarly, a *Kamba* is tied to each end of the *Kadamkani* and each *Kamba* has a length of 10 fathoms.

Operation of the Net

The net is launched from two boats, unlike the other gill-nets which are generally shot from a single boat. The *Korunna Nethal Vala* is generally employed by the Mappila fishermen in the months of *Chingam, Kanni* and *Tulam*. The fish is caught by detecting the shoal. While launching the gill-nets, the fishermen have to study the nature of the wind and current but in this case they do not generally attach much importance to the occurrence of winds and currents. However, they attach great importance to the migration of the fish when the *Korunna Nethal Vala* is shot. For instance, if the fish migrate from north to south, the *Vaikettu* of the net is dropped on the south and when the fish travel from south to north, the *Vaikettu* is shot first to the north. The net is employed either in the day or at night. Two Muslim dominated villages of Tanur and Puthiyangadi have replaced the traditional *Nethal Vala* by the nylon net. This net is used early in the morning at about 5 as well as before the sunset when the sky is red. The Mappilas say that when the sky is red just before the sunrise and sunset the *Vella Nethal* migrate towards the net and thus they are caught in abundance. Conversely the *Kari Nethal* is hauled abundantly in day time.

Mathur (1978) observes: "The depth of the Arabian sea varies from one region to another. The height and width of the gill-net in Kerala varies from one place to another. The sea is more deep in southern parts of Kerala than in the north. For instance, the gill-nets with 300 meshes width are used in Tanur, but nets which have 350 meshes width are used in Blangad about 50 miles south of Tanur. In Cannanore they have only 250 meshes height or width."

Blangad Fishing Village, Thrissur District

Blangad fishing village (Blangad Kadappuram) is situated at Manathala revenue village, Chavakkadu taluk, Thrissur district. The other beaches or *Kadappurams* are Puthan Kadappuram, Puthiya Kadappuram and Edakkazhyoor. But our study is confined to Blangad Kadappuram. The demographic structure of these fishing villages are as follows:

Name of Fishing Village	*Estimated Number of Households*		
	Hindus	*Muslims*	*Total*
Blangad	350	150	500
Puthiya Kadappuram	100	400	500
Puthan Kadappuram	100	350	450
Edakkazhiyoor	150	400	550

Traditionally, dugout wooden boats (*Maravanchi*) were used by the Mukkuva and Mappila fishermen in Blangad. These boats were made from Anjali, Tampakam and other tree trunks. They were replaced in 1987 by *Kettuvallam,* made of wooden planks and coconut husk. This technology of manufacturing boats was borrowed from the fishermen of Alleppey.

The Mukkuva fishermen of Blangad adapted small fibre covered boats during 1990s in which four men go out for fishing. They have also adapted *Valiya Kettuvallam* in which 20 fishermen go out for fishing with nylon *Tangu Vala (Valanjupidikkuna* net – encircling net) and *Chuda Vala,* for catching small fishes like *Chuda, Nethal, Para, Chemmin, Ayila* and *Mathi*. Earlier they used smaller wooden dugout boats in which only two fishing labourers used to go out for fishing. Mechanisation was adopted during the 1980s with Yamaha outboard engines of 9.9 HP, 25 HP and 40 HP. The *Beppukar* or hook-and-line fishermen migrate to Blangad during the peak seasons from far off

Kanyakumari and employ the modern hook-and-line technique of drift hook and lines for catching larger fishes like *Ayikkora, Tirandi, Avoli, Sravu,* etc. It was observed in February 2007, that the Tamil speaking Latin Christians use coconut bunches after removing tender coconuts for catching *Kanava/Koonthal*.

Both the Hindu and Muslim fishermen of Blangad own over 40 fibre *Kettuvallams*. In this village about 80 per cent of the Mukkuvas and 20 per cent of the Mappila fisherfolk use fibre *Kettuvallam*.

Hook-and-line fishing in Blangad

Over 150 Latin Christian fishermen from Colachel, Kanyakumari, district temporarily migrated to Blangad and started fishing thereby employing hook and lines and small fibre boats in which five men can work. The main catches are *Ayikkora, Kanava, Chura, Sravu, Kolli* and Catfish. The boats are fitted with 9.9 HP motors and nylon *Tangis* and steel hooks are used. The fishing boats which went out on 12 February 2007, got poor catches due to the appearance of a strong *Nir* (current). Moreover, a powerful south-western current and north-western wind appeared. With the help of hook-and-line the Latin Christians were able to catch *Kora, Palamin, Kanava, Olamin* and *Mothamin*. Only one unit got Rs 2 lakhs worth of fish on that day.

Kanava Fishing

Some Latin Christians from Cholechal of Kanyakumari district, Tamil Nadu, have temporarily settled in Blangad, Thrissur district, for employing hook-and-line fishing. They employ a very interesting technique for catching *Kanava* (cat fish) with the help of coconut bunches without coconuts. A number of coconut bunches are tied together to a gunny bag filled with sand and lowered to the sea bottom. Then, it is fed in the GPS. After a few days the fishermen visit this area with the help of GPS and catch Kanava fish. It is interesting to note here that no special baits are used in this technique but coconut bunches are the only baits.

According to our informant, Antony from Colachel, about 150 hook-and-line fishing units were engaged in fishing with coconut bunches technique in Blangad for two months from January to February 2007, and each unit was getting fish worth Rs 10,000 to Rs 20,000. Apart from the above technique, the migrant Latin Christians use also *Otta Cunda* (hook-and-line) for catching bigger fishes. They got larger fishes like *Chuda, Kanava, Sravu, Ayikkora, Kolli,* Catfish, *Kora* and *Para* worth from Rs 20,000 to 25,000 per an expedition by *Otta Cunda*. A single fibre boat was used for operating this technique. On 12 February 2007, about 150 *Choondakkar* (hook-and-line fishermen) went out for fishing using drift technique. They went out for hook-and-line fishing upto 40 to 50 *Maaru* employing *Otta Cunda* technique consisting of 200 to 300 metre of plastic line and an artificial fish into which three to six steel hooks were pierced. The boat was fitted with an outboard engine. Antony however told us that his unit got fish worth only Rs 2,000 on that day. The price of Kanava fish per kg was Rs 150 at that time.

According to Salam (31) of Blangad, the Mappila fishermen of this village are known as Puislam while the *Chundakkar* are known as *Beppukar*. It was found that Tamil-speaking Latin Christians of Puthantura, Valiyatura and Marthandam visit Blangad from October to May with hook-and-line for fishing larger fish like *Sravu*. The fishing labourers are benefited by them. Some fishermen carry GPS, a modern equipment for getting direction and also to locate the fish shoal in rocky places. Fishes like *Kanava, Para, Arakka, Chembal, Avoli* and *Sravu* generally inhabit these rocky places or *Chithal puttu*. With the help of the GPS the fishermen can locate the fishing grounds and fish shoal, apart from getting directions.

Boats

The Mukkuva fishermen of Blangad have adapted small fibre covered boats which are manufactured in the boat building yards of Trivandrum, Beypore and Cochin. We observed in July, 2006 that the Mukkuva fishermen employed small fibre *Vanchis* and *Ozukkuvala* for catching fishes like *Champan Sravu, Choora, Kuduka*, etc. These fishes are generally caught from August 15th. In Puthan Kadappuram, three kilometres to the north of Blangad, six Leyland fibre boats with a crew of forty men each are employed for catching *Mathi* and *Ayila* with the help of *Tangu Vala*. Side by side with the modern fibre boats the Mukkuva fishermen of Blangad employ the traditional wooden boats and nets. These bigger boats made of Anjali trees fitted with Yamaha engines having 9.9 HP, 25 HP and 40 HP motors are operated in Blangad with a crew of 20 to 30 men. During the period of ban on trawling, *Cheriya Kora, Nethal, Cheriya Chala* (small *Mathi*), and different species of *Chemmin* and *Para* are caught with Nylon *Chala Vala* and *Tangu Vala*.

The fishermen in the past used to row the wooden dugout boats with the help of rudders, oars and sails which helped them to keep in good health. Nowadays they go for fishing in mechanised boats and therefore suffer from cancer, cholesterol and heart diseases. They have to put in less labour today but they get good catches due to mechanisation of fishing. However, the expenditure for fuel (petrol and kerosene) is very high. The licensed 9.9 HP motor boats have sanction for 300 litres of kerosene in a month, but actually they get only 150 litres which is not enough for the operation of a Yamaha engine for two days. They require a minimum of 450 litres of kerosene for operating a 25 HP motor from 6 a.m. to 6 p.m.

There are six fibre *Valiya Vanchis* at Blangad in each of which 40 men go for fishing. There are only two traditional *Valiya Vanchis* without fibre covering at present at Blangad. They are Nagaraja and Nagayakshi. The cost of a mechanised *Valiya Vanchi* is Rs 8 lakhs as against Rs 1.5 lakhs for a small fibre covered boat with net. The Mukkuvas did not go out for fishing on 5 July 2006 due to bad weather. They said that the wind blowing at Blangad on that day was *Thekkan Kattu* (southerly wind), which is dangerous for fishing.

Nagaraja and Nagayakshi are operated at Blangad during the monsoon for midwater fishing, particularly *Ayila* and *Mathi*. These boats went out for fishing on 15 July 2006 along with 60 other units, but there was a very poor catch. The average catch for each unit was only Rs 1,000. The fishes caught were *Chemmin, Kora* and *Paral* (small kind of fish). Our informant Mohanan attributed *Kachan Kattu* and *Vadakkan Nir* for the day's poor catch. On the same day, a fishing boat was capsized at Munambu, sending one of the crew to the sea bottom. Unfortunately his body could not be traced. The deceased was married to a lady from Blangad a year ago. They had no issue. The widow attributed the cause of death to his fate (*Vidhi*).

Advantages of Fibre Boat

According to our informants the following are the advantages in adapting the fibre boats. Majority of the Mukkuva fishermen of

Blangad have adapted fibre covered boats, but not the fibre glass *Vallams* and trawlers for the following reasons:

1. Less fuel consumption than traditional wooden boats.
2. Fibre boats do not capsize easily. They can float on the top of the water and can go upto 25 kms in an hour as against 15 kms of mechanised wooden boats.
3. A fibre glass *Vanchi* fitted with a Leyland engine can travel for fishing up to any distance.
4. Rowing with rudders and oars are not required for the fibre *Vanchis*.
5. When a fibre *Vanchi* is operated with an encircling net (*Tangu Vala*), a fisherman sitting on the *Aniyam* draws the foot rope with the help of the engine. When the foot rope is drawn the net is entangled and fish is caught depending upon the occurrence of *Nir* and *Kattu*.
6. A fisherman sits on the *Amaram Padi* and detects the fish shoal. He then jumps into the sea and drives the fish into the net by beating the water. Within five minutes the net is entangled. It takes 1½ hours for launching the *Tangu Vala*.

Nooruddin, resident of Pulikakkal house, Blangad, owns a small mechanised fibre *Vanchi* and *Ozukkuvala*. The mechanised fibre *Vanchi* employs four fishermen. He is also working as a fishing labourer in the same unit. The total cost of his unit came to Rs 2.5 lakhs at that time. His boat is a *Kettuvallam* manufactured locally. Lighters are used for transporting the catch from a *Valiya Kettuvallam. Chakara* generally appears in Kerala during June–July and a bumper catch of *Chemmin* is expected then. Nooruddin operates his unit from August to May for catching sardine and mackerel. The *Ozukkuvala* is also called *Randamnumber Vala.* An *Ozukkuvala* with the sixth count of yarn is used for fishing *Chempan, Sravu* and *Choora*. On the other hand, the *Ozukkuvala* made of 12th count of yarn (*Pantrandamnumber*) is used for catching larger fish. Nooruddin has a GPS which he uses during his fishing expeditions.

The following are the different types of boats used in Blangad:

1. *Maravanchi:* crew of five men (dugout) go for fishing without fibre covering.

2. *Cheriya* fibre covered *Kettuvallam:* crew of four men.
3. *Valiya Kettuvallam:* crew of 20 men for fishing without fibre covering.
4. *Valiya* fibre covered *Kettuvallam:* 30 men, launched from Puthankadappuram.

A number of fishing nets including *Ozukkuvala* were destroyed by *Kadal Panni/Kadal Makri* at Blangad. This is a kind of frog which is inimical to the *Ozukkuvala*. Red crabs also destroy the *Ozukkuvala*. *Loopu Vala* (Nylon) popularly called *Ayila Vala* is also used for catching *Avoli* and Njara. The fishermen of Blangad use *Cheriya Kasavu* as *Era* (bait) for catching fish. Sometimes Mathi is also used as bait for catching Kanavan. The *Tangu Vallam* uses the *Valanjipidikkal* technique. A crew of 20 men is required for fishing with a *Tangu Vallam* (*Valiya Kettuvallam*), which is a mechanised boat. The fisherman who sits on the *Aniyam* is called *Aniyakaran* whereas the *Amarakaran* (shrank) drags the net. The *Aniyakaran* directs the crew for operating the boat and launching the net by showing his hand. He also directs one of the crew members to jump into the sea and encircle the boat with net. The fisherman who jumps into the water should be an expert in swimming. When the fishermen go with *Ozukkuvala* at 2 p.m. they normally return at 6 p.m. They also carry the ice for preserving the fish.

The *Aniyakaran* gets 1.50 per cent in addition to his wages while *Amarakaran* gets 1.25 per cent in addition to his wages, which are paid to the fishing labourers. *Chattakutty*, swimmer, also gets an addition of 1.50 per cent to his wages. The engine driver also gets 1.50 per cent of the total wages. The fisherman who beats the *Padi* also gets an additional 1.25 per cent of his wages. The total wages that the crew is eligible to get is 40 per cent of the catch (*Panis*).

Ramesh, our informant, told us that the bigger canoes, Nagaraja and Nagayakshi, could not go out for fishing in the first week of March 2007 due to shortage of fish harvest in the sea. He told us that for the past one month, the *Ozukkuvala* with fibre boats were employed and the fishermen got *Ayikkora* and *Chempan*. The *Vadakkan Nir* and *Vadakkan Kattu* appeared in Blangad on 7 March 2007 and so there was a poor catch. During the last one week of the season, the *Chundakkar*, the migrant Latin Christians, went out for fishing employing *Otta Cunda*. They got good catches of *Kanava* or *Koonthal*, which fetched from Rs 10,000 to Rs 15,000 per day. The hook and lines are operated from a single fibre *Vallam*.

On 10 July 2006 at 1 p.m., a cyclone visited Blangad when we were in a small thatched shed near the landing station. We were forced to shut the doors made of plated coconut leaves. Ramesh (31), a Hindu Mukkuva, told us that the character of the ocean was very unpredictable and the fishermen could expect their death at any moment. The ocean usually becomes violent when a south western wind and rain occur, which had happened that day. At Puthiya Kadappuram many houses and sea walls were destroyed on that day and no less than 50 tiled and thatched houses were destroyed by a cyclone at Blangad Kadappuram in the last week of July 2006.

Ramesh added that the GPS was widely used in Blangad for locating their hamlet, houses, the fish shoal, distance from the landing station, and the latitude (north/south) and longitude (east/west). Ramesh used GPS for *Chundapani* and for catching fish with *Ozukkuvala*. There were 25 fishermen in Blangad who had been using GPS for the last five years. In other fishing stations GPS was widely used for the past 10 years.

Kettuvallam

Usually thirty fishermen go out for fishing in a *Valiya Kettuvallam.* When a *Valiya Kettuvallam* goes on a fishing expedition two lighters are used for transporting the catch to the landing station. The owner of the fishing craft and gear meets the cost of the fuel and food. After deducting the fuel and food charges he gets 60 per cent and fishing labourers get 40 per cent of the net catch.

Cheriya Tanguvallam/Murivallam

The *Cheriya Tanguvallam* requires four men for fishing as against eight in a *Cheriya Vallam*. The double net (*Pothan Vala*) is used in the *Cheriya Tanguvallam*. The *Valiya Kettuvallam* uses the *Tangu Vala* for catching *Ayila* and different varieties of *Chemmin*. On the other hand, the *Cheriya Kettuvallam* is employed with *Pothan Vala* for catching *Chemmin*. Both these *Kettuvallams* are mechanised. The *Tangu Vala* and *Chuda Vala* are launched from *Valiya Kettuvallam*. Despite the ban on trawling during June, July and August, it was found that fibre boats and *Ozukkuvala* were used for fishing with outboard engine having more than 10 HP.

In view of the imposition of the ban on trawling even the non fibre boats with outboard engines above 10 HP as well as trawlers were prohibited from fishing. However, fibre covered boats fitted with 9.9 HP motors and *Ozukkuvala* were used in Blangad during the period of ban.

Two *Kettuvallams* were destroyed by Tsunami in Blangad in December 2004. Krishnan Kutty, owner of one of the boats, received Rs 5,000 towards compensation. The owner of the other boat was a Latin Christian fisherman from Colachel, Kanyakumari district, Tamil Nadu. Although he was insured and his next of kin applied for Rs 50,000 as compensation, his family did not get the amount. Another fisherman also died when his boat capsized from Puthan Kadappuram.

The introduction of modern technology like *Cheriya Tanguvallam* boats has helped the fisherfolk of Blangad to improve their lot but one cannot say they are economically better off today. Almost all fishermen of Blangad have retained their traditional techniques and fishing equipment.

Valiya Kettuvallam

As many as 15 fishermen in Blangad use GPS for catching different species of fish. In the neighbouring hamlet of Puthan Kadappuram both the Hindu and Mappila fishermen have adapted fibre covered *Valiya Vallam* with Leyland engines having 110 HP. Ramesh informed us that GPS (Global Positioning System) helps him to locate *Kanava, Para*, *Avoli* and *Sravu* when these fish migrate to rocky bottoms. The GPS is used even in small fibre covered *Vanchis*. As the fishermen get only 50 per cent of the sanctioned quantity of kerosene for their mechanised crafts, they are required to buy the fuel at the open market price. The fisherfolk of Blangad employ *Ozukkuvala* or gill-nets from the middle of August to May.

Kettuvallam uses the *Valanjipidikkal* technique with *Tangu Vala*. While the *Tangu Vala* is in operation, the *Chattakkaran* who sits on the *Aniyam* swims in the water. He also shows the direction while swimming for hauling the net. The engine driver (Serang) brings down the net. The bonus for different expenses in the fishing operation is governed by custom. The trainers get 0.50 to 0.75 per cent of the catch. If an agent (*Korunavan*) collects fish from a *Kalli*, he gets a fixed amount from the concerned company. The commission agent gets 10 per cent of the catch. Certain fish marketing companies like AKR, BMC, etc., in Chavakkadu buy fish through the fish traders at auctions. "Friends" is one of the biggest export companies in Cochin specialising in Kanava (*Koonthal*). This fish fetches Rs 350 per kg. The three-member crew among the *Chundakkar* are themselves the fishing labourers and owners of the equipment.

There are six mechanised *Valiya Kettuvallam* boats at the Puthan Kadappuram in which 40 men go out for fishing with a *Tangu Vala*. The *Tangu Vala* (boat seine net) is used in Leyland boats for catching *Ayila, Vatta, Chura, Para* and *Mathi*.

Kettuvallam and Modern Innovation

The *Ottathadi Vallam* or *Maravanchi* is the traditional dugout boat manufactured locally with *Ayini* wood. This *Vallam* has been replaced by *Kettuvallam*. Earlier a crew of eight to ten men used to go out for fishing in an *Ottathadi Vallam*, but now fibre boats, in which four men go out for fishing, have replaced *Kettuvallams*. Two *Cheriya Kettuvallams* are used for operating the *Pothan Vala* (double net). There are more than 450 households at Puthan Kadappuram of which 400 belong to the Hindu Mukkuvas. The Puthiya Kadappuram has 400 families belonging to the Mappila fishing community, whereas there are hardly 100 Hindu fishing families.

The Matsyafed used to advance loans at four per cent interest to fishing labourers' upto Rs 10,000 with 30 per cent subsidy for five years against security of land. The Matsyafed also advances loans up to Rs 1 lakh to a crew of eight men. The housing grant has been increased from Rs 25,000 to Rs 40,000, but this grant is paid in three instalments against the pledging of the title deeds. Majority of the fishermen cannot avail of the housing grants as they are landless.

Over 100 units went out for fishing on 10 July 2006 at 5 a.m. Some returned to the landing station at 12 noon and others at 4 p.m.

On that day the *Vadakkan Nir* flowed upto 7 a.m. and the *Tekkan Nir* thereafter. When the *Vadakkan Nir* flowed *Chemmin* was available. The combination of the *Vadakkan Kattu* and the *Tekkan Nir* resulted in the disappearance of *Chemmin*. However, *Kora* and *Karikadi* were caught. About 250 units of Mukkuva fishermen from the neighbouring villages of Nattika, Vadanapally, Azhikode and Talikulam came for fishing at Blangad. Each unit required a minimum of Rs 10,000 for meeting their fuel and food charges. They went upto a distance of 20 *Maarus*.

About 200 Muslim and Hindu fishermen are engaged in fish marketing at Blangad. Due to the introduction of ice, the export of dry fish is not in practice. The fibre boat was introduced in Blangad 10 years ago. The *Ozukkuvala* is employed from September to May for catching the larger fishes like *Sravu, Ayikkora* and *Tirandi*. Sometimes poor catch is reported due to the operation of foreign trawlers in the deeper sea.

Kadappuram Fishing Village

The Kadappuram Panchayat is situated three kilometres to the south of Blangad. It is another exclusive fishing village. Moidutty of Kadappuram (Kadappuram Panchayat, Chavakkadu Taluk) owns a fishing unit with a crew of eight men and several nets including two *Ozukkuvalas*, one *Aaramnumber* (6th count) for catching small fish and the other *Pantrandamnumber* (12th count) for catching larger fish. He also owns three *Chemmin Vala* and a *Maravanchi* (wooden boat). Two Yamaha engines fitted to his fibre boat cost him Rs 1,25,000. He also purchased two *Murivala* for Rs one lakh and a fibre boat for Rs 1,15,000. Although he is an expatriate from the Gulf, he had to pledge his wife's ornaments for taking a loan of Rs. 2 lakhs. He says that fishermen do not commit suicide like the farmers. Since he believes that the soul is given by *Allah* and therefore he can alone take it back. Moidutty went with one of his fishing units for fishing on 10 July 2006. The catch obtained was valued at Rs. 4,000. He had spent Rs. 3,000 for fuel (kerosene, petrol and engine oil) and Rs. 250 for food, making the total outlay to Rs. 3,250. The unit employs both Hindu and Muslim fishermen and each member got Rs. 100. The *Tekkan Nir* and *Tekkan Kattu* occurred in the sea on that day.

Narayanan of Kadappuram fishing village is the owner of a *Kettuvallam*, a *Pothan Vala* (double net) and a *Mara Vanchi* (dug-out boat). The *Pothan Vala* cost him Rs 10,000 and the *Kettuvallam* Rs 40,000. Two Yamaha engines were fitted to his crafts at a cost of Rs 80,000. Narayanan's unit leaves for fishing at 5 a.m. and returns by 5 p.m. The total fuel consumption for an expedition is: 150 litres of kerosene, 10 litres of petrol and 5 litres of engine oil. The total daily expenditure comes to Rs 50,000 including food charges. After deducting the expenditure the owner of the fishing unit gets 60 per cent of the catch and the fishing labourers get the remaining 40 per cent.

According to Ramesh, the mesh size of the trawl net is smaller than that of the *Pothan Vala*. The trawl net is dragged along with the bottom mud. It is used for catching demersal fish like Chemmin. The spawning period of *Chemmin* is in the monsoon season and spawning generally takes place in the *Chali* (mud). Trawling is banned during the spawing period because if permitted it would destroy the *Chemmin*. On the other hand, the *Pothan Vala* is launched at 1.50 metre above the bottom sea. This helps the fishermen to catch prawns since prawns start laying eggs during the monsoon that is immediately after the rains. On the other hand, when the water from rivers and backwaters enter from the east to the sea the fishes like Kanava migrate to the sea shore which are easily caught by the fishermen. The owners of the *Valiya Kettuvallam* complain that the ban on trawling is a violation of natural justice but this is not so since they catch many other kinds of fish with the help of *Pothan Vala*. The spawning period of the *Mathi* and *Ayila* is also in the monsoon. The appearance of bubbles in the sea is an indication that it is the spawning period.

There are five inboard *Valiya Vallams* at Puthan Kadappuram, of which three are owned by Muslim and two by Hindu fishermen. Blangad and Puthan Kadappuram have no trawlers but at Chettuva (Munakkakadavu boat jetty) the Muslim and Hindu fisherfolk have adapted trawlers. Chettuva is only six kilometres north of Blangad. A large number of fishermen have migrated from Kanyakumari and Quilon to Blangad and are using trawlers for fishing *Chemmin*. According to our local sources, the catch of *Chemmin* is very poor. The overwhelming majority of the trawlers are owned by Latin

Christian fishermen of Colachel and Kanyakumari (Tamil Nadu) and Quilon district (Kerala). They employ trawl nets or double nets. *Chemmin* is their main catch. Seven to ten men go out for fishing in a trawler. At Puthan Kadappuram, 3 kms away from Blangad, there are no trawlers.

The Mukkuvas of Blangad still depend upon their traditional techniques, knowledge of winds, currents and tides of the ocean, for fishing. They have adapted modern nylon nets and fibre covered plank-built boats but they continue to use the traditional *Kettuvallam* in which 30 to 35 men go out for fishing. The total cost of a *Kettuvallam* and *Tangu Vala* comes to about Rs 20 lakhs.

Kappad Fishing Village, Calicut District

The Mukkuvas of Kappad where Vasco da Gama landed in 1498, still followed the traditional techniques of fishing at the time of our fieldwork. They had not yet adapted trawlers and trawl nets but operated their traditional boats and nets. However, a few fishermen had become the owners of fibre covered canoes and *Tangu Vala*. It appears that Mukkuvas of Kappad are slower in adapting modern technological innovations as compared to even their counterparts Puthiyappa and the Mappila fisherfolk.

Mappila Bay, Cannanore District

Mappila Bay, a fishing harbour of Cannanore, is situated at Ayikkara beach or Kadappuram. Thayil Kadappuram is situated in Thayil ward of Cannanore Municipality. Azhikode, which is in the same Panchayat, is 10 kms away from the Mappila Bay.

The Mappila Bay fishing harbour was constructed as part of the Indo-Norwegian Project at Ayikkara about 15 years ago. Infrastructure includes ice plants and a boat manufacturing unit.

Our field work in the Mappila Bay, Thayil Kadappuram and Azhikode has brought home the fact that the Mappilas and Mukkuva Hindu fishermen of the area have adopted modern technological innovations. A Muslim fisherman who worked in an Arab country, returned to his native Mappila Bay and invested his savings in three trawlers and trawl nets.

The Mukkuva fishermen of the Mappila Bay did not venture out for fishing on 11 August 2006 due to the occurrence of *Vadakkan Nir* and *Vadakkan Kattu,* which generate violent waves, particularly, *Vadakkan Tiramala*. If the *Vadakkan Tiramala* occurs continuously it is called *Kachan Tiramala*. Generally, whenever a northerly wind is accompanied with northerly current, the sea would be very rough and there will be poor catches. Therefore nobody can go for fishing. However, some small boats with the help of *Tangis Vala* went out for fishing at 9 a.m. employing traditional techniques. *Tangis Vala,* known also as *Mala Vala,* is employed to catch small fish.

Dugout canoes were used in Cannanore in former times with cotton nets for fishing. The Mukkuvas of Cannanore have adopted modern technological innovations to a limited extent. They have adopted small fibre boats (traditional *Odam*s covered with fibre), in which two men are employed for fishing.

The floating gill-nets are used at Cannanore in small fibre boats (*Odam*) with a crew of two men for fishing larger species like *Ayikkora, Tirandi* and *Sravu*. The *Odams* (small fibre boats) are now fitted with outboard engines having 9.9 HP motor. A spare engine is also kept in the boat for use in an emergency. Two engines are employed for launching a gill-net and also for safe return after fishing expeditions. Normally, a single engine which requires eight litres of kerosene for an hour is used in fishing expeditions upto a distance of eight kilometres. The engine is switched off after launching the net. The fishermen carry kerosene for operating upto 50 to 100 fathoms. The use of encircling technique requires more kerosene. If the catch is nil, the cost of the fuel will be met by the owner of the unit and marked as loss for the expedition.

When the Mukkuva fishermen launch their *Ozukkuvala* they have to put in more labour since foreign trawlers often pass over the *Ozukkuvala*. Although they have to operate beyond 35 to 40 *Maarus* they often operate within 18 *Maarus,* violating the maritime laws.

Mohammad Kunju purchased an *Odam* (boat) and two trawl nets for Rs 39,000. This amount was raised as a loan from the Fishermen Cooperative Society at Ettikulam. The loan has since been cleared off.

Forty per cent of the catch is given to labourers and he takes 60 per cent as the owner after deducting the expenses for food and fuel. The total operating expenses for an expedition is round Rs 100. Kunju's fishing unit does not get regular catches. Sometimes he gets bumper catches and sometimes no catch at all. His fishing labourers get eight per cent as bonus above their wages when there is a bumper catch. They account their catches once in two weeks. Kunju worked in the boat for 15 years from 1972 to 1987 before buying the *Odam* and trawl nets. Whenever there was a poor catch he used to go with his hook-and-line during October, November and December for catching shark. In those days cotton lines were used for the rod or *Chunda* and he used to get cat fish. The cotton rope is now replaced by a plastic one but the hooks are the same. The Muslim Beppukar of Talassery still follow the hook-and-line fishing by using plastic dummy fish as bait on their *Otta Cunda.* There are also many Beppukar at Tanur Kadappuram.

Valiya Fibre Vallam and Modernisation of Mappila Bay

The Mukkuvas of the Mappila Bay and Thayil have adapted *Valiya* fibre *Vallam* and also fibre covered small *Vallams*. The *Valiya* fibre *Vallam* is fully covered with fibre glass excepting the *Padi* (cross bar), which is made of wood. The engine fitted at the *Amaram*. This boat costs Rs 1,80,000 and a crew of ten men can operate it. Two *Vallams* are required for a fishing expedition and two lighters carry them. The lighters also bring the catches to the landing stations. Altogether thirty men form the crew when two fibre glass *Valiya Vallams* are operated. The Ring Vala is used in such an expedition and it has a length of 200 and a width of 20 *Maarus*.

The fibre glass *Vallam* helps the fishermen to catch pelagic fishes like *Mathi, Ayila* and small *Pulappu min*. The cost of the boat depends upon its length. The fibre glass *Vallam* was introduced in Mangalore between 1975 and 1980, but it appeared at the Mappila Bay only in 2001. There are also wooden *Vallams* fitted with Leyland engines in which 50 to 70 men go out for fishing. These have been turned into trawling fishing boats. The following are the facilities available in the trawler boat:

Underground	—	Store boxes for ice, fish and engine
At Aniyam	—	There is a space for storing fish and space for ice
Storing place	—	Cabin
Serang	—	Operator
Engine driver	—	Driver
Crew of 5 men	—	Fishing labourers with specific functions

Abdul Salam, a Mappila fisherman who returned from the Arabian Gulf, owns three trawlers, viz., Sainco, Saimakku and Salmania. Each trawler costs Rs 30 lakhs. Before his departure for the Arabian Gulf, he along with three others, owned a trawler, but it was sold out due to heavy debts. Abdul worked as a fish trader for eight years in the Gulf. Sainco was purchased in 2001 and its Captain is Krishnan, a Thiya, with long experience in fishing. This trawler is often hired by the Coast Guard for patrolling. It is made of wood and iron and has a store made of fibrewood. It has a wooden platform. It is fitted with a Turbo engine of 122 HP and is operated by a crew of seven. A trawl net is used for catching various species of fish including *Chemmin*. Sainco is fitted with GPS, echosounder and wireless and travels to the borders of Karnataka and Tamil Nadu. A trawler requires 200 litres of diesel and 40 tons of ice (cost Rs 100 per ton) for its operation for a day. Out of the crew the captain alone belongs to the Thiya community and the rest six fishermen are Muslims. The cost of a trawl net is Rs 15,000 and its accessories like ropes and sinkers come to Rs 50,000. The floats cost Rs 10,000.

The owner of the Sainco trawler gets 65 per cent of the catch; the remaining 35 per cent is divided among the crew of seven. The owner (*Muthalali*) bears the expenses for fuel and food. Before the imposition of ban on trawling, three days' operation of Sainco fetched fish worth Rupees two lakhs. After deducting the fuel and food charges of Rs 50,000 and the commission of agents amounting to Rs 2,300, the remaining Rs 47,000 was divided among the crew of seven, each getting Rs 6,500 for three days' work. In other words, each fisherman got Rs 2,166 per day. *Kanava* (cat fish), *Ayila, Karimin* and *Puthiyappala Kora* the main catches from the expedition.

Technology of a Trawler

There are two separate stores in front of a trawl boat. One is meant for fish and the other for storing ice. There is also a separate compartment for keeping the 122 HP Turbo engine. In another compartment 10 to 15 nets and ropes are kept. All these stores are covered with wooden planks. The net is dragged with the help of engines.

According to our informants, unless and until a trawler fishing unit gets catches worth Rs 2 lakhs, it will incur a heavy loss in an expedition lasting from three to four days. If a unit gets Rs 2 lakhs worth of catches, a sum of Rs 50,000 will be incurred for fuel and food, and Rs 20,000 towards the commission of the agent. Of the remaining, Rs 1,30,000, i.e., 65 per cent, will go to the *Muthalali* and 35 per cent to the fishing labourers. If they do not get any catch the owner will incur a loss of Rs 50,000 as an amount of Rs 1,500 to 2,000 has to be paid to the fishing labourers for an expedition.

The Serang or captain of the boat controls the fishing expedition. The other functions of the crew are shooting, separation of the fish and hauling of the net. After removing the net from the boat, fish is separated and cleaned. Then the catch is stored in the boat.

The bottom fish can be located with the help of GPS in trawlers. In the past the fishermen went upto from 15 to 20 *Maarus* only in their wooden boats. Today they can go to Mangalore, Neeleshwaram and other far off places. The echosounder also helps to identify the flowing of fish according to the currents and winds.

In the event of the sea becoming very violent the officials of the Fishery Department and Marine Police hire the Sainco boat owned by Abdul Salam. In this case the Fishery Department and Matsyafed meet the fuel and food charges. Food is prepared in the boat itself.

Peethambaran Nair, Pallipurathu Veedu, Pappanassery (PO), Cannanore, is the son of Kunjiraman Nair, Aroli Veedu, a Panchayat member and a freedom fighter. His mother, 80 year old Ammalu Amma, is getting the freedom fighter's pension since Kunjiraman Nair's death.

Peethambaran Nair has a daughter and a son. Pappanassery is situated 11 kms from Cannanore. It is near the Valapatanam River. After his high school education, he was selected for training in fishery by the Department of Fishery, Government of Kerala on a stipend of Rs 100 per month for nine months in 1970. He became a member of the Pappanassery Fishermen's Cooperative Society. The Society had two boats (KLA 245 Kelvin engines and KLA 318 Varna engines), and a traditional net (*Vakkuvala*) was used for training. The net was subsequently replaced by a plastic *Vala* and trawl nets. After his training for one year in 1972 he joined Ettikulam Muhammad, a shareholder of a fishing unit, as one of a five man crew. Muhammad's boat was a mechanised one having a length of 30 feet. The trawl net was made of plastic.

Peethambaran Nair belongs to the Maniyani community (Laterite stone workers – OBC) of Cannanore district which produced a DGP and a High Court Judge. He purchased the fibre covered boat and trawl net from his master, Mohammad Kunju, for Rs 60,000. He goes out for fishing in his trawler and is its captain. Nair lives in his mother's house with his wife at Tanakkad, Kannur (Cannanore). His brother Mohanan works with him as a fishing labourer. Nair's younger brother is working as a constable in the Police Department. He got a special award from the Chief Minister for his meritorial services as a Coast Guard.

Peethambaran Nair holds that the ban on trawling is necessary during the spawning period from June 15th to August 15th as it increases the productivity of prawn, sardine and mackerel. He also feels that trawling at night should not be permitted. His brother who is a Coast Guard caught hold of a trawler when its owner violated the law. The owner of the trawler filed a writ petition before the High Court of Kerala, but he failed to get a favourable order.

Out of Peethambaran's crew of seven, six have ownership of the net. The unit owned by Peethambaran Nair goes out for fishing upto 50 *Maarus* using the trawl net. The boat was not employed during the period of ban but from 16th August, 2006 it was again operated for fishing. The unit carried the ice, rice, vegetables and coconuts. One of the members of the crew cooked food for the crew. As many as 24 trawlers came from Quilon and Cochin and they were engaged in fishing in Cannanore, during the pre-trawling ban and post-trawling ban periods.

There are also three trawlers at the Mappila Bay owned by three men called Babu, Unni and Sivarajan. In each of these trawlers, trawl net is used by a crew of seven. A rich middle man owns seven trawlers. He is also a commission agent and exporter. Another exporter owns four fishing trawlers. A Hajiyar owns four fishing trawlers. It must be mentioned here that the Latin Christians from Quilon visit Cannanore for fishing by employing traditional wooden boats.

Abdul Jabar, a Mappila fisherman of the Mappila Bay, bought a boat and a net in 1980 for Rs 5,000. He went out for fishing with fishing labourers for 18 years. As the owner he got 50 per cent of the catch and his fishing labourers also got 50 per cent after deducting fuel and food charges. Now he owns a small fibre boat, one sardine net and another mackerel net. Apart from sardine and mackerel, small species of fish are also caught. Each net costs Rs 5,000 and the fibrc boat Rs 20,000. Even now he goes out for fishing although he is quite old. Sometimes he alone goes out for fishing and catches pelagic fish. He is very happy since he is not indebted to anyone and he has a steady source of income. He used to earn Rs 1,000 a day earlier, but now gets only from Rs 200 to 300 per day.

Farook, a Mappila who returned from the Arabian Gulf region, purchased a fibre boat with outboard engine of 9.9 HP by taking a loan of Rs 1,92,000 from Matsyafed. He has repaid the loan excepting Rs 25,000. He also bought a trawl net for Rs 75,000 which was insured. The loan was sanctioned through the Fishermen Cooperative Society which is controlled by the Marxist Party. His *Ozukkuvala* was once destroyed by a foreign launcher. He claimed compensation from the insurance company. The net was insured for Rs 3,000 but the fisherman stopped payment of the premium. Muslims do not go out for fishing on Fridays while the Hindus do not undertake fishing on *Sankranthi* days.

At the age of 20 Farook became a shareholder in a small boat with a crew of four. He got an *Ayilachala Vala* for Rs 25,000 and also purchased an old boat for the unit. The *Ayilachala Vala* was used in this *Toni*. He worked as fishing labourer in this boat along with three others.

Mukkuvas of Thayil

Soman, a Mukkuva, owns 10 cents of land at Thayil. The Mukkuvas are patrilineal. Soman lives in a rented house with his two unmarried sons and two daughters. Although he lives in a rented house he could not avail of housing grants due to its cumbersome rules and regulations.

Soman does not own any fishing gear and craft. He gets employment as a fishing labourer whenever the encircling technique is used in *Valiya Vallam* with *Valakkuna Vala.* The equipment in this fishing unit are owned by a group of 40 fishermen but unfortunately majority of them do not have any share in the unit. The group took a loan of Rs 30 lakhs for the purchase of the *Valiya Vallam* and other fishing equipment from the Canara Bank and Grahmina Bank through a Fishermen's Cooperative Society. Soman works as a labourer in the same fishing unit. His son, Ratheesh, after passing his matriculation, has also been working as a fishing labourer for the past eight years. The cost of the fishing gear and fishing craft is as follows:

Fibre boat	—	Rs. 1 lakhs
Mackerel net	—	Rs. 10,000
Two Suzuki outboard engines	—	Rs. 65,000 × 2 = Rs 1,30,000

The fishing labourers get 40 per cent of the catch and the remaining 60 per cent goes to the owner. The owner meets the cost of the fuel and food. Generally, fishing is undertaken from 6 a.m. to 2 p.m., for which 60 litres of kerosene is required.

According to Dineshan (31) of Thayil Kadappuram, nobody could go out for fishing on 11 August 2006 due to the occurrence of *Vadakkan Nir* and *Vadakkan Kattu* that resulted in turbulence in the ocean and unpredictable weather conditions. However, a few traditional fishermen were able to go out for fishing with their small fibre boats and *Malan Vala* for catching small species of fish. Dineshan also went out as one of the five for fishing in one of these units with its two small fibre boats and a *Malan Vala.*

Rafi, a successful Muslim fisherman of Thayil Kadappuram, owns two small fibre boats and an encircling net. Rafi purchased 10 cents of land for Rs 15,000 in Thayil Kadappuram and is now constructing a new house. He also works as a fish trader since the past two years. In 1990 he purchased a 9.9 HP motor from Matsyafed through the Cooperative Society for Rs 53,000. He has repaid the loan. Ten fishing labourers work for him.

About 50 Latin Christian families are found in Thayil Kadappuram. Mogavirar and Bovis-Mogeyar are not found at Thayil. Trawl nets and gill-nets are employed from a *Cheriya Vallam* fitted with an outboard engine for catching *Ayikkora*. Two engines are used for launching the net.

Johnson worked as a fishing labourer in Saudi Arabia for six years. He returned home and started his traditional occupation of fishing. Johnson became a shareholder in a fishing unit with a *Valiya Vallam* which has got an inboard engine (40 HP) and two lighters. This unit has another boat with two 25 HP engines. The *Valiya Vallam* employs a crew of 20 to 25 men for fishing with *Pothan Vala* and *Ayila Vala* for catching *Chemmin* and *Mathi*. The lighters have 5 to 10 fishing labourers. Thus, this fishing unit has a crew of 40 to 60 including both fishing labourers and owners.

In the past, *Mathikolli Vala* was used for fishing during nights. The net was launched against the *Nir* and fish is caught by adjusting the floats. The fishermen adjust their techniques according to the occurrence of the *Nir*. They go out for fishing when the *Nir* and *Kattu* become *Patham* (not very rough).

Some fishermen in Cannanore use nets for catching mid-water fish like prawn, sardine, mackerel, *Avoli* and *Mullan*. They use the *Ozukkuvala* or floating gill-net according to the flow of the *Nir* and catch *Ayikkora* and *Nemmin*.

Valans

The Valans of Cannanore district are found in the fishing villages of Valipatanam, Pazhayangadi and Cheruvathur. They employ the *Kuttivala* with small meshes for catching small prawns in rivers, backwaters and estuaries. Prawns are available in Azimugam (back water). *Kuttis* (Pegs) are erected in water and the *Kuttivalas* are tied to them according to the high and low tides. Some Valans say that growing of *Kadal kadu* would increase the fish resources.

Modernisation of Hook-and-Line Fishing

The Mappila fishermen of Cannanore district in general and those of Talassery in particular follow hook-and-line for catching the larger fish like *Sravu* and *Ayikkora*. Hook-and-lines have been modernised by incorporating innovations like plastic fish with Kambi snoods. Small fishes are tied to the Kambi snoods as baits.

Stake Nets

The stake net is widely used in Cochin and Valapatanam by the Valans. This is a conical bag net and is called *Kuttivala* or *Unnivala*. It is usually shot against the tides. The net is spread and held in a vertical position by two stakes or sticks. It is particularly used for catching prawns. The net is generally hauled up at the turn of the tide. This method is followed in the backwaters of Kerala throughout the year whenever the tides are feeble. The size of the net varies widely. Some of the largest ones along the coast of Kerala are 300 to 500 feet in length with a mouth that is 150 to 200 feet in diameter.

The net is set in such a manner that the mouth is left open against the current. However, the edge of the coir wings and the opening of the bag are kept high and open by the device of floats and sinkers at regular intervals. The net is positioned vertically by attaching weights at the lower end (foot rope) and by tying wooden floats to the upper edge or head rope. The floats and weights are balanced in such a way that the net is held in a vertical position almost as if it is suspended. Stake nets are also operated without floats and sinkers. They are just tied to stakes or fixed on pairs of stakes driven into the bottom.

Cooperatives

Loans are given to the fishermen through the Fishermen Cooperative Societies. In Cannanore, the Fisherman Cooperatives are controlled by different political parties. There are fishermen cooperative societies organised by CPIM, Congress and BJP. The flags of these political parties are generally hoisted in their boats when they go out for fishing. There is also a Cooperative Society run by the RSS in Cannanore.

Technology and Cosmology

The character of the ocean is very uncertain due to various factors. Nobody can predict its behaviour. When the *Puram Nir* and *Vadakkan Kattu* is followed by *Adinir,* the boat and net are likely to flow upto 100 to 150 kms. This generally occurs after the *Velutha Vaavu* (Full Moon), particularly in the month of *Karkkidakam*. After the 28th of *Karkkidakam* all the currents flow as a single *Nir*. After *Karkkidakam*, a single *Nir* generally flows for 10 to 15 days. In other words, a particular *Nir* begins on either full moon or new moon and continues upto the next face of the moon. If the *Nir* occurs during the full moon it continues upto the new moon. After 15th of *Chingam,* the sea becomes very peaceful and all fishermen go out for fishing.

It was full moon on 11 August 2006 when we were in Thrissur for our fieldwork. For the past three days *Veliyerakkam* had taken place. It was the turn of the *Veliyettam* till 12th night. Usually *Veliyerakkam* starts after 12th night and continues upto 6 a.m. The *Veliyettam* again begins at 6 a.m. In this way the *Veliyettam* continues for 15 days until the new moon. On the *Karkkidaka Vaavu* day fishermen do not go out for fishing since the sea becomes very violent, causing boats to be capsized. If the fishermen fail to carry sufficient kerosene they might find it very difficult to return to their fishing stations. Sometimes the government seeks the help of the Navy the rescue to fishermen in dire necessities.

After the *Vellutha Vaavu* (full moon), *Veliyettam* occurred and so fishermen did not go for fishing on that day. Generally fishing boats do not operate against *Ozukkuvala*. If they go for fishing during nights they operate beyond 30 to 40 *Maarus*. They cannot anchor the trawlers near the Continental Shelf which would destroy their nets (*Ozukkuvala*). If they are good fishermen they sometimes tie the *Ozukkuvala* launched by the traditional fishermen to their trawlers. Otherwise the *Vala* is likely to be destroyed causing heavy loss.

Our informant Manu belonging to the Mukkuva community of Thrissur says: "I have been in Cannanore for the past eight years employing *Ozukkuvala* of different kinds. The *Ozukkuvala* having yarn count 6, is meant for catching small and medium species of fish. The *Ozukkuvala* costs Rs 1.5 lakhs. But it must be pointed out that after the month of *Tulam,* the *Ozukkuvala* with a larger mesh is used."

"In earlier days *Kamba Vala* or shore seine net with the help of 30 to 40 fishermen was used. The fishermen belonging to Thrissur do not venture to deep sea waters. Due to mechanisation, instead of a crew of four men, only three men are employed in fishing. I want separate laws for each technique of fishing to be enacted and fishing should be prohibited at nights."

Fish migrate according to the occurrence of different currents and winds. That is why there is scarcity of fish in some fishing grounds. A particular species of fish may be available at Cannanore today, but the shoal may swim to Kozhikode tomorrow. On the following day, the same species of fish appears at Ponnani and on the third day it may reappear at Chavakkadu and Munambu. Again if the current flows towards the north it will appear in Cannanore and Neeleswaram. In fact, on 11 August 2006, the *Nir* was *Karanir* (easterly current). Since this current flows from east to west and the wind was *Vadakkan Kattu* the sea became very rough, preventing fishermen from going out for fishing. When the *Vadakkan Kattu* and *Vadakkan Nir* occur violently the sea becomes very violent and the nets are also likely to be destroyed. But if *Vadakkan Nir* occurs alone there will be little catch of fish. Similarly when the *Thekkan Kattu* and *Tekkan Nir* occur violently there will be no catch at all. If four types of currents occur simultaneously the nets are likely to be torn and it would take another three days to repair them. The Mukkuvas of Cannanore have traditional knowledge of eight types of currents, eight types of winds and eight types of waves, like the Mappila fisherfolk of Tanur. They adjust their technology according to the behaviour of the ocean, habitat of fish and climatic condition.

Kasargod Fishing Village, Kasargod District

The Hindu fishermen (Mukkuvas) of Kasargod have adapted wooden *Valiya Vallam* fitted with 20 to 40 HP motors in which 25 to 30 fishermen go out for fishing. These *Vallams* are not covered with fibre. In other words, the Mukkuvas of Kasargod have not adapted the latest technological innovations like the *Valiya fibre Vallam,* trawlers and trawl nets.

Kasaba Kadappuram

The Mukkuvas, Mogeyar, Bovis-Mogeyar and Mogavirar are fishermen belonging to Kasargod. According to our informant the Mogeyar and Mogavirar are one and the same community. There is no intermarriage between Mukkuvas, Bovis-Mogeyar and Mogavirar. Valans are also found in Neeleshwaram, 35 kms from the Kasaba beach. The fishermen stay along the coast of Chandragiri river and inhabit several hamlets or Kadappurams, viz., Kizhur, Kottikulam,

Bekal, Kanjangad, Kumbla, Uppala, Kasargod and Manjeshwaram. The fishermen inhabiting the above fishing villages would be much benefited, if the proposed Kasargod fishing harbour is taken up soon.

Odam

Traditionally *Odam* (dugout canoes) without mechanisation were used with a crew from eight to ten fishermen to operate them. Besides small dugouts called *Cheriya Toni* were also used. Now the *Odam* has been converted into *Valiya Toni* with inboard motor of 40 HP, and from 12 to 15 men go out for fishing. There are 16 *Kallis* and 17 *Padis* in an *Odam*. The estimated cost of an *Odam* is Rs 3 lakhs. Even small boats are now fibre covered with Yamaha engine upto 10 HP. These boats are used as lighters for transshipment of catches from the big boats or *Valiya Toni* or *Valiya Vallam*. A *Valiya Vallam* has 10 *Kallis* and 11 *Padis*. The boats are mainly made of the straight trunks of *Ayini*, Mango and Malaysian woods. Even a small fibre covered boat costs about Rs 35,000 to 40,000 and a *Valiya Vallam* with a 40 HP motor engine costs no less than Rs 30 lakhs. A crew of five to six men go out for fishing in the small fibre covered boats. They cost Rs 2 lakhs and can operate upto 10 *Maarus* from October to May. In a Leyland inboard *Vallam*, a motor engine of 40 to 50 HP is fitted.

Cost of Engines (Suzuki/Yamaha)

9.9 HP	=	Rs. 67,500
12 HP	=	Rs. 1,05,188
40 HP	=	Rs. 1,22,000

Boats using gill-nets are fitted with a 9.9 HP motor (Suzuki or Yamaha). The Mukkuvas of Kasargod launch their *Rani Vala* with the help of four fibre boats each fitted with a 10 HP motor. There is a crew of four in such a fishing unit. *Rani Vala* is one of the encircling nets. There are 17 fishing units among the Mukkuvas of Kasargod.

The *Odam* of the Kasargod Kadapuram has 16 *Kallis* (divisions) and 12 *Padis*. These are made of *Ayini* wood and fitted with 40 HP Motors (Yamaha) costing Rs 30 lakhs. The crew number from 12 to 15 men. Two lighters are used for transporting the catch from the *Odam* to the beach. The fish is transported in lorries from the auctioning centre for marketing through wholesale merchants and commission agents in Cannanore, Talassery and even far off Trivandrum. The commission agents or *Hawalakkar* charge 10 to 14 per cent for loans advanced to the fishermen.

Since the establishment of ice factories in the fishing villages, the traders are able to send the fish to different villages of North Malabar properly iced. There are three ice factories at Kasargod, of which two are privately owned and the third one is run by the government. In the past fish was salted for marketing.

A fibre covered boat is generally operated by a crew of five to six men and it costs from Rs 2 to 3 lakhs. In the traditional non-mechanised *Odam* eight men were employed. A mechanised *Odam* requires a crew of 12 to 15 persons for fishing.

Bovis-Mogeyar

The Bovis-Mogeyar (Manjeeswaram Panchayat, Kasargod Taluk, Kasargod district) speak Malayalam and Kannada. They are predominantly found in Kunjathur, Uppala, Muttam and Mangalpady in Kasargod Taluk. They have their own traditional boats and nets. Their traditional boat is called *Pathi* and is a wooden dugout. *Toni* is its equivalent in Malayalam. About 500 families of the Mappila fisherfolk are reported to be living in Manjeeswaram. More than 100 Mappila fishermen hailing from Manjeeswaram are working in the Arabian Gulf countries as fishing labourers. Many of them have invested their savings in houses and lands. It must be pointed out here that the Mappila fisherfolk working in foreign countries in white-collar jobs are better off compared to their kinsmen who are fishing labourers.

The Bovis-Mogeyar use their *Pathi* (length and breadth 13 Kols × 5 Kols) for fishing. This boat is made of either Mango, *Ayini* or even Teak. In the past a boat used to cost from Rs 4,000 to Rs 5,000, but today it costs not less than Rs 50,000. Fibre covered plank-built boats have been recently introduced. Their *Pathis* are being gradually replaced by fibre covered boats keeping the length of 13 Kols and employing 13 people for fishing. Small fibre boats are operated with a crew of two to three men. There are seven such fibre boats at Kunjathur fitted with 8 HP motors. This mechanisation took place in 1985.

The Bovis-Mogeyars have adopted small and big fibre boats fitted with motors of 9.9 HP and 15 HP. However, they still use their traditional boats with a crew of two to three men. These *Pathis* are eight Kols long and are even today employed for fishing. Of the three fibre boats (*Valiya Toni*) used by the Bovis-Mogeyar, one is a fibre glass boat.

Three mechanised *Pathis* are employed for fishing by the Bovis-Mogeyar and each has a crew of 10 men. A *Pathi* fitted with a 25 HP outboard Yamaha engine requires 30 litres of kerosene for working 12 hours, which is usually from 5.30 a.m. to 6.30 p.m.

Ownership of Nets and Boats

Balakrishnan belonging to the Bovis-Mogeyar community of Kunjathur owns two mechanised boats, one Sardine net and one *Ayilachala Vala*. The cost of the net and boat is given below:

2 boats	—	Rs. 60,000
Chala Vala (Mathi Vala)	—	Rs. 20,000
Ayila Vala	—	Rs. 20,000
2 Engines (Yamaha) 9.9 HP	—	Rs. 1,25,000

Each boat has a five man crew for fishing. On 27 August 2006 one unit went out for fishing and caught *Katla* fish worth Rs 4,000. After deducting the charges for fuel (Rs. 400) and food, the catch was divided between the owners of the crafts and the fishing labourers. Each member of the crew employed in this operation got Rs 515. Three shares were given to the owner of the craft and Balakrishnan thus got Rs 1,545. On 29 August 2006 Balakrishnan's unit went out for fishing with an *Ayila Vala* and a mechanised boat to a distance of 20 *Maarus,* from 4 a.m. to 10 a.m. They incurred an expenditure of Rs 400 towards fuel but no food charges were given, as the crew returned within six hours. A loss of Rs 100 was incurred. The catch was sold directly to the traders as they did not take any loan from the commission agent. Fisher women sold this catch of *Katla* fish at the rate of Rs 100 per kg. Apart from Balakrishnan's unit, three other units went out for fishing on that day, but they could not get any catch due to unfavourable *Kattu, Nir* and *Adinir*.

Balakrishnan had taken a loan of Rs 25,000 from the Matsyafed through a cooperative society for buying an engine which costs Rs 60,000 at the bank interest of 14 per cent. Additionally eight per cent of the catch was paid as commission of which 4 per cent went to the owners of the boat and 4 per cent to the society as service charge. Since he cleared the amount within two years 22 per cent of the loan was given as a subsidy. Now, subsidy is no longer given. The Bovis-Mogeyars have not adapted trawlers. One hundred and ten fishermen of the village have joined the *Sampadya Padhathi* of the Matsyafed. The Bovis-Mogeyars believe that *Vadakkan Nir* and *Vadakkan Kattu* are favourable for good catch if there is no occurrence of *Adinir*.

A.K. Ahammed, son of Seethi, took a loan of Rs 61,000 from a moneylender for the purchase of an engine for his boat. Although he paid the amount, it was adjusted towards the interest. He says that the amount of Rs 61,000 is still outstanding against him. He has taken another loan of Rs 40,000 for purchasing a plywood fibre boat but he could not go out for fishing for three months. He sold out his boat for Rs 15,000.

Dasan, a Mogavira, went out for fishing but he got only five-and-a-half boxes (35 kgs per box) of fish.

Income and Expenditure

150 litre of kerosene oil	=	150 × 30	=	Rs. 4500
Oil	=	5 litre	=	Rs. 200
Petrol			=	Rs. 600
Total income			=	Rs. 1,400
Total loss			=	Rs. 4,200

There is a Jama Masjid at Kasargod where Ahammed offers his prayers five times daily. He donated two boxes out of the 100 boxes of fish that he caught to the mosque.

The *Cheriya Vallam* or small fibre boat, of the Bovis-Mogeyars has nine *Padis* and eight *Kallis*. The *Odam* has 10 *Kallis* and 11 *Padis*. Generally two boats are employed for catching fish. There is a fibre glass *Vallam* in this village with a 40 HP motor. There are two other boats with 9.9 HP and 25 HP motors respectively. 150 litres of kerosene is required for operating a boat for 12 hours and the local fishermen buy the oil from the open market from Rs 25–30 per litre. Although *Chakara* appeared in Kasargod they could not go out for fishing due to the ban on trawling during June-July.

Net

The Bovis-Mogeyars of Kasaba have discarded all their traditional cotton nets for catching different species of fish and now use nylon or plastic nets. The main nets used in the Kasaba beach are *Cheriya* trawl net, *Veesa Vala, Rani Vala, Malan Vala* and Ring net. The *Rani Vala* has a length of 200 *Maarus*, width of 30 *Maarus* and a weight 500 kgs. Its mesh size is 18 mm. It is launched upto a depth of seven feet. The launching of the *Rani Vala* requires four *Odams* and each *Odam* has a crew of 10 fishermen. This net is operated upto 30 *Maarus,* employing the encircling or *Valachupidikkuna* technique. The Bovis-Mogeyars of Kasaba go out for fishing upto 10 *Maarus*. The *Rani Vala* cannot be operated with a motor above 10 HP. Moreover, here the Azimugam is located 4 kms away from the landing station. One Babu, son of Raman of this Kadappuram died when his boat capsized in July 2006. Ten fishing units went out on 27 August 2006 when we were in the field. Each unit had four boats and 10 fishermen (total 40 men). These 40 members were jointly employed for launching the *Rani Vala*. The *Rani Vala* was operated with four boats fitted with 10 HP engines even when the ban on trawling was effective. The cost of a *Rani Vala* is Rs 2 lakhs. The forty men forming the crew had taken loans from Rs 1 to 5 lakhs for purchasing the net, engine and boat but they are finding it very difficult to repay the loan.

The Mukkuvas of Kasaba took to fishing when the ban on trawling was effective upto a distance of 10 *Maarus*. The total kerosene consumption for a fibre boat from 6 a.m. to 6 p.m. is 250 litres but they get only 150 litres per month at the subsidised rate of Rs 9.50. Kerosene costs from Rs 25 to 30 per litre in the open market. The kerosene rationed to the fishermen is hardly sufficient for two days. Sometimes the local fishermen go out for fishing at 6 a.m. and return around 2 p.m. They carry their food and ice. On 27 August 2006, a unit went out for fishing incurring an expenditure of Rs 8,000 towards fuel charges. Their catch of mackerel was worth Rs 5,500.

The *Vadakkan Kattu* and *Vadakkan Nir* occurred on that day and the fishermen expected a good catch but due to the onset of *Adinir,* the catch was very poor.

According to Jayan, storekeeper of Matsyafed, Kasargod, *Rani Vala, Ayilachala Vala (Kanathavala), Chala Vala*, Sardine net (gill-net) and *Ozukkuvala* are used for fishing at Kasaba. The nets are launched from *Pathi* (small fibre boats with a crew of five men). The Mukkuvas go out for fishing in the evening and return the next morning. Their nets have various counts of yarn such as *Ettaamnumber Vala* (8th count) and *Panthradamnumber Vala* (12th count). These Mukkuvas have not adapted trawlers and trawl nets. They still use their traditional *Ozhukkuvala* (gill-nets – with bigger mesh size) for catching *Sravu* and *Ayikkora*.

The yarn used for the mackerel net is synthetic plastic fibre. The mesh size varies from 20 to 54 mm. The net has a weight of 22 to 25 kgs and its length is 80 *Maarus*. Three *Murivalas* make a full mackerel net. One *Murivala* equals to 100 *Kanni.*

Chala Vala/Mathi Vala/Tangis Net/Sardine Net/Surrounding Nets

The following are the technological aspects of sardine net or *Chala Vala*.

1. Nylon sardine net
2. Monofilament

   Mesh size–1 × 2/36 +38 mm

   Mesh depth–200 *Kannis*

   Length 1 kilo–30 *Maarus*

   Weight 20 kgs (20 × 30 *Maarus*).

Small sardines are caught using this net but *Chemmin* cannot be caught with it.

The *Mathi Vala*, which is used at the Kunjathur village has a weight of 20 kgs and length of 600 *Maarus*. Sardine is caught using this from September to January and mackerel from February to June. Prawn (*Chemmin*) is caught during June, July, August and September. Different species of *Chemmin* like *Naran Chemmin,*

Puvalan and *Vella Chemmin* are caught from June 15 to July 31st. The fishermen of Kunjathur could not go out for fishing due to the roughness of the sea on 29 August 2007 when we were in the field. The Bovis-Mogeyar and Mogavirar welcomed the ban on trawling for 45 days from June 15th to July 31st on account of the spawning period. The *Ayila Vala* is operated upto a distance of 15 *Maarus*, although they support trawling ban from June 15th to July 31st. The fishermen of Kunjathur condemned the Supreme Court verdict extending the ban upto 15th August. The main reason put forward by the traditional fishermen is that July–August is the *Chakara* season and they expect bumper catches of prawn of various species at that time. The fishermen could not go out for fishing at Kunjathur due to the ban on trawling with *Rani Valas* and boats (*Pathis*) fitted with engines of 25 HP. Under the Supreme Court verdict the fishermen are prohibited from employing boats fitted with engines above 10 HP whereas the *Pathis* of Kunjathur are fitted with 25 HP motors. The fishermen held that the Apex Court had no knowledge of the *Chakara* season and the ecosystem of Kerala.

The Mappilas and Bovis-Mogeyars of Manjeswaram have adopted fibre boats and *Rani Vala*. The cost of a fibre boat is Rs 3 lakhs and that of a *Rani Vala* also Rs 3 lakhs. The *Valiya Pathi* (big boat) with a 13 Kols length and a crew of 10 men is operated from 5.30 a.m. to 6 p.m. Such a boat consumes 150 litres of kerosene in 12 hours. It is said that trawlers visit Kunjathur from Mangalore affecting the catch of the traditional fishermen. The local fishermen had to pledge their landed property of five to fifteen cents for purchasing fishing gear and fishing crafts. They borrow money from the Matsyafed through the cooperatives and nationalised banks.

A number of fishermen including Bovis-Mogeyar, Mogavirar and Mappilas are employed as fishing labourers in Saudi Arabia. Our informant Rajeendran worked as a fishing labourer in Saudi Arabia for two years and returned home as the prospects there were bleak. The plight of Tilak and Praveen of Kunjathur was the same even after spending Rs 75,000 each for Visa. In addition Tilak invested a sum of Rs 35,000 besides borrowing Rupees two lakhs from a Muslim trader for repairing his net and boat to work again at Kunjathur. Thirty fishermen took a loan of Rs 2 lakhs from a Muslim trader who is also a commission agent, for purchasing a *Rani Vala* 15 years ago on the condition that they would sell their catch to him only. Two per cent of the catch is set apart towards his commission. After deducting the expenses of the fuel, food and commission, the sale proceeds of the catch is distributed among the 30 fishermen. There is a Bagavathy temple at Kunjathur with ten subsidiary deities including *Gulikan* and *Naganmar* and each fishermen's family is bound to contribute Rs 1,000 annually for its maintenance.

Gill-Net (Ozukkuvala)

The gill-net or *Ozukkuvala* is used in Kasargod. There are three types of gill-nets according to the count of the yarn, namely 6th, 8th, 12th (85 mm, 90 mm and 95 mm). The mesh depth is 100 and 10 *Kannis*. Its length is 5 *Maarus* and total weight is from 200 to 250 kgs. The *Ozukkuvala* has a length of 400 to 500 *Maarus* and its depth is six *Maarus*.

The *Ozukkuvala* used for catching *Ayikkora* from September to March is operated from a fibre-boat (*Pathi*). In the past a single wooden boat was used, but nowadays two small dugout fibre boats are used for launching a gill-net. A crew of five men go out for fishing in the evening and return the next day morning. This is applicable in the case of a boat using *Ettamnumber* (8th count) and *Pantrandamnumber* (12th count) net. The Mukkuvas of Kasargod have not adopted trawlers and trawl nets. Larger fishes like *Sravu*, *Ayikkora* and *Tuna* are caught with gill-nets.

Ayilachala Vala

The mesh size of the *Ayilachala Vala* (net for catching mackerels and sardines) is 0.20/52/54 mm; mesh depth is 100 mm and its weight varies from 20 to 25 kgs. Its length is 80 *Maarus*, and it is locally called *Tangis*.

Rani Vala

The *Rani Vala* is used in a traditional wooden boat. It is manufactured by the net factory run by the Matsyafed in Cochin. This net is not used at the Mappila Bay and Talassery, but it is found at Puthiyangadi, near Ezumala. In Kasargod and Kumbla this net is widely used. It has a weight of 550 kgs and its total cost comes to Rs 2 lakhs, excluding accessories. The cost of accessories like floats, sinkers, ropes, etc., comes to Rs 50,000. This net is manufactured from monofilament yarn.

The *Rani Vala* is launched from four fibre boats with the help of as many as 40 fishermen. A fishing unit went out with this net in four boats on 28 August 2006 from 6 a.m. to 2 p.m., incurring a fuel charge of Rs 8,000 for its operation. The catch was worth only Rs 5,500 as the Mathi fetched Rs 140 per box (30 kgs per box). Unfortunately killer fish like *Kadikkuna Min* (*Makri*) and *Vadakkan Kattu* appeared in the sea and destroyed the net. The catch was very poor due to the occurrence of *Adinir* and migration of the fish to the bottom. The fishermen incurred a loss of Rs 3,000 in the operation. These fishermen purchased 9.9 HP motors through their cooperatives. At the time of our fieldwork we were informed that the Matsyafed used to give 22 per cent subsidy for the purchase of engines from 10 HP to 40 HP motors. The subsidy was 18 per cent initially but afterwards it was increased to 22 per cent. There was no subsidy for cash purchase.

Formerly, when a fisherman bought an 8 HP engine for Rs 60,000 he got a subsidy of Rs 10,000. Similarly, when he bought a net costing Rs 8,000 he got a subsidy of Rs 2,000. The subsidy for the purchase of motors has been discontinued these days.

The cost of a *Rani Vala* is Rs 375 per kg. The fishermen can buy upto 22 kgs of *Rani Vala* ($22 \times 375 = 8{,}250$). They get subsidy upto Rs 6,000 reducing the actual price to Rs 2,250.

The following are the conditions for the purchase of fishing crafts and gear through the fishermen cooperatives:

1. The beneficiary should belong to the fishing community
2. He must be a member of the society.
3. He must own at least a wooden craft and a net.
4. He must own a motor below 10 HP.

The Mukkuvas of Kasargod told us that the ban on trawling from June 15th to July 31st and from August 1st to 15th has affected their livelihood. During this period only outboard engines with 10 HP motor or below that capacity can be employed. The Government of Kerala failed to appreciate the problems confronting the poor fishermen of the state who have been much affected by the Supreme Court verdict.

Although free ration was sanctioned during the period of ban on trawling, many of the fishermen failed to get it due to lack of coordination on the part of the officials in charge of its distribution. Generally the fishermen go out upto 10 *Maarus* with *Rani Vala,* employing the *Valachilakkunna* (encircling) technique for catching fish.

For each fish, a separate cotton *Vala* was used in the past, but nowadays nylon and monofilament and *Rani Vala* are used. Due to the shortage of manpower, *Pathi* boats (dugouts) are manned by a crew of four to five men. Monofilament Vala, plastic *Avoli Vala, Kanatha Vala (Ayila Vala)* and *Kettu Vala (Mathi Vala)* are launched from small wooden boats. There are five such units at Kasargod which have not adapted modern fibre boats, trawlers and trawl nets. They still use kerosene for their outboard engines.

The Bovis-Mogeyars and Mukkuvas work together in the same fishing unit and their womenfolk are engaged in selling fish in the markets of the town. The Bovis-Mogeyars employ *Pathi* boats (dugouts) in Kasargod.

The Bovis-Mogeyars of Kasargod, like other traditional fishermen, were engaged in fishing employing their traditional techniques and equipment. This was the situation before the introduction of outboard motors and nylon nets. With the introduction of mechanised boats the Bovis-Mogeyars are able to go to distant fishing grounds, and hence the traditional characteristics of marine fishing have undergone drastic changes.

We have already mentioned elsewhere that *Vallam* or *Odam* were used traditionally in this region. *Toni* is another term for *Odam*. All over Kerala *Vanji* means a boat. The traditional fishing crafts fitted with outboard engines are found in the southern districts of the state and also in the district of Kasargod. Motors from 10 to 40 HP capacities are used for fishing. It must be pointed out here that the type of boat the fishermen use in Kerala, depends upon their financial resources and the ecology and behaviour of the ocean irrespective of caste and community affiliations.

During our survey of the villages of Kumbla, Uppala and Neeleshwaram, we found that *Cheriya Vallam* (small boat) and *Valiya Vallam* (big boat) were commonly used. The difference between these two types of crafts is their size and capacities. A *Valiya Vallam,* according to Elizabeth Zacheria (1996), is 6 to 6.5 metres long (20–21 feet or 8–8.5 Kols, 1 Kol = 2.5 feet), while the *Cheriya Vallam* is about 1.2 to 1.5 metres long (4 to 5 feet or 1.5 to 2 Kols).

A *Cheriya Vallam* goes upto a distance of 20 *Maarus* with the help of a crew of four men, whereas a *Valiya Vallam* can go beyond 40 *Maarus* with eight men. Even today the Hindu fisherfolk of Kasargod prefer the *Cheriya Vallam* or dugout for shore seine fishing. The *Valiya Vallams* are plank-built boats. Normally dugouts are suitably sized and planks are attached, particularly on the sides, prow, and stern, with coir thread (*Choodi*). The planks are sewed together, with copper nails so that they do not get corroded by the sea water.

Nowadays, the Bovis-Mogeyars and Mogavirars use mostly *Valiya Vallams* which are plank-built boats. However, older informants told us that they prefer small boats or *Cheriya Vallams* since they are less prone to leakage. On the other hand the younger generation prefers the bigger plank-built boats of various kinds, as, according to they, they are very resilient and can be employed at any time and in almost any condition, whereas dugouts have their own limitations. Moreover, it is difficult to repair the dugouts while the damaged planks can be replaced without trouble. The life of a dugout, if damaged, is greatly reduced, they said.

Fishing crafts generally ply in pairs. The catches are generally shared according to the established norms between the owner of the fishing crafts and gear and fishing labourers. With the introduction of mechanisation, *Valiya Vallams* are generally fitted with inboard Leyland engines and therefore they can go out for fishing quite far into the sea, particularly during the calm weather. These boats carry a sufficient stock of food, water and fuel if their expeditions are prolonged. The leader of the crew of a fishing expedition used to meet the cost of the fuel in olden times since he happened to be the owner of the fishing gear as well. Nowadays the expenditure for fuel and food is shared between the owner of the unit and the crew members.

Earlier the *Cheriya Vallam* had no outboard engines, which prevented the fishermen from going out too far. With the catches being drastically reduced, the fishermen using these boats are forced to go to distant parts of the sea. Thus, small boats are now being fitted with 3 HP motors whereas in the past they were rowed manually. When the mechanised small boats reach the fishing grounds the motor is switched off and oars are used to row to the destination. A *Cheriya Vallam* with an outboard motor can go upto 25 to 30 *Maarus* in calm weather.

The Bovis-Mogeyars and Mogavirars have not adapted the trawler boats and trawl nets. The trawlers are larger and fully mechanised boats and generally remain at sea for a week or so. These boats can go far out to the sea upto a distance of 1,000 *Maarus* or more, and have the capacity to carry more fish. Most of the *Valiya Vallams* have two lighters or fibre boats for transporting the catches to the landing station. Boat seine or ring net shaped like bags are dragged along the bottom of the sea. Trawling is resented by the traditional fishermen as they believe that this kind of continuous fishing, using purse seine nets, would sweep the seabed, harming the marine ecology and depleting the fish life.

Zachariah (1996) has discussed the impact and results of the introduction of modern technologies in the fisheries sector of the Bovis-Mogeyar and other communities. There are many reasons for the successes and failures, and differences between the various groups, which are mentioned by her. The role of the Bovis-Mogeyar temple organisation had a significant impact in shaping the attitudes of their members and in guiding their adaptations of these technologies "This role is quite akin to some of the successful cooperatives in other parts of the world. The set-up and adaptation that their temple organisation made in a situation with changed demands is comparable to that of the Latin Christian and Muslim religious bodies. Some of their members who are fishermen have also successfully adapted to modernisation with the help and support of their respective religious bodies. The combined or group ventures into economic and technological fields are supported by the collective actions of the temple organisation in various ways which is not feasible at individual levels. These include getting bank loans, lobbying and canvassing with government personnel too. The strong and well structured organisation, where the elders are respected by the community in general helped. The combined impact of all these is the break from the control of the middlemen." The various ways in which the temple organisation diversified to provide support to its members are outlined by Zachariah.

Kalamokku Private Fishing Harbour, Ernakulam District

Kalamokku is a private harbour situated in Cochin used by the Muslim, Mukkuva (Hindu), and Latin Christian fisherfolk. The Latin Christians adapted modern technology as far back as 1961. The Mukkuvas lagged far behind them. The Latin Christians are very adventurous and are prepared to take any risky operations at sea although they had received license for mechanised *Vallams*. They also have financial support from their churches for adapting modern technological innovations like mechanised boats and nylon nets. The Hindus took three decades for adapting outboard engines, modern nylon nets like nylon sardine and mackerel nets, and boat-seine nets. The Muslims of Kalamokku are also very quick in adapting mechanised boats and nylon nets. Most of the Hindus sold the *Vallams* as they could not afford to purchase kerosene for their operation from the open market where the ruling price is Rs 25 to 30 per litre as against subsidised price of Rs 9.50.

The Mukkuvas are economically very backward as there is no financial support from their own kins as they are also poor. The Hindu fisherfolk are highly indebted and they find it very difficult to repay their loans. Kalamokku has more than 100 boats. The Hindu fishermen often fail to undertake fishing expeditions due to the high fuel charges. Because of this they are afraid of taking loans amounting to Rs 15 to 20 lakhs from financial institutions to become owners of fishing units.

There is also a private harbour at Murikkinpadam near Vypin from which both traditional and small trawler boats operate. They go upto a distance of 100 to 120 *Maarus* for eight days incurring an expenditure of Rs 1,800 for diesel and Rs 100 per day per person for food. Leyland engines of 112 HP are fitted to their boats. For an inboard *Vallam* the catch usually varies from Rs 75,000 to Rs 1 lakh. The Vallams also carry GPS, echosounder, wireless and ice. Five fishermen who died in the Tsunami hailed from Edavanakkadu Kadappuram, 10 kms north of Vypin.

Tumpoli Fishing Village, Alleppey District

The Latin Christians, a numerically dominant community in Tumboli (Alleppey district) and the Arayans of this village, who constitute about 25 per cent of the total population of the fishermen, have adapted *Valiya* fibre *Kettuvallams* fitted with a Leyland engine. A fibre *Valiya Kettuvallam* employs fifty to seventy fishing labourers. It is accompanied by two fibre covered lighters with a crew of three men in each for collecting and transporting the catch to the landing station where auction takes place by the commission agent or a trader.

The commission agent is the man who has advanced the loan for the purchase of fishing crafts and gear and he is the sole authority to auction the catch. The *Valiya Kettuvallam* is generally owned by a group of seven to ten fishermen who are related to each other. For instance, Vikrant, one of the nine *Valiya* fibre *Kettuvallams* in Tumboli is owned by seven brothers. It cost them Rs 25 lakhs. Fifty to seventy fishing labourers are generally engaged in a fishing expedition in this vessel with a *Tangu Vala*. For instance, fifty fishermen went out for fishing in the *Valiya Kettuvallam* on 8 September 2006 at 4.30 p.m. and returned to the landing station next day at 5 p.m. They went upto a distance of 20 *Maarus*. The fuel consumption in this expedition was:

Diesel	=	200 litres
Oil	=	5 litres
Petrol	=	10 litres

The owner of the fishing equipment got 50 per cent and the fishing labourers 40 per cent of the catch respectively after deducting the expenses for fuel and food. The unit that went out for fishing on that day got 50 baskets of *Mathi* (1 basket = 40 kgs). It may be interesting to point out here that at 1 p.m. on this day the retail price of *Mathi* was Rs 3.50 per kg and at 6 p.m. the price came down to Rs 1.75. The total catch was auctioned at the rate of Rs 1.50 per kg, though the market price of a kilogram of *Mathi* on that day ranged between Rs 15 and Rs 20. In August 2006, the price of *Mathi* in Tumboli was between Rs 700 and Rs 1,500 per basket of 40 kgs.

Azheekkal Fishing Village, Quilon District

The Arayans of Azheekkal (Alappat Panchayat, Quilon district) have adapted big boats fitted with Leyland inboard engines and also outboard *Vallams*. There are about 100 outboard *Vallams* and 17 *Valiya Kettuvallams* with inboard Leyland engines at Azheekkal.

The Arayans of Azheekkal owned only the *Ottathadi Vallams* (dugout *Vallams*) in the past. But now the *Ottathadi* has been replaced by the *Valiya Kettuvallam* with inboard engines and the medium sized *Kettuvallam* with outboard engines. In those older days the *Ottathadi Vallam* was operated with the help of oars, rudders and sails while cotton nets were mainly used. The cotton nets are now replaced by nylon ones.

The *Valiya* fibre *Kettuvallam* was introduced at Azheekkal about 20 years ago. Forty to fifty persons go out for fishing in this boat. These were made of Anjali wood but no fibre was coated; instead sardine ghee was used. On the outer board Yamaha engines of 8 to 40 HP motors were fitted while the inboard *Vallam* (wooden *Kettuvallam*) covered with fibre was used.

A *Valiya Kettuvallam* called Mooladanam was owned by Sarangabani of Pandaraturuthu, Azheekkal. He bought this boat for Rs 27 lakhs in 2004. He also purchased a *Tangu Vala* for Rs 7 lakh and accessories (floats, ropes and sinkers) for Rs 1.5 lakhs. The Matsyafed advanced loans to a crew of 15 men in this unit as a group loan. This unit has been in operation for the past one and a half years. Apart from the loan taken from Matsyafed the unit has also taken loans from the following agencies:

Cooperative bank	=	Rs. 7 lakhs
Landmortgage bank	=	Rs. 3, 45,000
Commission agent (Tarakan)	=	Rs. 6 ½ lakhs
Individual loans	=	5 persons

The Mooladanam boat went out for fishing upto a distance of 42 fathoms in Azeekkal on 12 September 2006 on from 6 a.m. to 5 p.m., with a crew of 35 persons. Following are some of the items of expenditure incurred for the fishing expedition:

Fuel (150 litres)	=	Rs. 6000
Food and Tea	=	Rs. 1000
Commission agent	=	Rs. 2500
Expenditure for lighters	=	Rs. 11,000

Sakthikulangara Fishing Village, Quilon District

Traditional fishing technology and modern technological innovations co-exist at Sakthikulangara, a fishing village in Quilon district. In contrast with the situation found at Poovar, small and big *Vallams* are still employed. The size of the crew found at Sakthikulangara is eight to ten as against three in Poovar. Most of the gill-nets operated from small *Vallams* are either boat seine nets or drift nets. In contrast to Poovar the gill-nets used at Sakthikulangara are made of nylon. The fishermen of Sakthikulangara do not own any *Catamaran*. But several *Catamaran* are operated from Sakthikulangara which are owned by the Latin Christians of Kanyakumari.

Many of the *Catamarans* have been mechanised with outboard engines. The Latin Christians and the Nulayan Hindu fishermen employ two types of mechanised boats having 18 to 25 feet in length. Trawlers have been introduced here under the Indo-Norwegian project. The majority of the boats are trawlers fitted with Leyland engines having the capacity from 90 to 120 HP. These are operated by a crew of five to seven fishermen. The trawlers can go out from 10 kms to 50 kms.

Murickan (1991) had written: "Out of 3,500 fishing units operated in 1979 in this fishing village, the majority were small trawlers employed for catching shrimps and prawns. There was a decline in the number of fishing trawlers during 1980s. During peak seasons 1,000 to 1,500 used to operate while only one tenth of the total numbers operated daily during the lean seasons. Similarly there has been decline in the operation of *Valiya Vallam* owned by Latin Christians who are a dominant community in Sakthikulangara beach."

Prior to 1952, the Latin Christians were bonded labourers and they lived in thatched huts. They were kept in ignorance and subordination. The women were the main bread winners among the poor families and they worked as coir workers, fish vendors and net makers. Today over 2000 trawler boats operate daily from the Sakthikulangara harbour. The Latin Christians own over 97 per cent of these trawlers.

The Impact of Indo Norwegian Project (INP)

In 1952 the Indo-Norwegian project was started at Neendakara with the main objective of improving the living standard of the fishermen and simultaneously increasing the yield of fish. The Norwegian project was established to modernise technology but the local fishermen opposed the modernisation. They feared that any type of modern innovation might endanger their traditional life. The introduction of artificial bait for hook-and-line fishing was opposed by the fishermen in the past.

The traditional fishing bonded labourers who worked in trawlers have become the owners of the boats and nets. The majority of the trawler-boat owners and merchants recruited their crew not from Sakthikulangara but from outside.

According to Murickan (1991), the Norwegians left Neendakara in 1965 handing over the project to the Kerala Fishermen's Corporation. The economic and technological change brought about in Sakthikulangara can only be comprehended when we relate it to the modernisation of technology with its linkages of fish harvesting, processing and marketing. The traditional fishermen were bound to resist any change and were incapable of understanding the broader fish trading process. The relatively more adventurous traders, educated merchants and entrepreneurs gave boost to the prawn industry. Once they understood how profitable the prawn business could be they mobilised the fishermen to form cooperatives and purchase more crafts to increase production. Murickan observes in this connection: "a critical consequence of the INP project was the encouragement of a new pattern of prawn industry. The INP intended to increase fish yield and improve life standard of fishermen. In general it improved the fish landing above the sustainable level. In general it improved the standard of the life of the fisherfolk of Sakthikulangara. But it also created wide disparities among those involved in the prawn business. Some made huge profits, amassed assets, boat property, improved houses. This became possible only at the cost of vast majority of the fisherfolk who depended solidly on sea for their livelihood. The wealth created by INP did not become a social asset beneficial to the whole community and it merely resulted in the accumulation of wealth by a few being in a particular area."

The operation of mechanised trawlers was prohibited from June 15th to August 15th by the Government. This has created an unexpected situation in the mechanised sector of the fishing industry when the big capitalists sold their trawlers and shifted to other profitable business.

The ordinary fishermen who bought some of these boats have become the owners of mechanised trawlers for the first time. But it is impossible for them to pay back the loans to the credit institutions and they have landed in a severe financial crisis. This has threatened the security and development of the fishing economy of the state.

Murickan (1991) observes: "The use of heavy trawlers in the in-shore waters (where traditional fishermen are usually concentrated) has on several occasions resulted in conflicts between the owners of mechanised trawlers and traditional fishermen. While competing for shrimps the unrestricted use of trawlers often damages the nets of the traditional fishermen and upsets the sea bottom. The ill-feeling thus generated has led to organised confrontation between them. The traditional fishermen have many a time attacked all those associated with mechanised boats and even burnt the purse seine boats. In January 1984 two boats operating from Cochin were destroyed. They belonged to the Chairman of the Kerala Congress (J) and to the Secretary of the Indian Trawlers' Association. Another purse seine boat 'Mahalakshmi' worth Rs 5 lakhs and net worth Rs 2.5 lakhs were burnt down by fishermen at Cochin in September 1984. Several other incidents of conflict between the traditional fishermen and owners of mechanised trawlers have taken place during the last decade because of stiff competition."

Taking a serious view of such problems, the Government of Kerala passed the Marine Fishing Regulation Act in 1980. The Act was intended to protect the interests of traditional fishermen working on country crafts, *Catamarans* and canoes. It also aimed at the conservation of fish so as to regulate fishing activities on scientific lines. Through this act, the state government was empowered to regulate, restrict or prohibit fishing activities of any class of fishing vessels in any of the specified areas of the territorial waters of Kerala.

Statutory orders that followed the Act stipulated that the entire coastal waters upto 20 fathoms depth of the coast from Edava to Manjeswaram are reserved for the exclusive use of traditional fishermen. Mechanised boats were not permitted to fish in these coastal waters. Orders were also issued relating to the registration of all motorised and traditional crafts, the use of license by all motorised crafts for fishing in a particular area and the prohibition of purse seine fishing in territorial waters up to 22 kms. A special cell was created in January 1984 to enforce the Fishing Regulation Act, with a small squad and three patrol boats.

Murickan (1991) held that "in 1981, the traditional Fishermen's Federation launched an agitation for the complete ban on trawling during the months of June, July and August because of the social and biological implications of fishing during these months. The social implications pertained to the multifaceted problems of fishermen's social existence (including starvation) because of stagnation and resultant poor catch. The biological implications stemmed from over-fishing throughout the year since it led to the killing of juveniles in the under-water. The continuous trawling also destroyed the bio-mass at the sea bottom. In response to this agitation, the Fisheries Director prohibited in May 1981 mechanised fishing during these months. However the mechanised boat owners demanded that they could not remain without fishing. The consideration of substantial loss in foreign exchange also prevailed. Hence the Fisheries Director reversed the order and permitted trawling during the months of June, July and August. The matter was ultimately taken to court by the Fishermen's Federation for final judgment. In June 1984 it was declared that trawling must be banned during these months all over the state except in Sakthikulangara. The judgment was in favour of the boat owners of Sakthikulangara. However, the court felt the need for further investigation of this matter and referred it to the divisional bench."

In August 1981, the court also questioned the validity of the order on grounds that only a Government Secretary can issue such orders based on the Act. No person including the Fisheries Director has the requisite authority to pass such orders. Thus all orders were stayed till further action. A committee was formed of several experts in fishing and related fields to study the conservation aspects of marine resources during certain seasons. This committee was commonly known as the Babu Paul Commission since he was its head. The report was submitted in July 1982. Opinions on the banning of trawlers were divided among the government members – the majority being against the ban. Some of the unanimous recommendations of the Committee were:

1. Declaration of fish sanctuary at important bar mouths, viz., Neendakara, Cochin, Chavakkadu and Beypore.
2. Strict enforcement of mesh size regulation.
3. No more shrimp trawling boats to be licensed.
4. Export ban on tiny prawns.
5. Registration of all fishing crafts.
6. Prevention of water pollution.

Some of these recommendations have been enforced successfully. However, the problem of the ban on trawling continued to persist even after the committee's recommendations. The traditional fishermen kept on agitating for the conservation of those fish species which helped them earn their livelihood.

In April 1983, the purse seine owners took the matter to court in order to get the prohibition distance reduced for their fishing. The court ordered 10 kilometres limit as against 22 kilometres. The angry fishermen stormed the court at Cochin, disrupting the proceedings. The earlier order specifying a 22 kilometre line was stayed in order to bring the new order specifying the 10 kilometre boundary. The stay was later vacated in June 1984 and purse seiners were allowed to fish only beyond the 22 kilometre line. The boat owners continued to petition against this and even challenged the validity of the basic Marine Fishing Regulation Act. In September 1984, the Act was declared an order and was upheld, but the orders following the Act were discredited and struck down by the court.

In the wake of the agitation in 1984 the Government of Kerala constituted the Kalawar Commission consisting of three outside experts to study the issue. The report of the Commission confirmed most of the contentions of the Federation regarding the depletion of marine resources and stressed the need for regulating the fishing activities. During the period of 1969–82 Kerala had developed four fishing ports with a fleet of 3,500 trawlers, 1,000 mechanised gill-netters and 100 purse seiners. The Commission recommended the reduction in the number of trawlers to 1,145 and the strict implementation of the Kerala Marine Fisheries Regulation Act. The recommendations also included a complete ban on purse seiners within the territorial waters of 22 kilometres and trawlers within 20 kilometres. Further, the mesh-size should be raised to 35 mm. Since the government had not implemented the recommendations, the Federation launched a third agitation in 1985 which lasted for four and a half months. Hundreds of fishermen together with priests and nuns fasted and held demonstrations all over Kerala, pressing for their demands.

The opinions regarding the subject of unlimited fishing in inshore waters have varied according to the interests of both the mechanised and traditional sectors. Even though both the sectors have largely depended upon this fishery resource yet a long-term perspective for its conservation has not been viewed as a matter of common concern. The profit oriented powerful industrialists belonging to the mechanised sector consider conservation issue merely a favour to the traditional sector. The state policy has consistently favoured the interests of the mechanised sector. The powerful capitalist owners of mechanised boats, freezing plants and exporting companies have substantially influenced the state policy during various stages.

Tangasseri Fishing Village, Quilon District

According to John Jocob, a Latin Christian, his unit goes out for fishing at 4 a.m. and his fibre covered *Vallam* is operated by four fishing labourers. On 21 October 2006 his unit went out for fishing using a *Chala Vala* (sardine net) with the mesh size 28 mm. The unit got a good catch of sardine. The unit went upto a distance of 20 km. This fishing village has a number of outboard boats with 9.9 HP. There are more than 900 such fibre covered boats in this village. Sometimes they go out for fishing at 2 a.m. and return next day morning at 8 a.m. They go upto Varkkala. On that day about 400 fibre *Vallams* went out for catching sardine which are concentrated just above the bottom of the sea. The sardine net costs Rs 320 per kg and its total weight is 75 kgs. A crew of five men go out for fishing in a fibre *Vallam*. The cost of a fibre *Vallam* comes to Rs 70,000. A *Tattumadi* net was used in this *Vallam* during August and September with a mesh size of 6 mms to 50 mms. Larger and small fishes like *Kanava*, *Korichala* and *Avoli* are caught by employing the technique of *Valichilakkal* in which fishes are frightened and driven into the net.

Ayila Vala

The mesh size of *Ayila Vala* (mackerel net) is 64 mms and its cost 300 per kg. The total weight of the net is 80 kgs. The *Ayila Vala* unit generally leaves for fishing at 5 p.m. and returns the next day at 2 a.m. It goes upto a distance of 75 kms during October–December. There are also fibre covered boats in which four men go out for fishing. Each labourer works only for six months. The fishing labourers get their equipment at the rate of four per cent interest from Matsyafed upto a limit of Rs 3 lakhs. The Matsyafed takes charge of auctioning the catch at 15 per cent commission after deducting fuel and food charges. After deducting the commission, the sale proceeds are divided on the basis of 50:50 between the owner and the labourers.

There are no trawlers owned by the local fishermen at Tangasseri. There are also no *Valiya Vallams* here with inboard engines. *Disco Vala* is used for catching *Chemmin* during July–August. The mesh size of the *Disco Vala* is 42 mms. Its cost per kg is Rs 550 and 50 kgs are required for catching *Chemmin*. The local fishermen go out for fishing upto a distance of 20 kms. Apart from *Disco Vala*, there is a *Rani Vala* which is used for catching *Chemmin*. The ban on trawling has not affected the fishermen of Wadi. In fact they welcomed the ban. They say that the spawning period of *Chemmin* is from November to December and they wanted a total ban on trawling for these three months to protect the life of the fish. The traditional fishermen use only outboard engines with 9.9 HP. There are no *Valiya Vallams* with a crew of 40 to 50 in this village. Some Latin Christians from Virinjam, Cochin and Muttam visit Wadi for fishing. There are over 3,000 Latin Christians in this village.

Joseph (63) says that the local fishermen are not interested in doing hard work. He owns a *Chala Vala, Ayila Vala, Rani Vala, Disco Vala* and *Kariyala Vala. Kariyala Vala* and *Chala Vala*, which are quite similar, are used for catching *Parava, Chura, Nemmin* and *Chemmin*. Sometimes trawl nets are also used for catching demersal and pelagic fish.

From Wadi as many as 200 fishermen, particularly Latin Christians, have gone for fishing to Dubai and other (Arabian Gulf countries. They are employed as fishing labourers and workers in other sectors. Joseph's son is working in a construction company getting a salary of Rs 20,000 per month.

According to Joseph who started fishing at the age of 15, when the *Kattu* occurs from north to south it is called *Kachan Kattu*. When the wind travels from west to east it is called *Konda Kattu*. Whenever there is a *Vada Nir* they expect a good catch. When a *Nir* comes from south and travels to north also they get a good catch. However when a *Nir* occurs from west to east there will be a poor catch. When a *Nir* originates from south-west there will be a strong wind and the sea will be very violent. The *Nir* that travels from north to west brings a violent wind. When eggs are found in the sea water they do not get any catch. Joseph said that the nets used by them are now manufactured by foreign companies and they are imported. The mesh size of the *Kariyala Vala* is 32 mms. A minimum of 50 kgs is required for a *Vala*. There are no GPS, echosounder and wireless in his unit. Joseph reported that he had earlier worked in Bahrain and in a chemical factory at Bombay. He speaks English and Hindi. Joseph had got married when he was 22. He has six children and all of them are employed.

Poovar Fishing Village, Trivandrum District

The Poovar fishing village is situated 32 kms south of Trivandrum (Neyyattinkara taluk of Trivandrum district). It has the lowest level of fishing technology although transport and communication are fairly good here. The Neyyar River flows through the eastern border of the fishing village and joins the Arabian sea. The Neyyar dam has been built over the river in 1990s.

Nulayans (Hindu), Marakkan (Muslim) and Latin Christian fisherfolk inhabit Poovar. All the fishermen use traditional crafts and gear. They employ *Catamarans* and small dugout canoes for the operation of nets like *Thattumadi* and *Karamadi* (shore seine net).

Three to four logs of wood are tied together with cross bar at both ends for making a *Catamaran*. Small dugout canoes and big plank-built boats are also employed in this village.

Catamarans are employed for off-shore fishing and small dugout for the shore seine nets. The gill-nets, bottom-set nets and *Tattumadi* are operated from the *Catamaran*. Hook-and-lines are also operated from the *Catamarans*. The size of the mesh is adjusted to the type of species like sardines, mackerels and prawns. Bottom-set nets are operated for catching demersal fishes like skates, rays and sharks. Two *Catamarans* each having three fishing labourers are employed for launching the *Tattumadi,* which are bell-shaped boat seine nets. The ends of two coir ropes are tied to the net and they are pulled from the *Catamaran* after having driven the fish shoal into it. Hook-and-line fishing is also employed along the shore and deep waters.

Two *Karamadis* (shore seine nets) are employed by 30 to 40 fishermen from the shore. One end of the net tied to the fishing gear is held on the shore while the net is launched into the sea from the canoe. The other end of the net is brought to the shore about 100 to 200 metres away from the former rope. When the net gets the 'U' shape, it is pulled ashore by carefully coiling the ends of ropes on the ground. The fishes caught are mainly pelagic.

Vizhinjam Fishing Village, Trivandrum District

The main craft used at Vizhinjam, in the district of Trivandrum, for fishing is the *Catamaran*. There are mechanised and traditional *Catamarans* here. The *Catamaran* is the traditional fishing craft used in Vizhinjam that is mainly used by the Latin Christians and Hindu Nulayans. Four to five logs of lightwood are lashed together with coir ropes to make the *Catamaran*. The craft is rowed with the help of a bamboo pole. As part of Tsunami rehabilitation programme as many as 50 *Catamarans* have been mechanised at Vizhinjam. A *Valiya Vanchi* fitted with an inboard Leyland engine can employ from 40 to 50 fishermen in launching a *Tangu Vala*. The Hindu fishermen here do not have financial resources to adapt modern technological innovations, as a result of which they do not own fishing boats and trawlers.

Impact of Technological Innovations and Overfishing

The major fishing communities inhabiting the southern districts of Kerala are the Latin Christians, Muslims and the Hindu fisherfolk. The Hindus belong to the Arayan and Nulayan communities. The population of the Muslim fishermen in this region was very low at the time of our fieldwork. The technological innovations introduced in this region have not benefited all these fishing communities uniformly. Our study has revealed that the Latin Christians have reaped the lions' share of benefits of these modernisations, followed by the Marakkan, a Muslim fishing community of Trivandrum.

The Hindu fishermen, particularly the Nulayans and Arayans, have failed to respond to the modern innovations as much as expected. They continue to register a lower rate of response to the technological innovations. In some fishing villages like Tumboli (Alleppey), Poovar (Trivandrum) and Vizhinjam, the Nulayans have even retrogressed to worse conditions than they were before the introduction of modernisation and mechanisation. In fact these technological innovations have even affected them adversely. Many of them have got into heavy debts and have even lost their fishing equipment.

The Latin Christians of Quilon and Trivandrum catch more varieties of fish with modern technological innovations, whereas the Arayans and Nulayans catch lesser varieties since they continue to follow the traditional technology. While the Latin Christian fisherfolk have both the means to transport their catch to far off and more lucrative markets and to preserve their larger catches, the Hindu fishermen, particularly the Nulayans and Arayans, are forced to sell their catch to the greedy commission agents at very low prices. They were even forced to mortgage their landed properties and fishing gear and crafts to repay their heavy debts. This has resulted in a lot of communal conflict between the traditional Hindu fisherfolk and the Latin Christians and these situations are being exploited by various political parties.

The Bovis-Mogeyar and Mogavirar of Kasargod have responded to modern innovations introduced in the neighbouring fishing centres of Mangalore and Malpe of Karnataka state. This indirect exposure to the technological innovations has been welcomed by the Bovis-Mogeyar. This is also true in the case of Mukkuvas of

Puthiyappa (Kozhikode district) to whom the technological innovations were introduced directly. They have also been benefited by the construction of Puthiyappa fishing harbour. The Mukkuvas of the Mappila Bay (Cannanore district) have also responded to mechanisation and technological innovations to a great extent compared to their kinsmen of Kasargod and Arayans of south Kerala.

New technologies have helped the fishermen to garner more varieties and quantities of fish. The demand for marine products in the world market has also increased several times. Above all, there has been a 25-fold increase in the number of fishermen engaged in actual fishing. This is due to addition from the non-fishing communities. The new technological innovations like GPS, echosounder and wireless help fishing even during the monsoon periods. However there has been a drastic fall in marine catch due to overfishing.

The Bovis-Mogeyar of Kasargod hold the view that there is an unwritten law that no fishing should be undertaken during the spawning seasons. These are the seasons that permit marine life to recuperate and regenerate itself. According to our older informants belonging to Bovis-Mogeyar, Mogavirar and Mukkuvas of Malabar and the Mappila fisherfolk, even when though the sea is teeming with life at that time, they are unable to undertake fishing expeditions as it is very rough. They say it is Nature's way of ensuring preservation of marine life. Modern technologies are outbidding Nature. During the period of south-west monsoon from June to August almost all species of fish spawn. They migrate to the sea bottom and traditional fishermen suspend fishing during the spawning season.

Unfortunately, the age-old traditional marine laws are being violated and ignored by the new-comers who are ignorant of the marine ecology and environment. The members of the non-fishing communities who have invested in the fishery sector have no respect for Nature's law. Additionally, when the traditional fishing is abandoned during the spawning period in their territorial waters, foreign trawlers are still allowed by the Government of India to fish in the deep seas. This has resulted in the depletion of the fish harvests further here. As a consequence of the overfishing by the foreign trawlers, the traditional fishermen suffer the most since their fishing grounds are deprived of their resources. The mechanised trawlers can sweep the sea bottoms for fishing but the traditional fishermen are unable to reach such depths in their traditional boats. Above all, continuous fishing by mechanised crafts have led to market fluctuations locally, affecting the economy of the traditional fisherfolk further whose catches are much smaller in scale compared to the huge outputs of trawlers.

In view of the above facts, an agitation has been launched by the traditional fisherfolk in Kerala. For the first time the state government was forced to declare a ban on fishing by mechanised crafts in 1997 during the spawning period to protect the interests of the traditional fishermen, to conserve fish wealth, to regulate fishing and to effectively control the clandestine fishing operations in the open sea.

The Marine Zone Act of 1988 prevents the poaching of fishes in the exclusive economic zone of India by foreign trawlers without license as stipulated under the Act. According to this Act, fishing is prohibited in any marine zone of India upto the depth of 40 fathoms (1 fathom = 6 feet) and 22 nautical miles from the shore. Despite this Act the Government of India has given permission to foreign trawlers to fish on the ground that the permission to the foreign trawlers applies beyond the 22 nautical mile limit. The Hindu fisherfolk, particularly the Mogavira, Bovis-Mogeyar and Mukkuvas of Malabar, have protested against fishing in India's economic zone.

The traditional fishermen with mechanised boats fitted with outboard engines beyond 10 HP motor are prohibited from fishing during the period of ban on trawling while the foreign trawlers are allowed to fish without any restraint at any time. The foreign trawlers dump their catches in the local markets which result in the crashing of the prices of fish. Moreover, there is a poor harvest during the lean period owing to the depletion of fish resources due to overfishing. On the whole, the traditional fisherfolk of Kerala are badly affected by overfishing as every one is influenced by the catch of the foreign trawlers.

Our study has revealed that the fishing communities of Kerala have been unable to adapt the new technologies uniformly on account of variations in ecological conditions, cultural ecology and traditional belief systems. If all the Hindu fishing communities have failed to adapt the modern technological innovations, we can hypothesise that it is due to the same religious belief systems, cultural factors and world view that inhibit them from responding to modern innovations. To a great extent, this seems to be true. The caste structure and social ranking in the caste hierarchy play a very important role here.

The low status of fishing communities like Mukkuvas and Arayans has a deep-rooted impact on the psyche of their members. The low social position of the fisherfolk in the caste hierarchy seems to have played a psychological factor in responding to change. These fisherfolk have further the disadvantage of having no outside or inside influence to alter their negative attitude. This is not found among the Bovis-Mogeyar and Mogavira of Kasargod, Mukkuvas of Puthiyappa and the Mappila fisherfolk of Kerala. The temple organisation of the Bovis-Mogeyar of Kasargod looks after the interests of their members by even advancing loans to the needy fishermen to purchase modern equipment on very low interests. This has helped the community to respond to modern innovations to a great extent.

Impact of Overfishing

Kurien and Thankappan Achari (2006) observe: "Overfishing not only implied a fall in the fish harvest but led to a very skewed distribution of the benefits and costs in the fish economy. This in turn came to attain larger socio-political implications which today plague the state. It is customary to distinguish between two types of overfishing: economic and biological. Economic overfishing occurs when marginal costs of an additional unit of fishing effort are higher than marginal revenues. The economy experiences loss (even though total fish catch may still increase) because of a misallocation of capital and labour which might have produced higher economic yields in alternative activities. Biological overfishing occurs when the marginal yield of an additional unit of fishing effort is negative. At such a level of effort the fish population stock is prevented from generating its maximum sustainable yield."

These scholars have expressed the view that there are five factors contributing to overfishing. They are: (a) the free access nature of the fishery, (b) the employment of inappropriate technology, (c) the demand of various varieties of fish in local and foreign markets, (d) financial subsidies given by the government which have encouraged investment, and (e) population explosion among the coastal communities.

Open Access to the Sea

Social factors like untouchability of the fishing communities in the past prevented free entry of outsiders and capital into the fishery sector of Kerala. But the introduction of mechanised boats and the great demand for prawn in the foreign markets changed the scenario to a great extent. The mercantile class of Kerala took the initiative to break the social barriers by setting up fish processing units and increasing the export of prawns. This gave them free access to the sea with their mechanised boats for which no licence or registration was necessary. Consequently the post-1966 period witnessed a considerable influx of non-fishermen owners into the fishing sector with their mechanised trawlers. Between 1966 and 1985 the number of trawlers increased by leaps and bounds. At Sakthikulangara itself more than 2,000 trawlers are operating today.

Unsuitable Technology

The traditional fishing technologies were time-tested to suit a particular ecological system and behaviour of fish. Fishing nets were used according to their special mesh size for catching specific species of fish. "Trawling (the method of scraping the sea bottom with a bell shaped net to catch demersal fish) and purse-seining (the method of quickly encircling whole shoals of pelagic fish) were two such techniques introduced after the decade of the sixties. Both these techniques were very capital-intensive and initially raised the labour productivity. In the short-run harvesting costs were low and given the high prices of certain species of fish, the profits to owners were very high. This led to a rapid increase in numbers and the extensive use of these techniques. This contributed very significantly to overfishing by destroying the sea-bottom eco-niche (trawling) and by indiscriminate and non-selective fishing of whole shoals of pelagic fishes (purse-seining)" (Kurien and Achari, 2006).

Rise in Demand for Prawns in International Markets

There is a great demand for prawns in the international markets, especially in the US and Japan and this boom induced many fishermen and non-fishermen to introduce trawlers into Kerala. These demand pull factors are outside the control of the local economy and it is also difficult to insulate the fishery resources from being harvested in response to them. Kurien and Achari say: "In 1971–72 prawn prices reached Rs 1810 per ton. Between then and 1984–85 it increased nearly sevenfold while the prices of oil sardines and mackerels rose by 184 and 213 per cent respectively. In the case of the domestically consumed fish species – oil sardines and mackerels – there is evidence to show that the increased prices were the result of the inability to enhance the harvest in keeping with the growing demand for fish from the local population. Purse-seiners were first introduced in Kerala in 1976. Until then oil sardines and mackerels were caught exclusively by fishermen using traditional crafts and gear" (Kurien, 1978).

State Subsidies

The state government of Kerala took active interest in promoting the adoption of modernisation and technological innovations. The state promoted capital-intensive, infrastructural facilities like harbours and landing centres, besides providing training facility and subsidy to the owners of modern crafts. No less than 25 per cent of the cost of a boat and 50 per cent of the cost of an engine were provided as grants, and the remainder was treated as a loan to be repaid in 64 instalments spread over a period of eight years at seven per cent interest. The state policy encouraged the traditional fishermen to purchase outboard engines with subsidies at the rate of Rs 3,000 per engine and Rs 2,600 per craft and gear. The rapid growth of outboard engines in Kerala is due to these incentives.

Population Explosion

The population explosion encouraged the traditional fishermen to invest more capital in fishing equipment within the limited area of the coastal waters. The increase in population of the fishermen within the coastal area and also increase in the investment of a large number of fishermen in fishing assets have led to overfishing. Kurien and Achari (2006) have stated: "The pressure exerted by increasing numbers of fishermen using increasing amounts of fishing equipment within the limited area of the coastal waters has this effect .... By 1985 the population increased by 65 per cent to 1,34,000, increasing the fishermen population density in the coastal sea area to 10.6 per sq km. This reduced the average coastal commons per fisherman to nine hectares as against 30 hectares for all of India."

The increase in the number of fishermen led to the increase in their fishing assets as well. More important are the increases in the quality and the quantity of the fishing gear. However, there was a declining tendency in productivity due to overfishing by the traditional fishermen.

Impact of Trawling Ban

A committee was constituted by the Government of India to study and report on the impact of the closed marine fishing season in India, under the chairmanship of the Director, Central Marine Fisheries Research Institute, Cochin, with members from the Department of Fisheries from the maritime states of Gujarat, Maharashtra, Karnataka, Kerala, Tamil Nadu, Andhra Pradesh and West Bengal. The committee held three sittings during 2004–2005 and after holding discussions and examining all aspects of the issues involved, prepared the report.

Recommendations

The major observations and recommendations of the Committee are as follows:

- The duration and period of the closed season varied in the maritime states. There was no uniform closed fishing season observed either in the East Coast or the West Coast.
- The type of fishing craft/fishing banned also varied from state to state. In some states all fishing activities were banned while in some others only trawling or mechanised fishing was banned. The committee observed that such loose implementation of the closed season negates the positive benefits which otherwise would have accrued if the spirit of the ban had been understood and implemented.
- The closed season in its present operational regime in many maritime states, has not positively impacted on the fishery

resource in ensuring sustainability and resource resilience, but has helped to some extent in allowing a respite for the resources for partial recovery and the benthic fauna to recover from the adverse impact of continuous trawling.

- It was observed that the spawning season of major species of fish and shellfish on both coasts are prolonged with one or more peaks. There were only a few species, which had their peak spawning season coinciding with the closed season. Based on scientific information, it was also observed that it is impossible to have a closed season coinciding with the peak breeding of all commercial species on either coasts of India. However, the committee strongly felt that a closed season is very essential for a recovery of the fish stock as well as the biota as several studies have shown positive impacts when closed seasons are followed.
- The Committee recommended a mandatory closed season for the West Coast of India from 15th June to 31st July every year (45 days) both days inclusive.
- Similarly, the Committee recommended a mandatory closed season for the East Coast of India from 15th April to 31st May every year (45 days) both days inclusive.
- The Committee recommended that only sustenance fishery using traditional non-motorised or motorised with OBM/IBM of less than 10 HP vessels should be permitted for fishing during the closed season.
- It was proposed that there should be no further increase in traditional motorised OBM/IBM crafts for fishing during the closed season in any state.
- It was recommended that licensing of all fishing vessels, both traditional and motorised/mechanised, should be mandatory and only such vessels permitted to operate in the seas.
- All fishing activities beyond the 12 nautical miles in the Special Economic Zone should be banned during the closed season and Government of India must ensure compliance and prevent poaching by vessels of other countries.
- It was also felt that the maritime states must consider additional closed seasons for stressed local species to meet special conservational needs.

- The Committee recommended that destructive and resource unfriendly gears such as Ring seines, Surface Pelagic Trawls, Monofilament Disco nets, destructive fishing practices such as dynamite-fishing must not be allowed under any circumstances during any part of the year.
- The Committee recommended that the minimum legal size for mesh in any type of net should be 35 mm. No fishing net with mesh size below 35 mm should be permitted to be operated from any vessel any time in the seas. The 35 mm mesh size regulation is applicable to the cod-end of trawl nets also during the regular fishing season.
- The Committee recommended formation of a national policy advisory group at the Ministry consisting of scientists, resource managers, policy specialists and stakeholders for an informed and knowledge based marine fisheries governance.
- There should be a mechanism for effective monitoring and surveillance with the help of Coast Guard/massive patrol for effective implementation of regulations and to prevent poaching.
- Majority of the traditional fishermen employ outboard engines with the capacity of 9.9 HP, 25 HP and 40 HP motor using kerosene for fishing. Although the government sanctioned 300 litres of kerosene on control rate of Rs 9.50 for 9.9 HP, the fishermen get only 150 litres of kerosene. They buy the kerosene from open market at the rate of Rs 25–30. As a matter of fact the lion's share of their income is spent towards fuel charges. State government should supply at least 3,000 litres of kerosene for 9.9 HP motor for one month. The quantity of kerosene should be increased according to the horse power of the motor used.
- The inboard vessels/canoes employ 110 to 120 HP motors using diesel in which 50 to 70 men go out for fishing. They incur an expenditure of Rs 5,000 to Rs 10,000 for 24 hours operation although the government gives subsidy of Rs 1.50 per litre limiting Rs 1,000. The government should supply the diesel oil giving maximum subsidy of 50 per cent so that the fishermen

can be liberated from the clutches of moneylenders and commission agents.

- The Matsyafed should be strengthened in order to extend their financial assistance to the fishermen according to their felt needs.
- The foreign trawlers should be prohibited from fishing in deep seas.
- There should be regulation of the operation of trawlers within the areas regulated by the government.

On the basis of the above recommendations, the Government of Kerala issued an order dated 12th June, 2006 prohibiting the use of bottom trawl in the sea along the entire coast line of the state within the area specified under G.O. (P) No. 136/g4/PWF&PD dated the 30th November, 1984, published as S.R.O. No. 1496/84 in the Kerala Gazette Extraordinary No. 1005 dated the 3rd December, 1984, for a period of 45 days with effect from the 15th day of June, 2006 to the 29th day of July, 2006 (both days inclusive).

The Hon'ble Supreme Court of India by order in WP (C) No. 393/2005 dated the 17th July, 2006, directed that there will be a ban for mechanised boats, mechanised fishing vessels and trawlers for fishing in the coastal waters of Maharashtra, Gujarat, Goa, Karnataka and Kerala upto the 15th August, 2006.

The Government of Kerala decided to ban all mechanised boats, mechanised fishing vessels and trawlers for fishing in the coastal waters and to extend the period of such ban upto the 15th August, 2006 as ordered by the Supreme Court.

In compliance with the Order of the Supreme Court dated 17th July, 2006, the Government of Kerala prohibited bottom trawl and the use of mechanised fishing vessels and trawlers for fishing in the coastal waters along the entire coast line of the state within the area specified for a effect from the 15th June, 2006 to the 15th August, 2006 (both days inclusive).

For the purpose of this notification, a mechanised boat/mechanised fishing vessel will mean a ship or boat fitted with mechanical means of propulsion and includes country crafts and canoes fitted with an inboard or outboard motor of the capacity above 10 BHP.

The Government of Kerala filed an appeal in the Supreme Court to modify the order dated 17 July 2006 and also to cancel the extended trawling ban period for 2006 in the interest of justice. But, the Court dismissed the review petition filed by the State Government.

The state government submitted before the Supreme Court that the extension of trawling ban from July 29th to August 15th has no scientific relationship with the basic concern of the Apex Court, viz., the conservation of marine life and protection of livelihood of traditional fishermen. On the contrary it was submitted that it would affect the livelihood of lakhs of fishermen and the entire economic activities related to the marine export, peeling industries, etc., would come to a standstill.

Following are some of the grounds for the appeal:

1. Most of the fisherfolk who are dependent on fishing are indebted to local moneylenders and various financial institutions. Such loans are raised hoping that they would have a decent income immediately after lifting of trawling bans, that too after a long period of unemployment and lack of income. The unexpected extension of ban would upset the individual calculations of the poor fishermen and thousands of allied workers, and create a social catastrophe.
2. Monsoon months being a lean period, the fish produced by the traditional fishermen fetches very good price when compared to other seasons. Thus the monsoon fishing has very strong bearing in setting right the economic conditions of the traditional fishermen of the state. By imposing a ban for the traditional sector also it will definitely deny them the one and only source of livelihood and create unemployment situations.
3. Imposing a ban for the crafts of more than 10 BHP will amount to the imposition of a total ban in the fishery sector of Kerala as these days country crafts are using outboard motors having more than 10 BHP.
4. The monsoon fishery provides fish to around 40 lakh fish consumers of the state, the cheapest protein available to the rural poor in Kerala. The ban will deter such availability.
5. The extension of trawling ban may cause the loss of highly priced prawns like 'Karikkadi' as they are abundant during the

extended period. The prawn variety 'Karikkadi' is the most sought after commodity for export during relevant period.

The government, after considering all the aspects relating to the matter, particularly the adverse effect of bottom trawling, banned it in the territorial waters of the state for a period of 45 days in the monsoon season starting on and from the 15th June 2006. In the meanwhile, Supreme Court of India issued an Order on 17 July 2006 in Writ Petition (C) No. 393 of 2005 filed by the Goa Environment Federation directing that there will be a ban for mechanised boats and trawlers for fishing in the coastal waters of Maharashtra, Gujarat, Goa, Karnataka and Kerala upto 15th August 2006. It was further ordered that, for the present purpose, a mechanised boat/mechanised fishing vessel will mean a ship or boat fitted with mechanical means of propulsion and includes country crafts and canoes fitted with an inboard or outboard motor of the capacity above 10 BHP.

Lakhs of fishermen are dependent on monsoon fishery activity, and they are daily wage earners. Since the fisherfolk are in absolute poverty during ban the government has been providing free ration to the fishermen families. The further extension has affected huge financial loss to the fisherfolk and the government. There are lakhs of traditional allied workers, viz., the village fish vendors, merchants, ice plant workers, mechanics, coolies, drivers, etc., totally dependent on fishing activity, who would continue to be in poverty. Though the Apex Court has permitted boats upto 10 BHP, in fact, such boats are not even used by traditional fishermen for monsoon fishing.

In spite of the Supreme Court order, inboard *Valiya Vallams* (fibre boats) with more than 10 HP were operated with the help of purse seine nets during the ban period by both Hindu and Muslim fishermen in some fishing villages of Malabar by employing *Peyithan Vala* for catching the pelagic fish.

5

Technology of Fishing and Change

The efficiency of fishing depends on the implements used for its technique. While natural materials were used in fishing gears since times immemorial till the mid-20th century, and they were replaced by synthetic ones. The use of the synthetic fibre has brought about technological innovations, enhancing efficiency and productivity. Cotton nets which were widely used in the past are nowadays being replaced by synthetic ones. Apart from resistance to salt water, synthetic fibres have several advantages over the natural materials. They are durable and stronger; they do not absorb water; there is no need for using any preservative; and above all their efficiency is better.

Fishing Accessories

Ropes

Several ropes are used in fishing boats and fishing gears. In purse-seine and trawl nets ropes are used to prevent the fish from escaping. This purpose is attained by using purse line in the former and cod-end in the latter. In seine nets they guide the fish into the net and help in lowering the gear. The ropes are made of (a) fibre,

(b) metal, and (c) a combination of the two. The fibre ropes are made of coir, hemp, and cotton.

Floats

The floats help in keeping the net at the required depth, obtaining proper shape during fishing operations, and maintaining vertical opening in gears like trawl nets. The floats used in fishing should be able to withstand pressure and remain highly buoyant. They must be light in weight, durable and should have means of additions. In the past low density materials like wood, cork, sponge, plastic and synthetic rubber were used as floats. Nowadays high-density materials like glass, aluminium and steel are used. The traditional wooden floats of the mackerel net are replaced by aluminium and sponge in nylon nets.

Wooden floats were used in traditional fishing, particularly for gill-nets. They are easily available and are relatively cheap. Cork is used in the surface gill-nets. They cannot stand rough handling. Sponge plastic floats are light, do not absorb water and do not rot. Aluminium floats are mainly used in trawls and also in gill-nets and seine nets.

Sinkers

Red bricks were formerly used as sinkers for the cotton mackerel nets. They have now been replaced by small granite stones. It is interesting to point out here that there is no change in the total number of sinkers used for the modern nylon mackerel nets for catching the demersal fish.

Anchors

Anchors are designed to hold the boat in any desired locality and prevent it from drifting at the mercy of the wind, tide and current. Anchors are very essential, particularly in the event of engine failures or radar breakdown in the mechanised boats, which are highly vulnerable. They are also handy in operating boats when the wind is unfavourable. Often the boats are anchored near the set gear. Stones and wood tied with stones are used as anchors.

Hook-and-Line Fishing

The Mappila fisherfolk of Kerala employ cotton lines (*Kaichunda*), long lines (*Ayiram Chunda*) and bigger hooks for catching larger fishes like sharks and *Ayikkora* (seer). Line fishing is undertaken from a single anchored dugout canoe. This technique is also prevalent among the Latin Christians and the Hindu fisherfolk of Kerala. A hook-and-line fishing unit employs a dugout canoe, five to seven sets of lines and a crew of four. The lines were manufactured from cotton yarns by the Mappilas themselves in the past. These cotton lines have now been replaced by synthetic fibres. The hook is made from tempered cast iron and it is galvanised to protect it from seawater as otherwise it becomes rusty and blunt.

Sea worms and small fish are used as baits for fishing skates and rays, catfish and prawns and also large species of fish like shark. The Mappilas used to go out to the sea upto a maximum depth of 20 fathoms in the earlier days but now they go much farther. The traditional hook-and-line fishing has also been replaced by modern fishing lines.

In the technique of line fishing fishes are lured by offering artificial or natural bait. The bait may be either with hook or without it. Hooks help to capture the fish by ripping. We found that hand-lines, set-lines, troll-lines and drift-lines are used by different fishing communities in Kerala. The hand-line is operated with a few hooks by a single man while in the set-line fishing several hooks are set in a fixed position at a regular interval with or without baits. In the southern districts of Kerala drift-lines are employed over larger areas. These lines are fastened to a fishing boat. They have considerable lengths and the snoods are generally longer.

In line fishing, the fishes are baited in such a manner that they are neither able to take it away nor can they escape once it is taken. Hand-line is the simplest form of fishing. However, the number of lines, which can be operated by a person, is limited. With the development of set-lines, it is possible to use more lines. From simple fishing lines capable of catching a single fish, have emerged the long lines, which have several hooks and extend to a considerable length. These are a kind of unwatched set-lines. The need to operate over a larger area using a movable fishing line led to the development of trawling.

The Mappilas used to employ the *Changala Chunda* or chain line with a different kind of strong hook for catching the sharks. This is three times stronger than the longer lines. Unlike the long-lines the snoods of the *Changala Chunda* were made of chain or thicker yarn. Earlier the *Beppu* (long-line) constituted a special form of set-lines consisting of about 1500 feet length of fishing twine made of cotton yarn. But today the *Beppu* lines have been replaced by nylon yarns. The traditional iron hooks have been replaced by modern hooks. (For details see the section on hook-and-line fishing operation in the book *Mappila Fisherfolk of Kerala*, 1978:144).

Fishing Gears

Gill-Nets

The gill-nets can be classified into three kinds: set gill-nets (entangling the fish shoal), floating gill-nets (Ozhukuvala) and drifting gill-nets (*Nittielakkal*). The set gill-nets are wall nets set at the bottom between stakes or anchors wherever the current is weak. The floating gill-nets are not set but only fixed between floats on the upper end and anchors on the lower end, so as to suspend them in the water. The floating gill-nets, when set to drift with the tides or current, are known as the drift nets. They are not permitted to drift alone, but are tied to a boat, which drifts the nets. The drift nets are particularly useful for fishing mackerels. They are very useful as large areas of water can be covered. They can be used to catch even scattered fish. Apart from the above three types there is a net which encircles the fish and they are given to the gear. All the traditional cotton gill-nets have been replaced by nylon gill-nets. The gill-nets (*Ozhukkuvala*) used in Kerala are: (1) *Ayilachala Vala* (the gill-net used for catching mackerel), (2) *Ozukkuvala* (the gill-net used for seers, eel and cat fish), (3) *Tirandivala* (gill-net used for skates and rays), (4) *Nethalvala* (gill-net used for white baits), (5) *Baminvala* (gill-net used for threadfins), and (6) *Sravuvala* (gill-net used for sharks). The traditional *Tirandivala* (Vakuvala) has been replaced by the modern nylon net. Similarly, the traditional *Nethalvala*, *Sravu*vala (made of hemp) and other cotton nets have been replaced by nylon nets.

Encircling Nets

The use of an encircling net is preferred in daytime for catching the mackerel, particularly during the months of *Makaram, Kumbam, Minam* and *Medam*. The fishermen beat the cross bars of their respective boats with the oars and rudders and also amplify their sounds to scare the fish and drive them into the nets. This technique has the advantage of preventing the fish from migrating to the bottom as well as to distant areas of the sea. The traditional cotton *Ayilachala Vala* has been replaced by nylon mackerel nets. Gill-nets are also used for catching skates, rays, sharks and *Nethal*. During the operation of the encircling nets the fish are surrounded from all sides and below. The encircling nets are of two types, viz., purse-seines (*Tanguvala*) and ring nets.

Seine Nets

The seine nets encompass a particular water area, generally coastal or shallow water. These nets are set with the help of lines, either from the shore or from a boat.

The seine nets are mainly of two kinds, viz., one with a bag and another without it. The latter type is not found in Kerala. The seine net with a bag may be classified into three types according to its method of operation, viz., (1) stake-nets, (2) shore/beach seine nets (*Kambavala*), and (3) boat-seine nets.

Stake Nets

The stake net (*Kuttivala*) is one of the smallest of its kind used in the backwaters and estuaries of Cochin by Valans and Patanans. This is a conical bag net and is called *Kuttivala* or *Unnivala*. It is usually shot against the tides. The net is spread and held in a vertical position by two stakes or sticks. It is particularly useful for catching prawns. The net is generally hauled up at the turn of the tide. This method is followed in the backwaters of Kerala throughout the year whenever the tides are feeble. The size of the net varies widely. Some of the largest ones along the coast of Kerala are 300–500 feet in length with a mouth of 150–200 feet in circumference. Stake nets are also used by the immigrant Valans of Kannur district.

The net is set in such a manner that the mouth is left open against the current. However, the edge of the coir wings and the openings of the bag are kept high and open by the device of floats and sinkers at regular intervals. The net is positioned vertically by attaching weights at the lower end (foot rope) and by tying wooden floats to the upper edge or head rope. The floats and weights are kept in balance in such a way that the net is held in a vertical position almost as if it is suspended. Stake nets are also operated without floats and sinkers. They are just tied to a stake or on pairs of stakes driven into the bottom.

Shore Seine Nets

The net operated from the shore is called shore seine net (*Kambavala*). The shore seines (*Karamadis* or *Kambavalas*) are shot in Kerala with the help of a boat in such a way that the Kamba (warp) of one wing (*Kadankani*) remains on the beach while the bag and other wings are taken out in a canoe and the drag line is finally brought back to the shore. It may be noted that the *Vanchi* (dugout canoe) moves in a semicircular manner releasing the warp or drag line, the coir wing reaching the shore, thus completing the shooting. Now both ends of the net are pulled, thereby encircling a particular area. As the operation advances both the ends come closer and the bag is pulled to the beach in order to complete the operation. It was observed that both the Hindu and the Muslim fishermen of Blangad employed shore seine nets for catching pelagic fish. As many as 30 to 50 fishermen are generally employed for dragging the shore seine net.

A *Vanchi* with a crew of seven is employed in shooting the net while twenty to twenty-five persons are employed for pulling the net. It is shot at a depth of half to one fathom.

The shore seine nets are used along the coast of the districts of Alleppey, Quilon and Trivandrum. About thirty five years ago the shore seine nets were introduced into some fishing villages of the districts of Malappuram and Trichur.

Boat Seine Nets

The seine net operated from a boat is called the boat seine net. This is one of the commonest types of nets employed for fishing prawns, sardines, mackerels, soles, etc., along the Kerala coast. Though there may be some minor regional variation a boat seine net is essentially composed of a bag-like part or *Vaikettu* (cod-end), *Mevala* (belly), two coir wings (*Kadankanis*), and two coir warps (*Kambas* or dragline). This net is usually operated by a four man crew from each of the two boats. It is generally shot at a depth of 10 to 20 fathoms.

The technological aspects of a few important boat seine nets (made of cotton, hemp and nylon) are discussed below. The traditional boat seine nets used in Tanur are: (i) *Muyyan Vala,* (ii) *Mathi Kolli,* (iii) *Ayila Kolli* or *Odakolli,* (iv) *Peyithuvala* or *Peyitham Vala,* (v) *Ettavala* or *Vakkuvala,* and (vi) *Avolivala*. Some of these nets, particularly the first three ones, have been almost completely discarded and replaced by nylon nets in most of the fishing villages of Malabar.

Mathikollivala/Chala Vala

The cotton *Mathikollivala* (sardine net) was used for fishing sardines in the past. In recent times the traditional *Mathikollivala* has been replaced by the nylon one. The nylon *Mathikollivala* is known as Silkuvala among the Mappilas of Tanur. The cost of a nylon sardine net is more than Rs 20,000. Its size is almost the same as that of the cotton *Mathikollivala*. It has a long conical bag net with a mesh size of 3/8" which increases to 3/4" at the mouth. The rule is that the size of the mesh should increase from the bottom to the upper part of the net. The nylon *Mathikollivala* is a long conical bag net with long and wide platforms and is designed to catch prawns and sardines. The *Mathikolli* (nylon-net) is still used in the villages of Tanur (Malappuram district) and Blangad (Trichur or Thrissur district).

Operation of the Mathikolli

The *Mathikolli* is launched from two fibre boats fitted with a Yamaha 9.9 HP engine. Two techniques are employed in this respect. These are called *Tattipeyyal* and *Iccapeyyal*. Six to ten boats (3 to 5 fishing units) are employed in the *Tattipeyyal* technique. Two of these boats are used for shooting the net while the rest are for beating. The boats carrying the 'beaters' drive the fish shoal into the wide gap of the net.

During 1970–72 there were as many as 25 dugout canoes around Ernakulam for beating, scaring and driving the fish into the nets. When the beaters detect the fish shoal they hail the *Valapayirrukar* (shooters of the net) by showing the rudder. This is called *Madivilikkal*. The *Mathikolli Vala* is one of the boat seine nets operated by the Mappila fisherfolk of Kerala even after the advent of nylon nets. The nylon *Mathikolli Vala* is used for catching almost all the species except larger fish like shark, skate, ray and seer. It is generally employed during the fishing season from August to April. Although the traditional cotton *Mathikolli Vala* has been replaced by nylon *Mathikolli Vala*, the fishermen have retained their traditional techniques. This is true in the case of the Muslim fishermen of Tanur and the Hindu Mukkuva fishermen of Blangad.

Peyitham Vala

The *Peyitham Vala* is one of the boat seine nets, which is in extensive use in the Malabar area and is the most popular and efficient gear. It is a nylon drag seine net. The net was operated in the past from two dugout canoes, which employed 4–8 fishing hands in each boat. The canoes are now replaced by plank-built fibre covered boats. The area of the operation is restricted to eight fathoms and the net is employed during the southwest monsoon in August–September for catching mainly soles. The major difference between the *Peyitham Vala* and other boat seine nets is that the former has a square bag net or cod-end. The *Peyitham Vala* is one of the boat seine nets and it is a mid-water trawl net. The *Peyitham Vala* is even now operated in Tanur and Blangad by the Mappila and the Hindu fisherfolks. As soon as the fury of the south-west monsoon is over the sole migrates from inshore to the mid and surface waters. The rapid increase in the temperature just after the monsoon may be the main reason for the appearance of sole in large shoals. Two small fibre *Vallams* or boats are used for fishing with the *Peyitham Vala* in Tanur and Blangad. The boat is propelled with the help of a Yamaha motor or Suzuki engine (9.9 HP) during the period of monsoon. The fishermen go out for fishing between 5 a.m. and 6 a.m. and return to the landing place at 1 p.m., consuming a minimum of 100 litres of kerosene, besides five litres each of petrol and engine oil. Thus, the total fuel charge varies from Rs 3,500 to Rs 4,000 for the fishing expedition for seven hours. Although the government has sanctioned 300 litres of kerosene for a mechanised boat per month the fishermen say that they get only 150 litres. The price fixed by the government is Rs 9.50 per litre but the fishermen are forced to buy kerosene from the open market at the rate of from Rs 25 to Rs 30 since the ration is only sufficient for two days.

According to our informant, the cost of a cotton *Peyitham Vala* was only Rs 500 whereas the same net made of synthetic fibre or nylon is priced between Rs 8,000 and Rs 10,000. Some Hindu fishermen of the village of Puthiyappa (Kozhikode district) went out for fishing on 17 July 2006 employing a *Peyitham Vala* with a crew of 10 in each unit. For the operation of a *Peyitham Vala* two fibre boats were used for catching *Ayila* (mackerel), *Mathi* (sardine) and small fishes. The *Potham Vala* or *Peyitham Vala* is also known as double net in Blangad. Writing on the technique of operation of the *Peyitham Vala* Mathur (1978) has observed: "The *Peyitham Vala* is carried in one of the two canoes and shot into a depth of about 10 fathoms. The two boats are brought face to face. The net is first dropped into the water followed by the *Kadankani* (wing). After the *Kamba* (drag lines) are attached to the ends of each boat, the boats drift by themselves to the length of the net. The boats are then rowed dragging the net behind them. After dragging the net for half an hour or so the steersman holds the Kamba or the drag line and pulls up the net slowly. The boats are propelled backwards towards the net to face each other while pulling up the *Kamba*. The coir net (*kadankani*) is first hauled up when the boats face each other and then the net itself. The fish caught is thereafter removed. They operate the net for 8–10 times in an expedition in this manner." The *Peyitham Vala* is widely used by the Mappila and the Hindu fisherfolk of Puthiyappa (Kozhikode), Blangad, Tanur, Mappila Bay (Kannur) and other villages. The *Peyitham Vala* is shot from fibre boats in these villages.

Midwater trawl (*Peyitham Vala*) is a trawl-net shot in the pelagic region either from one or two boats. "In view of the vertical and horizontal distribution of the fish in midwater, a midwater trawl is provided with a large vertical as well as horizontal mouth opening. To prevent the downward escape of the fish, the lower surface of the net extends as far forward as the upper surface and possibly beyond it. Midwater trawls are made relatively long, and have finely tapered funnel shaped bags to minimise the escape of fish through large meshes from the front of the net. The overall drag of the gear is reduced by various means like using lighter materials, smaller size of ropes and warps, absence of flappers, false bellies, etc. The shearing devices are also reduced in size and weight and are designed to provide a better lift/drag ratio" (Sreekrishna and Shenoy, 2001: 104).

Purse Seine Nets (Tanguvala)

Purse seine nets are encircling gears provided with a purse line at the bottom which, when pulled through the purse rings close the bottom of the gear, preventing the escape of the fish. It can be operated from one or two boats.

"Purse seining is one of the most important commercial fishing methods of the world. This method is used for the capture of small pelagic fishes for manufacture of fish meal as well as for catching fishes of high individual value like tunas. The principle of catching fish in a purse seine involves surrounding a shoal of fish by a long wall of webbing and subsequently pulling the bottom of the net by means of the purse line. The fish catch from this artificial pond of webbing is then removed either by bailing or pumping. The efficiency of the purse seines has gone up considerably with the introduction of synthetic twines and ropes, power block, improved method of fish detection by sonar, increased manoeuverability of the vessel by side thrusters, controlable pitch propellers, and improved bulk fish handling. The purse seines are used to catch fish like sardines, mackerels, tuna herring, pilchard, white bait, cat fishes, etc." (Sreekrishna and Shenoy, 2001: 91). Purse seine nets are operated either from a single boat, a pair of boats or combination of large and auxiliary boats.

The *Tanguvala* (purse seine net) is operated from a mechanised fibre boat (*Valiya Kettuvallam* covered with fibre and Leyland inboard engine), by the Mappila and the Hindu fisherfolk of Blangad, Puthiyappa and Beypore (Kozhikode district). Thirty to forty men go out for fishing in the *Valiya Kettuvallam*. It was observed at Blangad that both small *Kettuvallams* (small fibre boats) and big boats are used for fishing with the *Tanguvala*. The *Valiya Kettuvallam Tanguvala* and *Chudavala* are used for catching small fishes like *Chuda, Nethal* and *mullets* and larger fishes like *Etta* and *Ayikkora*. *Chemmin* (prawn) is also caught with the help of a *Tanguvala*. But both the Hindu and the Mappila fisherfolk complained that their nylon nets like *Tanguvala*, *Chalavala* (sardine net), *Ayila Vala* (mackerel net) are often destroyed by *Kadalpanni* or *Kadalmakry*. This is a kind of a crab that destroys the gill-nets, Purse seine nets and seine nets. The launching of the *Chalavala* (sardine net) and *Ayila Vala* (mackerel net) is done with the help of two small *Kettuvallams* (fibre boats) for which there is a crew of four men in each boat, whereas in the *Valiya Kettuvallam* no less than 30 to 40 fishermen are required for shooting the *Tanguvala*. The cost of a fishing unit with a *Valiya Kettuvallam* and *Tanguvala* comes to around Rs 8 lakhs, whereas the cost of a small fibre *Kettuvallam* with an *Ayila Vala* or *Chalavala* would be Rs 1–1.5 lakhs. The *Tanguvala* is also used in inboard Leyland boats for catching *Chemmin*, *Ayila*, *Chuda*, *Para* and *Mathi*. This net is used not only in the fishing villages of Malabar but also in those of Cochin and Travancore.

Ring Nets

The ring nets are shaped like a purse seine but unlike it the netting has different mesh sizes. It is a hybrid type between *lampara* and purse seine. These are generally launched from a small boat.

Trawl Nets

The trawl nets are dragged through the water from a fishing boat. Trawlers are employed at the bottom or surface water. They account for nearly 25 per cent of the world marine catch. In India trawlers are operated by about 70 per cent of the mechanised boats, contributing substantially to the foreign exchange. The trawl has been classified into three types, viz., (1) beam trawl, the net mouth of which is held open by a beam, (2) otter trawl, of which the net mouth is kept open horizontally by otter boards and (3) ball trawal, the net mouth of which is kept open by two boats pulling apart, without the use of either the otter boards or beams (Sreekrishna and Shenoy, 2001: 72).

The trawl is a bag-shaped net towed through the water, the mouth of which is kept open by a frame, beam, otter boards and floats and sinkers. Trawl also refers to dragging on the bottom. When the net is dragged its mouth remains open. Trawling as a modern fishing technique has been introduced in recent years in India and trawlers of different sizes are in operation in various parts of the country. The trawl nets were introduced on the Kerala coast for marine fishing by Norwegian technicians in the year 1958. The nets are used in Kerala for catching bottom fish, particularly prawns, and are operated from November to May, from mechanised trawlers. The Mappilas believe that the fish move to the bottom of the sea with the onset of summer.

Bottom trawls are conical shaped nets towed over the bottom for catching the demersal fish. The mouth of the net is kept open by means of either a beam, otter boards or two boats pulling apart.

Double Stick Net (Kuttivala)

These are winged bags of netting held between two sticks. They are more popular with the river fishermen (Valans). They are held in river currents and then encircled and thereupon the catch is lifted. This technique is widely practiced in the Valapattanam river (Cannanore district), by the Valans.

Lift Nets

The technique of fishing with the help of a lift net is by lifting the gear submerged at the required depth, either by hand or mechanically from the shore, or from a boat. The opening of the net faces upwards. In Cochin the lift nets (Chinese nets) are operated by the Valans. The Kudumbis of Vypin, a migrant community from Goa, traditionally engaged in the making of *Pappadam* and beaten rice, have adopted fishing with lift nets. The Chinese lift nets are operated along the Kerala backwaters and several of them are mechanised.

We have already mentioned that the gill-nets have a long wall of webbing hung vertically in water. Depending on the mode of operation they are either set or drift encircling gear. Set gill-nets are positioned in water by means of anchors, whereas drift gill-nets are permitted to drift with the current or from a boat to which they are attached. The encircling net surround the fishes and they are driven into it by some means like frightening, swimming, or driving. This method is known as drive-in-net.

On 10 July 2006 the sea was very rough at Blangad due to the occurrence of *Karanir* or *Karavallam*. The *Karakattu* (westerly wind) and *Vadakkankattu* (northerly wind) also blew in the morning. Consequently there was a storm and rain besides the occurrence of high tides due to the full moon. The appearance of *Chakara* was reported by the media and therefore a number of fishing units came to Blangad from Alleppey for catching *Chemmin* (prawn). The fishermen at Blangad believe that when the *Nir* flows from east to west the catch will be very poor, but when a current (*Nir*) flows from northwest to southeast there will be a good catch. Similarly, when there is *Temprayil Nir* (current from south-west) there will be good catches, whereas in the case of the *Karanir* (easterly current) the catch will be very poor. In other words, whenever a *Karanir* and *Karakattu* occur the fishermen do not get any catch at Blangad. The Mappila fisherfolk of Tanur and their counterparts at Blangad have collective traditional knowledge regarding various kinds of currents and winds and also different fishing grounds.

Fishing Crafts

The fishing boats can be classified on the basis of various criteria like depth of operation, duration of voyage, size of the boat, material used in their construction, type of propelling system and the actual operation. Sreekrishna and Shenoy (2001) have classified vessels based on the depth of operation into (1) inshore vessels – vessels operating up to 10 fathoms depth, (2) offshore vessels – vessels operating in the depth range of 10 to 40 fathoms, and (3) deep sea vessels – vessels operating beyond 40 fathoms depth. They have also classified the vessels based on their size and duration of the voyage. These are: (1) near water-vessels of length up to 12 m which mostly make daily trips or trips of short duration (small fibre boats); (2) middle water-vessels of length between 12 and 30 m having an endurance of seven days or more but not exceeding 25 days (inbuilt big canoes fitted with Leyland engine); and (3) distant water-vessels of size over 30 m length with endurance exceeding 25 days (trawler boats).

Big boats (*Valiya Vallam* or *Valiya Vanchi*) fitted with Leyland engines are operated with the help of a crew from 30 to 40 fishermen. They can be compared with the midwater vessels used for catching the pelagic fish. The distant water vessels can be compared with the trawler, which is employed for catching the demersal fish with the endurance of more than 25 days. When bigger boats (bigger fibre *Vallam*) are employed with a purse seine net (*Ring Vala*) a forty-man crew is employed for fishing. Two lighters are used to carry the catch from the fishing boat to the landing station. The fishing boats are either made of timber or of iron. In Cannanore district many of the Tiyyas, Muslims and Latin Christians work in fibre *Valiya Vallam* or fibre glass *Vallam* with a crew of sixty men, whereas in the case of trawlers only a crew of seven men is required. A fishing unit consisting of a trawler and net would cost from Rs 22 to 25 lakhs. A Leyland engine with 120 HP is fitted in the fibre *Valiya Vallam*. This is also used for encircling the fish. The Matsyafed gives loans upto Rs 10 lakhs for the purchase of equipment, charging 10 to 12 per cent interest. A *Valiya Vallam* requires 250 litres of diesel for fishing for 12 hours. On the other hand, small trawlers fitted with inboard engines require 80 to 90 litres of diesel for working 12 hours.

The main craft used in the district of Trivandrum for fishing is the *Catamaran*. *Catamarans* can be broadly classified into two categories: mechanised and non-mechanised. The *Catamaran* is one of the traditional types of fishing crafts used in Virinjam, (Trivandrum district) and Kollam district, mainly by the Latin Christians and Hindu Nulayans. Four to five logs of lightwood are lashed together with coir ropes to make the *Catamaran*. The craft is rowed with the help of a bamboo pole. As part of Tsunami rehabilitation programme as many as 50 *Catamarans* have been mechanised in Virinjam. The traditional *Kettuvallams* have been replaced by fibre built *Valiya Vanchi* in the districts of Alleppey and Kollam. This *Valiya Vanchi* fitted with an inboard Leyland engine, can employ 40 to 50 fishermen in launching a *Tanguvala*. More than 2,000 trawlers are employed by the Latin Christians at Neendakara and Sakthikulangara (Quilon district) for deep sea fishing. They go out for fishing from 10 to 15 days and return with catches worth Rs 2 to 3 lakhs. It must be mentioned here that while 2,000 trawlers are owned by the Latin Christians of this area only nine of them are owned by the Arayans of

Puthenthura. It is evident from this fact that the Hindu fishermen do not have financial resources to adapt modern technological innovations. They cannot afford to buy mechanised fishing boats and trawlers. The cost of a trawler with a nylon net would be nearly Rs 30 lakhs but only seven to ten fishermen can find employment in it.

The Mappila fishermen of Tanur used to employ in the past, dugouts and plank-built boats for fishing with the help of various kinds of cotton nets. They used to employ special cotton and hemp nets for catching different species of fish. These cotton and hemp nets have now been replaced by nylon nets. The hook-and-line fishermen (*Beppukar*) were slow to adapt nylon nets and plank-built boats after discarding their dugouts. They now employ fibre boats and nylon mackerel and sardine nets. The traditional *Ayiram Chunda* (thousand hook-and-lines) is no more to be found among them. Majority of the *Beppukar* of Tanur have their traditional crafts and nets but the dugouts are no longer used by them and the *Valakkar* (traditional fishermen). The Mappila fisherfolk of Tanur now use bigger boats (fibre *Valiya Vallams*). The cost of this Vallam is nearly Rs 8 lakhs. It can be operated with a crew of 25 to 30 men fitted with a Leyland engine. The fishermen of Tanur travel with their *Valiya Vallam* upto Ponnani and Beypore. Their dugouts have been replaced by small fibre Vallams which cost them from Rs 40,000 to Rs 65,000 for a pair.

A number of Muslim fishermen at Puthan Kadapuram near Blangad have become owners of trawlers. They started operating these trawlers as soon as the ban on trawling was lifted on 16th August 2006. The fishermen of the villages of Puthiyappa and Beypore (Kozhikode district) use numerous trawlers for deep-sea fishing. As many as 85 fibre glass Vallams are owned by the Mappila fishermen of Tanur but they have to leave for fishing from Ponnani and Beypore as these boats cannot be anchored at Tanur in the absence of a fishing harbour. In short, the Mappilas are very quick in adapting modernisation unlike the Hindu fisherfolk, particularly the Mukkuvas of Blangad and Arayans of Alleppey. Big fibre glass inboard boats are widely used in Malabar and Cochin. Forty to fifty fishermen are employed in a bigger canoe (*Valiya Vallam*). While the Muslim fishermen have adapted the *Valiya Vallams* and trawlers in Mappila Bay and Kannur district, the Hindu fisherfolk of the same areas have failed to adapt these modern crafts due to their financial stringency.

Boats in Malabar

The traditional boats in the districts of Cannanore and Kasargod are called *Vallams* or *Odams*. *Toni* is the name for a plank-built boat in Northern Malabar. A *Vanji* is generally used for fishing in the southern districts of Kerala. The traditional designs of these boats are still used in Tanur and Blangad though they are now made of fibre glass and fitted with motors having capacity of 10 to 30 HP. The kind of boat used by the fisherman depends upon his financial conditions and the ecology of the sea.

In the district of Kasargod, the fishing crafts employed are mainly the *Cheriya Vallam* (small boat) and *Valiya Vallam* (big boat). The bigger crafts are also known as *Odam*. The difference between these two kinds of crafts is their size and therefore their capacity. Small boats are about 1.2 and 1.5 metres (4 and 5 feet). The boat's capacity limits the catch of each fishing expedition.

A *Cheriya Vallam*, which carried a crew of four, was used to sail to depths of 20 *Maarus* (1 *Maaru* = 6 feet) while the *Valiya Vallam* with 8 to 10 men sailed to depths of 35 to 40 *Maarus*. However, mechanised boats are operated normally by thirty to forty fishing hands. The *Cheriya Vallams* were earlier all dugouts, that is a trunk of a tree of appropriate girth and length was scooped out and shaped as desired. The *Valiya Vallams* were planks lashed together especially on the sides, prow and stern with coir (*Choodi*). Additional wooden planks are attached with coir ropes by literally sewing the pieces together, and pinning them with copper nails (*Chembutharra*), which do not corrode in sea water. These are called *Valiya Kettuvallams* at Blangad.

The *Valiya Vallams* (bigger boats) in Malabar today are built of timber planks joined together. The final procedure is to smear the vessel with sardine oil, cashewnut oil and *Pantham* (a kind of resin obtained from the *Ledebouria robusta* and *Acanthus ilicifolius* trees). This is to prevent the deterioration of the timber which remains constantly in salt water and to make it water-tight. Fishing boats are treated regularly with this substance. The younger members of the fisherfolk, however, prefer the plank-built boats as these are very resilient and can be used at any time and in almost any condition.

Moreover, dugouts are more difficult to repair, whereas the damaged pieces in a plank-built boat can be replaced.

Fishing crafts are generally employed in pairs and the process by which the two boats diverge and then converge during a fishing operation is called *Peithu*. The catches are shared as per established norms between the crew of the boats and owner of the fishing unit. With the introduction of mechanisation, *Valiya Vallams* are generally fitted with outboard motors and these are therefore able to go out quite far into the sea, especially during calm weathers. These boats carry the fuel (petrol and kerosene) for the outboard motors needed for their prolonged operations and also carry the necessary food and drinking water for these periods. The owner of the fishing gear and craft bear the cost for food and fuel, which are deducted from the sale price of the total catch before the profits are shared. However, there is a tendency now in some of the units to bear these initial expenses among the members of the crew themselves.

Earlier few *Cheriya Vallams* used outboard motors. The reduction in the catches and the strenuous physical exertion in rowing have compelled these smaller crafts also to use outboard motors of less than 10 HP. Once the fishing grounds are reached the motors are switched off and then oars are used to row to the spot. This is done so as not to scare away the fish shoals with noise. It was reported that *Cheriya Vallams* fitted with outboard motors can go upto a depth of 25 *Maarus*. During the calms some fishermen have even gone as far as 30 *Maarus*.

As mentioned earlier, trawler boats are larger and fully mechanised units and are put out at sea for weeks together. These trawlers go on fishing expeditions in the sea to a depth of even 1000 *Maarus*, and have the capacity to carry more fish. Most of these and even some of the mechanised *Valiya Vallams*, use smaller crafts (lighters) to undertake shuttle services between boats and landing stations to offload the catches and bring in essential requirements. Trawlers (also called *Vallakkar* or *boatukar*) are despised by the traditional fishermen as they believe that continuous fishing, especially using purse seine and trawl nets, which according to them literally sweep the seabed, are harmful to the marine ecology and deplete the fish life.

The cost of the fishing craft depends on the quality of the timber used. Local carpenters are experts in making specialised boats but these days most boats are made in boat-building yards. The first such yard was set up at Mangalore in 1958. Now there is also a boat-building yard at Beypore near Kozhikode. Both these are government-owned establishments. There are also a few private yards at both these centres. The most expensive timber used for the construction of the boats is *Ayini/Aanyiniplavu* (Artocarpus hirsute) and *Cheeni* (Austiarie genus). The timber of both these trees are light and strong. Besides these two, *Maavu* (Mangifera indica) and *Ellavu* (Bombax malabaricum) are the other two commonly used timber trees for making boats. *Thek* or teak (Tectona grandis) and *Plaavu* or jack (Artocarpus integrifolia) are used for the construction of some parts of the boats like keel and stern. A *Cheriya Vallam* made of these woods costs as much as one lakh of rupees, while the cost of a *Valiya Vallam* ranges between Rs 2 and 2.5 lakhs. A bigger fibre-covered boat with inboard engine and a nylon net would cost from Rs 25 to 30 lakhs.

Fishing boats are generally purchased with loans taken from banks and the Matsyafed. Sometimes the fishermen have to sell/mortgage their gold and land to raise the necessary funds to buy the boats. Loans are also taken from middlemen at ruinous interest rates. The construction of a new boat is preceded by elaborate rituals often conducted by Brahmin priests. These rituals are becoming costly affairs these days, but their expenses are neither spared nor curtailed, especially among the Hindu fisherfolk.

A fully equipped wooden trawler boat about 15 metres long costs from 8 to 10 lakhs of rupees at the prevailing market price depending on the timber used for its construction and the equipment installed in it. The cost depends upon the length of the boat. These are bought from the boat-yards of Mangalore usually, and can employ a crew of about 12 to 18 men. There is provision to carry barrels to stock fuel (diesel and engine oil), kerosene stoves for cooking, drinking water, and stock of raw food in these boats. Food is cooked daily on board. In the case of a steel or iron trawler with inboard Leyland engine the cost would be Rs 30 lakhs. The initial investments to purchase the essentials like fuel and provisions are shared by the owners of the trawler boat. The funds for the purchase of these costly trawlers or fibre covered boats are raised by a number of friends. Sometimes the funds are raised as loans from banks and private moneylenders, Tarakans, Matsyafed and Temple Committees in the districts of Kasargod and Quilon, are the other financiers of the fishermen.

The launching of a boat is also accompanied with very elaborate rituals or *Pujas* in order to seek the blessings and protection of the goddess Bagavathy. Brahmin priests are sometimes engaged to conduct these rites.

Modern Equipment in Fishing

Bigger boats and trawlers have sophisticated instruments like GPS (Global Positioning System), echosounder and wireless for fishing expeditions. Several fishermen of Tanur, Blangad and Mappila Bay use GPS in their trawlers and *Valiya Vallam*. The merits of the GPS according to Ramesh, a Hindu fisherman, are as follows:

1. To have contact with the fishing hamlets by the fishermen during their expeditions,
2. To locate the rocky beds,
3. To find out the fish living the rocky places,
4. To identify the type of currents and winds,
5. To locate the fishing grounds, and
6. To find out the distance from the fishing grounds located at a distance of 25 kilometres from the landing station.

GPS

The GPS 126/128 is a powerful navigation tool that can guide us anywhere in the world. To better understand its operation and capabilities, it may be helpful to review the basic terms and concepts briefly given below. Other navigation and GPS definitions used in manuals are defined in the appropriate reference sections of the manual.

The following are the numbers of the GPS indicating the name of the fishing docks. When the fishing unit reaches a particular landing station the following numbers are shown in the GPS.

Kollam	*Neendakara*	*Kayamkulam*	*Kochi*	*Munambam*
08. 52. 216 076. 34. 744	08. 56. 045 076. 32. 123	09. 08. 352 076. 27. 665	09. 58. 098 076. 13. 150	10. 10. 570 076. 09. 570
Tottapally	*Mangalore*	*Cannanore*	*Azheekkal*	*Beypore*
09. 19. 142 076. 22. 700	12. 50. 750 074. 46. 896	11. 50. 535 075. 22. 172	11. 56. 320 075. 17. 465	11. 09. 469 75. 47. 074

Advantages

1. With the help of GPS the fishing expedition can reach a particular specified distance of the ocean by seeing the mark of the GPS.
2. After identifying a particular area of the ocean by seeing the identification mark in the GPS the fishermen can reach that place with the help of GPS.
3. GPS can also identify the type of current.
4. The compass number of the GPS helps the fishermen to reach the same place from where they went for fishing.

Almanac Data

Satellite Constellation Information (including location and health of satellites) is transmitted to the fishermen's receiver from every GPS satellite. Almanac data must be acquired before GPS navigation can begin.

Bearing:	The compass direction from fishermen's position to a destination.
Coarse Made Good (CMG):	The bearing from the 'active from' position (your starting point) to fishermen's present position.
Crosstrack Error (XTK):	The fishermen's distance are off a desired course in either direction.
Desired Track (DTK):	The compass course between the 'from' and 'to' waypoints
Differential GPS (DGPS):	An extension of the GPS system that uses land-based radio beacons to transmit position corrections to GPS receivers.

Estimated Time of Arrival (ETA):	The time of day of fishermen's arrival at a destination.
Estimated Time Enroute (ETE):	The time left to fishermen's destination at fishermen's present speed.
Grid:	Coordinate system that projects the earth on a flat surface, using square zones for position measurements. UTM/UPS and maidenhead formats are grid systems.
Ground Speed:	The velocity that the fishermen are travelling relative to a ground position.
Latitude:	The north/south measurement of position perpendicular to the earth's polar axis.
Longitude:	An east/west measurement of position in relation to the prime meridian, and imaginary circle that passes through the north and south poles.
Navigation:	The process of travelling from one place to another and knowing where fishermen are in relation to their desired course.
Position:	An exact, unique location based on a geographic coordinate system.
Track (TRK):	The direction of movement relative to a ground position.
Universal Transverse Mercator (UTM):	A grid coordinate system that projects global sections onto a flat surface to measure position in a specific zone.
Velocity Made Good (VMG):	The speed that the fishermen are trawling in the direction of the destination
Waypoint:	A specific location saved in the receiver's memory.

Navigation Basics

The GPS 126/128 provides a steering guidance and navigation information using degrees, a measurement measured in a clockwise direction from a north reference. North is described as 000°, east as 090°, south as 180° and west as 270°.

Functions of the Wireless

Trawling expeditions generally take place at different distances of the sea. But with the help of wireless the fishermen can communicate with the fishermen of the other boats which have the same wireless facilities. They can be contacted and requested to come to the fixed place. Sometimes, GPS is also used.

The fishermen working in a trawler can send wireless messages to their friends in other trawlers about the appearance of shoals in a particular fishing ground.

Echosounder

According to our informants the echosounder helps the fishermen to locate the fish shoal and also the distance in the sea. Sreekrishna and Shenoy (2001) say that the "echosounder is widely used by the fishermen. They hold that the most widely used fish-detection instrument is the echosounder. Though it was primarily developed for depth measurement, its efficiency has been improved with time to record even the very weak echo signals reflected by the fish. Both echosounders and fish-finders are essentially the same but for additional features in a fish finder like more powerful amplifiers to receive and amplify the weak fish echoes". Echosounders are used in trawlers and bigger canoes in the districts of Kannur and Kollam.

6

Economic Organisation and Change

Organisation of Fishing

Devoting ourselves mainly to a discussion of the economic organisation of the Mappila, Hindu and Latin Christian fisherfolk of Kerala, for the present, we shall consider the nature of ownership of fishing crafts and gears and the distribution of the share in the catch. We shall also investigate the relations which subsist among these fishing communities. The effects of economic development, modern technology, introduction of fishermen's cooperatives and other innovations will be analysed. An attempt is also made to analyse the rights, duties and obligations of the owners of fishing tools on the one hand and the fishing labourers on the other, vis-à-vis the fish merchants and those who actually follow fishing for their livelihood.

Ownership of Fishing Equipment

Hook-and-Line

As has been already stated in our work 'The Mappila Fisherfolk of Kerala' (1978), that fishing with hook and line had been the exclusive prerogative of the Beppukar (Muslim hook-and-line fishermen), a sub-caste of the Mappila fisherfolk. A fishing unit of this kind consists of six to seven sets of line, a dugout canoe and a crew from four to five men who are attached traditionally to it. This does not include provision for recruiting *Ottakkanmar* (fishing labourers). Most of the units are operated by a crew of four men. Over 90 per cent of the owners go out for fishing with their *Ottakkanmars*. They start fishing operations before sunrise and return to the shore at sundown. Sometimes they go out to the sea at about 3 a.m. and come back at 12 noon. Their traditional hook and lines (*Kaiccunda, Cangalaccunda* and *Ayiramcunda*) have how been replaced by the modern *Otta Cunda* (drift line) and fibre boat with outboard engine of 9.9 HP but still they go out for fishing in the early morning and return in the afternoon. They go out for fishing with *Otta Cunda* upto a distance of 25 to 30 *Maarus*. Apart from the Mappila fisherfolk, the Latin Christians of Trivandrum and Quilon district and the Mogeyar of Puthiyappa and Quilandy have adapted modern hook and line fishing. A crew of four to five hook and line fishermen own a fishing unit consisting of a fibre boat and modern hook and lines.

Boats and Nets

The Mappila fisherfolk of Tanur today own 900 mechanised fibre boats, 400 small non-mechanised boats and 85 *Valiya* Glass fibre *Vallam* (big canoe). These Mappilas do not own trawler and trawl nets because Tanur does not have harbour facilities.

In earlier times a fishing unit of *Valakkar* in Tanur consisted of one or more *Valas* (nets), two *Tonis* (plank-built boats), two *Cukkan* (rudders) and eight *Tandus* (oars), eight *Mokkus* (wooden pegs for fixing the oars in the gunwale), one or more *Kadankanis* (coir wings), one or more *Kambas* (warps) and an eight-man crew. A unit was owned either by a single individual or by a number of persons. The general pattern is the joint-ownership. A fisherman used to get a share in the ownership of the unit by contributing to any one of the fishing tools mentioned above. A person could acquire a small share in a net by supplying even a *Kadankani* (coir wing) or a *Kamba* (coir rope or warp). The smallest share that a fisherman could acquire in a fishing unit is *Mahani* (one-sixteenth of the unit) and the maximum, *Mukkalarakkal* (7/8th of the unit). In a fishing unit the shareholders were called *Mahanikkarans* (owners of 1/16th share), *Arakkal Pakutikkar* (owners of 1/8th share), *Kalpakutikkar* (owners of 1/4th share), *Kalerakkal Pakutikkar* (owners of 3/8th share), *Arapakutikkar* (owners of one half share), *Ara Arakkal Pakutikkar* (owners of 5/8th share), *Mukkal Pakutikkar* (owners of 3/4th share) and *Mukkal Arakkal Pakutikkar* (owners of 7/8th share). The owner of the entire unit was called *Valakkaran.*

The shareholders of a fishing unit were called *Pakutikkar* and the overwhelming majority of them used to work in their units as fishing labourers. Each *Pakutikkar* (shareholder) was, and is, responsible for providing not only the appropriate share of the fishing implements but also the *Ottakkanmar* (fishing labourers). For instance, an *Arakkal Pakutikkaran* provides an *Orral* (fishing labourer) if he cannot himself be present in the expedition as one of the members of the crew. This principle is still followed in the case of the other shareholders like *Ara Pakutikkar, Mukkal Pakutikkar,* etc. The former provides four fishing hands as against six labourers of the latter.

We had observed: "A particular fishing unit is generally managed by a *Karanavan* (headman). He is a man who is able to command the respect of his crew. The *Karanavan* is a man of good economic standing and usually has the maximum number of shares in the fishing unit, varying from one-fourth to three-fourth. He keeps an account of the daily catch and settles the account of each shareholder fortnightly, monthly or even quarterly. Formerly the account was settled only once a year on the Bakrid day. It is his (*Karanavan's*) duty to see that the unit is functioning properly without any difference of opinion among its constituents. He distributes the quantum of pre-arranged share to the members of the unit out of the catch and also to the *Ottakkanmar*. Most of the fishing labourers are illiterate and so there have been several allegations of defalcations and embezzlements. It is the *Karanavan's* responsibility to maintain the crew including the *Orakkanmar* during off-seasons. He may even take loans from fish merchants, moneylenders and fishmongers in order to provide for the members for his unit. The loans are generally secured against personal credit and on the understanding that the catch would be supplied at reduced rates" (Mathur, 1978).

Nylon Nets

The nylon *Mathikollivala* and nylon *Odakollivala* (boat-seine nets) introduced in Tanur in 1968 were owned by a number of shareholders who formed a fishing unit (*Peyithu*). Each shareholder had to contribute a fixed amount of cash or a section of the net to the extent of shares he wanted in the unit.

Distribution of the Catch

The catch (value of the catch) was divided into six shares if the number of the crew was four. The owner of the hook-and-line and boat used to get two shares and the crew got four equal shares. If the number of the crew was five, the catch was divided into seven shares and the crew got five shares and the owner two shares. The owner of the unit had to meet the expenses for refreshment and bait. Nowadays, the owner of the fishing unit meets the fuel and charges of mechanised fishing expeditions.

Mathur had written in 1978: "Among the *Valakkar* (those fishermen who use nets and boats for fishing) the value of the catch is generally divided between *Pakutikkars* and *Ottakkanmars*, after setting apart the share payable to the owner of the nylon nets, if any, and to the boats, after deducting the *Cayapaisa* (petty expense for refreshments) of the crew which varies from Rs 15 to 20 irrespective of the nature of the catch. The share for the nylon net is 15 to 20 per cent of the catch and the rent for the two boats is 12 per cent. After deducting these three items from the catch the remainder is divided among the shareholders. Each shareholder pays the wages of his *Ottakkanmar* from his own share according to the agreement. If the *Ottakkanmar* are recruited by the *Karanavan* they are paid by him from the total catch. Sometimes all *Ottakkanmar* are paid off before sharing the catch among the owners of the nets and boats. The payment to *Ottakkanmar* depends upon the nature of the contract they have entered into with the owners of the unit. Majority (six out of eight members) of the crew were formerly *Ottakkanmar*. Today the number of *Ottakkanmar* has become reduced to two or four out of a crew of eight. The bumper catch of prawns, particularly the *Naran Chemmin* (Penaeus indicus) during the past two years has helped a number of *Ottakkanmar* to clear off their debts and acquire shares in their respective units. The struggle to break off the shackles is going on but it is difficult to do so as the *Ottakkanmar* are unable to get financial assistance during off season (south-west monsoon) and emergencies."

The *Ottakkanmars* generally used to get from 5 to 12 per cent of the catch, either after setting apart shares for the nets and boats, or before doing so. If an *Ottakkan* was a steersman, he got 12 per cent, while his comrades got only from 5 to 10 per cent of the catch as the remuneration depended upon the skill and age. The *Ottakkanmars* aged between 16 and 18 got five per cent and those above 18 years got from 7.5 to 10 per cent of the catch.

Although the *Ottakkanmars* used to get their dues from the catch according to their agreement, their share was reduced proportionate to their debt. In other words, if an *Ottakkan* owed a debt of Rs 500 to his master, he got 2 per cent less than his usual share of the catch. On the other hand, if his loan had been cleared and reduced to a lesser amount, he got his full share. If an *Ottakkan's* debt was below Rs 250 he may get one per cent more than his share.

The owner of the unit even today gets a share larger than the crew and the steersman gets an additional share. Nobody raises any objection to this inequitable distribution. The catch is sold by auction. The highest bidder gets the catch. The large and medium sized fish like skates, rays, sharks, catfish and mackerel are auctioned and the prices are quoted per fixed numbers. Small fish like *Puvalan* (small prawns), soles, mullets, etc., are also auctioned. When a particular unit fails to get any catch, the members borrow fish from those units, which have got bumper catches. This is still practiced not infrequently by the Mappila fisherfolk. The loan is repaid conveniently. These obligations are part of the social fabric of the Mappila fishing community.

As soon as the catch is landed, it is apportioned by one of the most important men of the unit, generally the *Karanavan* or *Chukkankaran*. The owner of the boat is allowed his share; so also the owner of the nylon net. If the members of the crew including the *Chukkankaran* happen to be of the same unit and have equal number of shares in the unit all get their shares proportionately. If the catch is poor it is divided by the members amongst themselves in order to meet their personal expenses.

When a number of fishing units jointly undertake an expedition the nets and boats belonging to all may not be used at the same time. Some boats are used for shooting the net while others for beating and scaring the fish.

"In an expedition the beaters also receive the same share of the catch as other members of the crew do. The catch is divided equally among the number of units participating in the expedition. The *Chukkankaran* and the owners of the units are given bonus shares. The distribution of the catch of a joint expedition generally takes place smoothly. When there is a bumper harvest and the demand for fish particularly for *Naran Chemmin* is great the members of the units dispose of the catch at an unscheduled landing station without the knowledge of the owners. This irritates the owner of a particular unit as well as the members of the other units. If such misappropriation is detected the culprit gets a tough lashing. This kind of a breach of trust is sometimes indulged in by a unit which gets a bumper catch but does not like to distribute the same among other units which have hauled a poor catch. It is said that elders and other respectable people do not commit such frauds as they get equal share of the total catch" (Mathur, 1978).

There are occasions, however, when family tie-ups lead fishermen to cooperate in large-scale expeditions, for launching and hauling up boats, particularly during monsoon. As has been already pointed out, five to ten units participate in an expedition during south-west monsoon. Only able-bodied men take part in an expedition in this season. The catch is divided in equal shares among the units.

Mathur (1978) observes: "Three to ten units went out together for catching sardines though only a single nylon net was used. At the same time all the boats belonging to the units were used for beating and driving the fish into the net. Similarly, in the case of catching mackerel by entangling technique 6 to 10 units join hands together and undertake a joint expedition. In the case of catching sardines 25 to 100 fishermen participate in an expedition whereas in the case of mackerel 32 to 64 fishermen are required. The entangling technique is used for catching the mackerel. The fishermen encircle an immense stretch of sea and gradually close it with much splashing, leaving only the space near the nets. The men driving the fish shoal are generally very vigilant, since some of the fish make attempts to jump over as they are driven nearer. Should the crew of the unit haul up large fish like *Kakka Tirandi* (eagle rays) or sharks, the crew generally do not keep it for themselves but leave it to the *Karanavan* of the unit.

Sometimes such a prized catch is cut into pieces and distributed among the members of the crew and shareholders for domestic consumption. Another factor affecting the catch is the presence of children at the beach as soon as it is landed. They surround the boats and grab five to eight pieces of small fish. The childish pranks are not resented or objected to by the members of the crew. In this manner a boy grabs 7 to 10 fish, the cost of which comes to Rs 2 to 3. Children are thus allowed to play truant by parents who refuse to send them to school."

Another significant feature of the distribution of the catch is that the Mappila fisherfolk set apart a share for defraying the expenses connected with their rituals. A share is set apart for religious purposes from almost every good catch. The Hindu, Mappila and Latin Christian fishermen send Rs 10 to 20 to their respective shrines for performing fishing rituals from every catch.

The apportionment of the catch from cooperative fishing is primarily a matter of satisfying the claims of the labour and the major items of capital and equipment. The same proportion is given to an individual worker. The minor items of capital equipment owned by the various individuals of the unit receive no share at all.

There is reciprocity among the Mappila, Latin Christian and Hindu fishermen. The norm is that all the participants in a productive enterprise should receive their due share but the social considerations enjoin the fishermen that they should take into consideration the age, skill and share in the fishing unit. It appears from the above that the maxim 'each according to his labour' seems to be the governing principle in the distribution of the catch whenever there is a joint expedition.

Relation between Patrons and Clients

The relationship between the *Muthalalis* (patrons) and their *Ottakkanmar* (fishing labourers) is hereditary. This age-old bond has not been fully shaken off even after the introduction of modern nylon nets and crafts like *Valiya Vallam* and trawlers. There are *Ottakkanmar* or *Ottakkans* who have been attached to their masters for the last three to four decades. They are obliged to go out for fishing daily when their services are required among the Mappila fisherfolk, Hindu fishing communities and Latin Christians.

Writing on the interrelation between the owners of the unit and fishing labourers, Mathur (1978) had observed that "the *Ottakkanmar* among the Mappilas are recruited to fishing units on the *Valiya Perunal* day. The *Ottakkanmar* cannot leave their *Muthalalis* as and when they like. Once they are attached to a particular fishing unit they are obliged to work at least for a year. If an *Ottakkan* wants to change his *Muthalali* and join another, he has to announce his decision a year in advance. He should pay off all his debts to his patron before he joins a new unit. A patron, keen on taking a new recruit, pays off the debts incurred by him on the day of his recruitment, i.e., on Bakrid, the 10th Zilhijja, the last month of the lunar calendar of the Muslims. The violation of this customary rule is punishable by the local *Kadakodi* (caste panchayat). Under normal circumstances the hereditary relationship between the patrons and clients is not broken off. Even today the recruitment of the *Ottakkans* takes place on the day of Bakrid. The exchange of patrons and clients among the Mappila fisherfolk of Tanur is called *Pirinnupokal*. As has already been pointed out the *Ottakkans* are obliged to work under their masters till the next Bakrid day. Once they are recruited, and their masters have the reciprocatory obligation of giving them work and finance during off-seasons and life crises. The *Ottakkans* pay off their patrons' debts in a number of instalments as and when bumper catch is available." It is interesting to point out that although the fishing labourers are not recruited on the *Valiya Perunal* day, if the fishing labourers take loans from the owner of the fishing unit, they cannot leave their masters until they repay the loans. This is also true among the Hindu fishing communities and Latin Christian fisherfolk throughout the coastal villages of Kerala.

All the fishing labourers used to get a fixed percentage of the catch according to their skill and age. For instance, an able-bodied steersman got 10 to 12 per cent of the total catch or the share as his master. The *Tandukar* (Oarsman) used to get 9 to 10 per cent of the catch. The fishing labourers above 60 years and below 18 got only 5 to 7 per cent of the catch.

Labour Organisation

The labour is distributed among the Mappila, Hindu and Latin Christian fishermen of Kerala on the basis of sex, age, rank and, to a certain extent, on the principle of specialisation. It is organised for different kinds of work in accordance with certain norms for achieving maximum efficiency in fishing operations and getting bumper catches. Some of the important factors of labour supply are discussed below:

Division of Labour

Fishing, the principal avocation of the coastal people of Kerala, is based on sex and age. The division of labour is based on operations such as removal of entrails, heads and fins of fish by women and catching and marketing by men. In certain spheres of fishing operations like net-making, curing and icing and pealing of fish both men and women cooperate with each other. Touching of a fishing net or other implements by women during their periods is tabooed among the fisherfolk of Kerala. The repair of larger nets is however, usually undertaken by the menfolk. The Mappila women of Tanur do not perform the task of fishermongers, fish merchants or headload workers like the Bovis-Mogeyar of Kasargod and Christian counterparts of Quilon district.

Occupational Specialisation

Mathur (1978) observed that "there is a system of hiring labour among the Mappila fisherfolk. Specialists in particular types of work recruited by the *Valapeyithus* (fishing units) are given shares according to their skill and age. The fisherfolk divide a five-man crew in a *Toni* into a *Chukkankaran* and four *Tandukkarans* who are all proficient in fishing and seamanship. Sometimes, the *Karanavan* or head of the unit himself works as the *Chukkankaran*. The *Tandukarans* are called after each *Kalli*. Thus, the four *Tandukarans* are known as (1) *Tamankallikaran* or *Taman Tandukaran,* (2) *Kada Tandukaran* or *Kadakkallikkaran/Ida Tandukaran* or *Idakallikaran, (3) Nalla Tandukaran* or *Nalla Kallikaran,* and *(4) Mumba Tandukaran or Mumbakallikaran.* The members of the crew in a fishing unit have specialised functions. For instance, it is the prerogative of the *Chukkankaran* of the *Valapeyithu* to select the members of the crew irrespective of the fact that whether he is the owner or not. He chooses only physically strong and diligent fishermen. He is endowed with the faculty of detecting fish shoal and possesses cosmological knowledge. But this does not mean that he is supposed to possess any supernatural power. The *Tandukaran* is distinguished from the *Chukkankaran* by his expertise. Thus, the most specialised job among the crew of the *Valpeyithu* is that of a *Chukkankaran* who is required to have a good knowledge of wind, weather, current, tides and habitat and other relevant factors of fishing. He must know where to find the fish and must have "good eyes" to sight the different species. The *Chukkankaran* is in overall charge of the unit and fishing operations are conducted under his directions. The *Tandukarans* are subordinate to the *Chukkankaran* and assist him in rowing, paddling, shooting and hauling the net. The *Chukkankaran* is usually responsible for the maintenance of the boat and nets though not for its replacement unless he is the owner. Generally, it is the *Karanavan* or his son who is the *Chukkankaran* but in four of the seven fishing units owned by Kadarkutty, one of the richest fishermen, the *Chukkankaran* or even a *Tandukaran* in a unit is selected exclusively on the basis of his expertise in fishing operations. There is no short-cut to a Mappila fisherman to get the title of a *Chukkankaran* or *Tandukaran*. This requires long experience in shooting and hauling the net. The technological and sometimes even ritualistic knowledge is considered an added qualification in the elevation of the individual and kin groups in the social ladder. According to the Mappila fisherfolk a man who has not shown his prowess is not properly equipped with theoretical knowledge of the craft." The occupational specialisation is still prevalent even after the introduction of fibre and mechanised boats and big canoes and trawlers to a limited extent.

The *Chukkankaran* as the leader of the crew wields considerable influence during the fishing operation. His words are law at sea. His main work as the steersman is to direct the operations of the fishing unit. As has already been stated earlier, he gets the lion's share of the catch as compared to the *Tandukarans*. The organisation of a crew in a cooperative fishing unit is identical with that of an individually-owned unit except that the steersman receives no extra share in the catch for his position. Although he is in a position of authority, his real ability to command the members of the crew in an individually-owned fishing unit may be limited by the fact that all the members of the crew own equal share in the unit.

The *Chukkankaran* and *Tandukarans* among the traditional fishermen belonging to the Mappila, Hindu and Latin Christian fishermen have specific duties to perform during the fishing operations. They discharge their duties from their respective *Kallis* from *Amaram* (stern) to *Aniyam* (prow). For instance, the *Chukkankaran* who occupies his seat in the *Cukkankalli*, near the stern, directs the entire fishing operations. But this does not mean that he performs no other job during the shooting and hauling. When the net is launched he drops the floats and sinkers and brings the boat around the fish shoal. In the case of launching the hook-and-lines, it is he who hands over it to the *Tandukaran (Kada Tandukaran* or *Idakallikaran*) for its dropping. He also keeps the boat against the current and wind while the hook-and-lines are launched whereas the oarsmen are asked to shoot the lines according to the direction of these natural forces. It is believed that when the boat is kept against the wind and current the hook-and-lines do not go far off from the fish shoal as well from the boat itself. The *Mumbakallikaran* or *Mumba Tandukaran* hauls up the hook-and-lines when the fish is caught from the *Mumbakalli*. We have already mentioned that the *Chukkankaran* should have sufficient knowledge of the behaviour of the fish, particularly their migratory character. For instance, if the fish migrates from north to south the *Tandukkarans* may suggest that the net may be launched on to the north. The *Cukkankaran* detects the flow of the *Nir* (current) and if its direction is from *Tengarakara* (south-east), he directs the oarsmen to launch the net in the *Vattakara* directional angle (north-east). Sometimes, he rules out the launching of the net to the north as it does not lie afloat properly in consequence of which fish are not caught. In this connection Mohammedkutty, an expert *Chukkankaran*, maintained that when the fish migrate from north-west to south-east, the *Tandukarans* are able to conjecture the direction only. On the other hand, it is the duty of the *Chukkankaran* to ascertain the fact and proclaim the direction in which the fish travel. The *Chukkankaran* generally orders the *Tandukarans* to bring the two boats very close to each other and keep the prows of the boats facing north-east when a fish shoal is detected. The *Tandukarans* are then asked to launch the net. When the *Mathikolli* net is launched, its cod-end is shot half a furlong from the rear side of the stern (*Amaram*) of the boat. The distance between the nets and boats is generally about 20 to 30 yards in the case of boat seine nets like *Mathikolli*vala and *Odakollivala*. The *Chukkankaran* is an expert in calculating the distance. If the fish shoal does not come under the net, the *Chukkankaran* finds out the actual reason for it. In case he discovers that the defect is due to the higher level of the *Kadamkani* (coir net), he asks the *Mumbatandukaran* to raise the anchor. This results in the lowering of the *Kadamkani* to the sea bottom. Thus, when the coir net goes down to the sea bottom and touches the muddy layer the fish enter the net. It is only then that the *Chukkankaran* directs the *Mumbakallikaran* to launch the net.

Among the *Tandukarans* the *Tammankallikaran* is the most important oarsman. He performs his prescribed duties from the *Tammankalli*. He assists the steersman in dropping the sinkers and floats. When the net is hauled up, he raises the head rope. It is he who throws the net into the *Tammankalli* when the net is hauled up. The *Idakallikaran* or *Ida Tandukaran* is considered the "centre forward" in a fishing operation by the Mappilas. He helps the other members of the crew in shooting and hauling up the net. The catch is generally removed from the net and put into the *Kumbidikalli*. When the *Kumbidikalli* is full with the catch, it is kept in the *Idakkalli*. When the catch is transferred to the custody of *Idatandukaran*, or *Idakallikaran*, he is called *Kada Tandukaran*. The net is generally kept in the *Kumbidikalli*. The *Nallakallikaran* and *Idatandukaran* jointly shoot the net after lifting it from the *Kumbidipadi*. When the net is hauled up it is the *Nallakallikaran* who puts it back either on the *Kumbidipadi* (the crossbar of the *Kumbidikalli*) or into the *Kumbidikalli* itself. The *Nallakallikaran* is also responsible for releasing the mast and hauling it whenever the situation demands. The *Mumbakallikaran* is responsible for shooting the *Kadamkani* (coir warp), besides extending a helping hand to the *Nallatandukaran* in hauling up the net. During this operation the *Mumbatandukaran* lifts up the foot-rope and sinkers whereas the steersman and *Idatandukaran* raise the hold up and head rope. The *Mumbatandukaran* also shoots the iron anchor and stone sinkers when the net is launched. When the wind and current occur simultaneously from the same direction, say from the south, and the net is launched in the same direction, it is the *Mumbakallikaran* who immediately lowers the anchor in order to protect the net. In other words, if the net is launched in the direction of north during the occurrence of wind and current from the south, it is not necessary to use the anchor. Whenever it is to be employed it is the *Mumbatandukaran* who discharges this function very timely. The *Vittalakallikaran* has the duty of keeping in safe custody the floats, sinkers, coir rope, coir net and hammers during the fishing expedition. He also assists the other oarsmen in shooting and hauling the coir rope, coir net, sinkers and floats.

From the above description it is evident that all the oarsmen cooperate with each other under the leadership of the steersman during a fishing operation, although each of them has specific duties. Thus, the success of the fishing expedition depends upon the team work of the members of the crew. However, the steersman has the final say in the entire operation.

Training plays an important part in the making of specialists among the fisherfolk of Kerala. Children take part in fishing operations in which their parents and kinsfolks are engaged. From an early age they perform minor works such as draining water from boats, carrying small fishing tools to the boat, etc. Gradually, they are entrusted with more significant jobs. Sometimes, a father makes a model boat for his son to demonstrate the fishing techniques; occasionally his son is seated by his side during fishing operations or while he is making boats. Although personal factors such as aptitude and individual interests are important in learning the technique, it cannot be denied that there is always a conscientious attempt to persuade children to follow the hereditary profession. Elderly people seldom take part in fishing operations, though there is no particular age for retirement. They go out for fishing only as long as they are fit. The sick and infirm do not go out for catching fish. Thus, there is plenty of scope for every member of the family to take part in fishing operations and specialise in any particular branch according to his initiative and aptitude.

Mode of Cooperation

The mode and extent of cooperation in fishing is determined mainly by two factors namely, (1) technical requirements, and (2) nature of social convention.

The Mappila, Hindu and Latin Christian fishermen collaborate with each other in the technical sphere, particularly in the making of boats and nets. The supply of labour is generally supplemented by a large number of kinsfolk and neighbours, wherever such assistance is required, for certain types of works like bringing a tree trunk to the beach for making boats, or for undertaking fishing expeditions and also for repairing larger nylon nets. This cooperative spirit is evident among the Mappila, Hindu and Latin Christian fisherfolk in the techniques of catching fish by joint expeditions for employing shore seine nets and for launching boat seine nets and gill-nets from big canoes. A *Valiya Vallam* or *Valiya Vanchi* employs more than fifty labourers in the use of boat seine nets for catching sardine and mackerel. The *Tattepeyyal* (frightening technique) requires cooperation of five to ten units, i.e., 40 to 80 fishermen, in fishing sardines using the boat seine net. At Ernakulam as many as 20 units with 150 to 200 persons were found cooperating with each other in fishing by the *Tattipeyyal* technique during 1970–72. Majority of these fishermen acted as beaters in order to scare and drive the fish into the net. The operation of a shore seine net at Virinjam (Trivandrum district) and Blangad (Thrissur district) requires a minimum of 30 to 50 persons.

Cooperation in fishing is based on the ties of kinship and mutual understanding. The ownership of the fishing unit is also based on kinship ties. Majority of the fishing units in the fishing villages of Kerala are jointly owned by the members of one or two kin groups. In a joint venture like fishing each individual contributes his labour and skill to the success of the operation. The fisherfolk of Kerala are fully aware of the advantages of cooperation in fishing as it enhances the efficiency of the unit.

The ownership pattern of the fishing implements discussed earlier has demonstrated the fact that the majority of the *Valiya Vallams* are generally owned by the rich and men of rank. Most of the boats in the coastal fishing villages of Kerala are built of wooden planks covered by fibre. The traditional wooden boats had no ornamentation but the *Odams* (larger dugout canoes) of Calicut, Cannanore and some parts of Thrissur have figureheads and ornamented sterns. The craft, whether it be *Vallam* or *Toni*, is regarded as the most valuable piece of the owner's property. This is because of their larger sizes and greater efficiency in shooting larger nylon nets and cotton nets. The traditional wooden boats across the coastal fishing villages of Kerala have been mostly replaced by either medium size fibre boats with outboard engine or by *Valiya Vallam* or *Vanji* with inboard engine. In the Calicut, Cannanore and Quilon districts trawlers and trawl boats have also been introduced. The average cost of a nylon net varies from Rs 3 to 5 lakhs and mutual cooperation and assistance among the fisherfolk of Kerala, particularly among the kinsfolk and neighbours, is required. This cooperative spirit lies at the root of fishing which is their mainstay.

Cooperation among the shareholders of fishing units on the one hand and the members of the crew on the other is essential, particularly during the monsoon when they have to go out for fishing in the rough and violent sea. During this period the regular members of the crew are unable to launch a boat or haul it themselves. In spite of cooperation and mutual understanding of the members of the crew, however, sometimes it is impossible to save life and property on account of the fury of nature. Thus, the Mappila fishermen, Hindu fisherfolk and Latin Christians are always at the mercy of Nature. In spite of cooperation and assistance there have been occasions when the boats of the fisherfolk got capsized and destroyed on account of the violence of the south-west monsoon.

What we wrote in 1978 holds good even today with regard to the different institutional forms of cooperation in fishing.

> "Cooperative spirit is the striking characteristic feature of the fishing economy of the Mappila fisherfolk.
>
> A great deal of the fishing operations is performed by the mutual cooperation of different individuals belonging to same kingroup.
>
> The economy of fishing is related to the social and ritual life of the people. There is division of labour on the basis of age, sex and skill.
>
> Labour is governed by kingroup or kinship relation and the kins cooperate among themselves for any difficult task like repairing the fishing crafts and gears. Sometimes, all members of a particular kingroup assemble under the leadership of a *Karanavan* for joint fishing expeditions.

> The composition of labour required depends upon the mode of fishing as well as the nature of the catch.
>
> Casual labour is employed but they mainly consist of children and teenagers. Casual labour is employed only when the permanent members of the crew are absent on account of illness or life crisis.
>
> The efficiency of the labour of a fishing unit is governed by the personality of the *Karanavan*. Labour supply depends upon the economic and social status of the head of the fishing unit.
>
> Technological innovation has not lowered the requirement for labour supply.
>
> The local social groups, Beppukar and Valakkar, play an important part in the labour supply. Each group supplies labour with specialised knowledge of different techniques. Neighbours and fellow-villagers participate in fishing operations whatever be the kinship ties and social obligations. Capital resources of fishing unit and kinship ties have a considerable effect on the supply of labour."

A fishing unit employing an eight-man crew in the past was generally owned by an individual or by a number of them with one or two nets and boats, besides the necessary accessories like rudder, oars, etc. Joint ownership is the general pattern even now. A fisherman can get a share in the ownership of the unit either by contributing a net or a boat. The catch is divided between the owners of the fishing gear and the craft, and the fishing labourers. The share varies from one coastal region to another. Earlier in Tanur the catch was divided into six shares if the number of the crew was limited to four among the hook-and-line fishermen (*Beppukkar*). The owner of the hook-and-line used to get two shares and the members of the crew equal shares after deducting the expenses for bait and refreshment. Among the *Valakkar* (traditional fishermen who use exclusively nets and boats), the catch was divided between the owners of the fishing gears and fishing labourers after setting apart the share payable to the owner of the nylon nets and boats after deducting the daily expenses. A number of fishing labourers have been able to become shareholders in the fishing units during the past three decades or so. According to Mathur (1978), "the introduction of the nylon nets and modern techniques of catching fish in recent years have resulted in bumper catches, particularly of delicious prawn, locally called *Naran Chemmin*, and improvement in the economy of the Mappila fisherfolk in general and those of *Ottakkans* (fishing labourers) in particular. The *Ottakkan* or *Ottakkanmar* have been in the forefront of breaking off the traditional ties with their patrons in becoming owners of fishing units. The number of *Ottakkanmar* in fishing units has come down to two/four in a ten or eight-man crew. Rest of the members of the crew have become share-holders-cum-fishing labourers. They are mostly consanguine kins. Sometimes affinal kins also join as partners in a unit. It has been observed that both consanguineal and affinal kins work as *Ottakkanmar* in the same unit. The economy of the Mappila fisherfolk appears to have been founded upon the kinship relationship. Thus the most striking feature of ownership of the fishing units is that majority of them are jointly owned, particularly by consanguine kins."

The owner of a fishing unit used to get an additional share than the other members of the crew. The highest bidder used to get the catch at auctions. Nowadays the middleman (*Tarakan* or *Dallal*) who advances loans to the fishermen for the purchase of fishing nets and boats is the auctioneer and he fixes the price. Today the owner of the fishing gear has to meet the expenses for the fuel and food. After deducting the fuel and other incidental charges, the cash price is divided between the owner and the labourers. The fishing labourers in Malabar get 60 per cent and the owner gets 40 per cent after deducting the expenses for fuel and food. But in the district of Kollam the fishing labourers who work in trawlers get 50 per cent of the catch and the owner the same share after deducting the cost of fuel and food. In the case of *Valiya Vallam* in Cannanore (Kannur) district the fishing labourers get 60 per cent of the catch and the owner or *Muthalali* gets 40 per cent after deducting the fuel and food charges. A Thiyya, who is the captain in a trawler costing Rs 28 lakhs owned by a Muslim, says that 65 per cent of the catch goes to the owner and only 35 per cent is distributed among the crew of seven. The *Muthalali* meets the expenditure for diesel, fuel, and food. The captain said his unit went out for fishing with the trawler before the ban for three days and the unit got a catch worth Rs 2 lakhs as against the total expenditure of Rs 50,000 for fuel and food. Out of Rs 1,50,000, an amount of 47,500 was distributed among the seven labourers and each got Rs 6,500 for three days. The landing consisted of catfish, mackerel and *Puthiyapla Kora*.

In this connection, Kalpana Ram (1991) observes: "The perishable nature of fish as a commodity, particularly in the absence of facilities for freezing and preserving it, dictates those fishermen must sell their catch rapidly, and on any terms. Merchants, on the other hand, aim at ensuring as large a share of the market as possible, and manipulate relations of credit in order to ensure a steady supply. They therefore advance loans to owners of equipment on the understanding that a predetermined proportion of the sale proceeds accrues to them. The merchants thus not only ensure their supply, but they are also able to manipulate sale prices even during an official auction. The advantage to the fishermen is that there is no insistence on repayment of the principal. The merchant has a vested interest in the continuation of the relationship. Interest may be waived on days when the landings are small or nil, since it is a stipulated percentage of the catch." This is true in the case of the relationship between the owner of the fishing units and the middlemen or *Tarakan* in the coastal villages of Malabar, Cochin and Travancore.

Fishing is no more a purely subsistence occupation that depends on market conditions for its prosperity. The export market is gaining importance, as it is a major foreign exchange earner for the country. The supply of fresh fish to markets depends greatly on the availability of transport. As fish does not keep fresh for long in hot climates, especially in Kerala, speed is essential in the transport to the market. Alternatively, fish may be frozen or packed in ice or cured by other methods. Or, they may be processed and canned.

The economic condition of the fisherfolk is of the worst kind. They have no capital and earn their livelihood by their back-breaking labour and most often they have to borrow to sustain their families during the monsoons. The fish trade and its profits are grabbed by the middlemen who advance money to the poor fishermen during the off-season.

Capital, Saving and Credit

The capital of the fishermen is represented by boats, nets, rudders, oars, coir nets, ropes and masts for sailing, mechanised boats, trawlers and trawl nets. Majority of the fishing labourers are poor and do not have adequate savings to meet their occupational requirements. Most of the working capital has to be borrowed. All members of a larger fishing unit (*Valapeyittu*) are responsible for repayment of the loan. Sometimes, the headman of the unit enters into a contract with moneylenders on behalf of the crew of the unit, which, however, does not relieve the individual members from their responsibility.

Capital formation among the fishermen of Kerala, especially among the Hindu fisherfolk, is very poor. The Hindu fisherfolk are much poorer than their Christian and Muslim counterparts and their grinding poverty is one of the main factors responsible for the absence of capital formation. The Arayans of Alleppey and Quilon are also at the same level as they too have not responded to technological innovations. The most important problem before the fishermen's economy today is finding out a means by which the level of income and capital formation can be improved.

The wholesale commission agents, fish traders and businessmen, are the important sources of finance. The Mappila and Hindu fisherfolk offer their future catch as security; hence the middlemen (*Tarakan*) who finances for the purchase of nets and boats, stipulate that the best fish should be sold to, or marketed, through them. Loans are generally given as and when the need arises and the amount is not given in one lump. The demand for credit is more pressing at the beginning of the fishing season. Short-term or working capital is absolutely necessary for the maintenance/repair of boats and nets, purchase of yarn, ropes, floats and sinkers, and oil and grease in the case of mechanised boats. Thus, the financial requirements depend on the kind of the boats, the fishing method employed and such factors as accidental damages, losses, etc. The expenses of a mechanised boat are almost three to five times more than the traditional craft. Credit is required for marketing the catch after landing it. Loans are normally repaid by adjusting the sale proceeds. Each *Valapeyittu* (bigger fishing unit), therefore, keeps a running account with the creditor. Thus, the sale, production and marketing of fish are linked together. Over 75 per cent of the fishing units at Tanur and Blangad depend upon private individuals through whom the fish is marketed for funds. Thus, traders and big businessmen provide the major part of the short-term credit to the Mappila and Hindu fisherfolk. Other sources of credit include commercial banks, cooperative banks, Matsyafed, professional moneylenders, relatives and friends. Matsyafed grants loans and subsidies for buying mechanised boats through cooperative societies at the rate of 12 to 14 per cent of interest. The Matsyafed does not give the loan in cash but supplies the fishing gear and fishing crafts against the loan.

The members of a fishing unit undertake to repay the principal/interest before the end of the season whenever a loan is taken from a private source. But the creditors who are not fishmongers or entrepreneurs to give security in the absence of which a very high interest is demanded. The Mappila and Hindu fisherfolk pawn their jewellery with private banks which charge them interest from 12 to 20 per cent per annum.

Private individuals (*Tarakans*) who advance credit engage in the marketing of fish and insist on the fishermen selling their catch to or through them. Although no fixed interest is charged, it is not against the custom to realise 5 to 10 per cent commission on the sale proceeds. The rate of commission generally depends upon the integrity of the fishing unit and its relation with the creditor. There is a tendency on the part of the fishermen to over-borrow and increase their dependence on the middlemen. Sometimes the commission is charged throughout even if the debt is cleared off even before the end of the season. The rate of commission often fluctuates, as it is not fixed. This often results in the exploitation of the fishermen by the middlemen. Private traders are the main source through which the occupational requisites of fishing are bought by the fisherfolk of Kerala.

In the existing economic conditions the traditional fishermen are unable to play an important role in capital formation in Kerala. Majority of the Hindu and the Mappila fishermen operate within extremely narrow margins. Several traditional fishermen have adopted technological innovations in fishing, as a result of which quite a few have purchased rice lands and coconut gardens from the sale proceeds of their bumper catches.

Moneylenders and Credit

Moneylenders, who are usually the fish traders, are the most cruel exploiters of the fishing communities. They cash in on the fishermen's general ignorance, poverty and lack of sustenance. The fishermen are in a vicious circle. Having no land, building or other properties fit for hypothecation to secure a loan, they have no option but to approach the moneylenders for their help, since they ask for no securities. The relation between a fisherman and his *Tarakan/Dallal* (moneylender) is based on trust and age-old tradition. Despite the fact that the fisherman knows that he is mercilessly exploited, he has nothing but regard for his *Tarakan*. During the lean months and life crises he has no one to turn to and there is none to help him except his *Tarakan*.

The *Tarakan*, on the other hand, at times goes to the extent of borrowing to help a needy fisherman. The fishermen pay exorbitant interests to the middlemen ranging from 4 to 10 per cent of the proceeds of the catch. The entire catch is sold to the middlemen who buy it at their own price, which is about 25 per cent below the market values. The fishermen are further cheated in weight as well. As fish is a perishable commodity, and the fisherfolk are in a helpless condition, in the absence of their own transport and storage facilities, they are too eager to dispose off their catch at the earliest opportunity. There is none more alive to this situation than the *Tarakan*, who has the means and facilities and who can, therefore, dictate the price.

Firth (1946) states that the advancement of loans to the poor fishermen is primarily a matter of providing their subsistence and that of their families during the monsoon. During this lean period, when the fishermen have no other source of income, they have nobody to advance loans for their daily sustenance except their middlemen. Till this debt is cleared the fisherfolk maintain their contact with the middlemen. Should they wish to change their allegiance, they have first to clear off their debt, which they do by borrowing from another *Tarakan*. Since the fisherfolk are seldom able to repay their loans and hence they are always indebted to their creditors.

The *Tarakans* or their agents, wait on the beaches for the fishermen to land their catch. Failure to deliver the catch to the *Tarakan* is a breach of trust, which would even entail forfeiture of their boats, nets, etc., if they have any, besides, seizure of their properties or house and even prevention of their going out to fish in other men's boats. This is hardly relished by any fisherman.

Most of the fisherfolk in Kerala are under the clutches of the *Tarakan*. Some work on the boats of the *Tarakan* to pay off their debts, and are often under-paid. This kind of bonded labour is on the wane now and absent in many places; nevertheless, it is present in some places. A good number of fishermen who had land have lost their property to their creditors as loans are advanced by mortgaging the land and the title deeds are kept under their custody. The fisherfolk are unable to redeem the mortgage and consequently they lose their land too. Besides land, ornaments, boats and nets too have been lost by pawning.

Where fishermen work as bonded labour, their share in the catch is proportionate to their debt; the greater their debt, the lesser their share. Over and above this, they have to pay the interest of the loan too, reducing further their take-home amount. "A patron gets casual labour freely. The fishing labourers are unable to leave their masters on account of social etiquette and fishing norms" (Mathur, 1978: 244). There is no way for the fishermen to break this traditional bondage for as long as they are indebted to middlemen, they are tied up to them till they pay off the loan, which they are unable to do.

Capital being scarce among the Hindu fishing communities, there is pressing demand for funds, despite higher rates of interest. The interest rates are high because the security given is poor. The borrowers, in majority of the cases, can offer nothing but their personal security (Mathur, 1978). The credit system being such, what alternative is there for the poor fisherfolk than to turn to the middlemen and suffering exploitation practically all their lives?

The fishermen have genuine complaints and grievances against the banks and cooperatives and Matsyafed, for during the lean months when they have no income whatsoever, they are obliged to pay the interests on loans, the failure of which entails penalties in the form of increased rates of interest.

Another complaint against the institutionalised agencies is that they demand securities for advancing credit. In a community as poor as the fishing community, landholdings, if any, are flimsy and houses are but ramshackle. The official credit agencies do not accept such unworthy securities. If fishermen own boats and nets they might have only a share in it, in which case they cannot mortgage it. In such circumstances what they should do? The availability of credit, for both short and long terms, is very important to the fishing communities and so long as their need is pressing, their dependence on the traditional moneylenders will go on increasing for want of a better source to turn to.

The relationship between the fisherfolk and their middlemen or traders is strengthened every year during the monsoons. They feel obliged to give their catch to their *Tarakan* (middlemen) though they do not stand indebted to them. Of course, the fact remains that they have little choice in selling their catch except to the middlemen simply because they have themselves no means of transporting fish nor any facilities for its preservation. In crises and emergencies they have none but the middlemen to approach to. They can expect help from relatives and friends only to a limited extent, for their lot is no better than their own. The middleman is always there close at hand to get the fishermen out of any crisis. In fact, in all financial crisis, the fishermen actually expect the middlemen to help them out and the middlemen too feel obliged to do so. This traditional relationship is time-tested though the setting today points out towards a change. The fisherfolk's reliance on local moneylenders cannot be done away with, as he is a sure source of credit who does not demand security, does not impose penal interests, is within his reach, and does not insist on long-winding procedures. The institutionalised credit agencies who are plagued by red-tapism and other abhorrent practices can never be a substitute for the indigenous middlemen. The fishermen say that even "Matsyafed is a blade whereas the *Tarakan* is their god."

The technological revolution in fishing has encroached upon the traditional fisherman's catch and income, as prices are brought down by glut in the market. He is unable to reap the benefits of technological innovations. Added to this is the inflation, which makes the fisherman's lot even more miserable. To eke out a living he has to borrow and since banks and other official credit agencies leave him out of their purview, he has none but his middleman to help him and to give him the means to make both his ends meet, in which process he himself lands deeper into debts.

The fishing communities are, therefore, characterised by a population steeped in poverty and low standards of education. Economically they hold a very weak position and the few who have the means and genuine interest to improve their lot are faced with insurmountable problems. There is, therefore, a strong feeling of helplessness among the fisherfolk, especially among the younger generation.

The economy of the Hindu fisherfolk is almost entirely based on fishing. The marine resources exploited by the fishermen of Kerala are nearly identical all along the state's coastline. Farming and other occupations among the Hindu fisherfolk are quite insignificant. Some retail fish and a few have incomes from salaried employment. Coir-making, which could have provided them with a subsidiary source of income, is not easing their pecuniary conditions as few have their own coconut gardens and the owners of the factories engaged in this line prefer to recruit the Ezhava women. However, the womenfolk of all Hindu fishing communities are engaged actively in coir-production in the district of Trivandrum and biri-making in Blangad.

Although a number of fishermen, particularly Muslims from Tanur and Blangad, migrated to the Arabian Gulf countries and worked as fishermen, majority of them could not improve their lot. For instance, our informant Muhammad worked as a fisherman in Abu Dhabi for eight years in a fishing unit which had two fibre boats and one Ozhukkuvala. He was one of the five men crew. He used to stay in an air-conditioned room and paid Rs 1,000–1,200 towards the rent and electricity charges. He also remitted Rs 5,000 in some months to his family. But he could not construct even a small house although he worked for 8 years. He returned three years ago. Now he finds it difficult to pay for his visa. Again he has started working as one of the fishing labourers at Mappila Bay. This is also true with many of his friends who worked in Abu Dhabi, he says. However, a few of them invested their earnings from Arab countries and became owners of trawlers and big boats.

The Mappila fishermen still perform *Kutikalyanam* (special feast) or *Cayakalyanam* for raising capital. It may not be out of place to quote Mathur (1978). He observes: "The *Kutikalyanam* is generally celebrated by poor as well as wealthy fishermen. Money is accumulated to construct a *Pandal* and buy provisions. Friends, kin, fishermen from neighbouring villages, join and even members of non-fishing communities are invited to the feast. The number of people attending such a *Kutikalayanam* often goes up to 100 and beyond while the number of guests and invitees for feasts at wedding and circumcision, etc., may be more than this number. The guests and invitees bring presents, either in cash or kind or both, to the host, in return for the feast. The value of individual contribution depends upon the status of the guest and his friendship and kinship with the host. The host is obliged to make a matching contribution to the donor on a similar future occasion. The presentation articles with the names of their donors are announced loudly by an elderly person and usually an account of them is kept by an educated person for future reference.

The advantages derived from *Kutikalyanam/Cayakalyanam* in respect of capital mobilisation are worth mentioning here. Firstly, this enables a fisherman to meet the cost of his fishing equipment which would otherwise have been impossible on his part. Expectations of handsome contributions prompt a fisherman to spend all his savings and even take loans from others to make the feast or tea party a memorable one in his life by preparing various delicious dishes and decorating the *pandal* lavishly, etc. In the absence of the whole-hearted help of his kinsfolk, neighbours and villagers, the *Kutikalyanam/Cayakalyanam* will turn to be a dead loss.

The funds raised by the *Kutikalyanam/Cayakalyanam* are not spent on household items; they are invested in capital items like a boat or a net after repaying any outstanding loans taken for meeting the cost of feast/tea party. The host is amply rewarded with presents by numerous kinsfolk, friends, neighbours and villagers before the feast/tea party is partaken of and he is required to repay presents at his convenience over a long period of time.

The main disadvantage of this method of raising capital is as follows: In the first place a wealthy fisherman gives a feast/tea party by spending a lot of money for the sake of display of his wealth but the presents he receives may not be matching. Secondly the giver of the feast/tea party cannot expect returns on account of the poverty or death of those who take part in it. Sometimes individuals make valuable presents to keep up their prestige though the prospects of similar returns are dim."

Indebtedness

Our study has revealed that most of the fishermen in general and the Hindu fisherfolk in particular, are indebted for economic reasons. The owners of the fishing crafts and gears require loans for the purchase of boats and nets and accessories. Reports of losing and destroying nets by *Kadal Makri* are rampant. Therefore, most of the owners have to seek financial help from the *Muthalali* on hypothecation of their future catch. Above all, during the fishing season, owners take loans from private moneylenders to meet the running expenses which are not properly covered by the loan taken from the *Muthalalis* or from the commission agents.

Fishing largely depends on the caprices of the marine environment. Whenever the catch is very poor, the owners of crafts and gears are unable to repay the loans to the money-lenders or *Muthalalis.* The members of the crew also take loan from the owners of the fishing gears and so they are indebted to the owners. In consequence, both owners and crew resort to loans. The decline in income is the main reason for the indebtedness of the fisherfolk of Kerala. Owners need large loans for the investment of their fishing occupation whereas the crew require smaller loans to meet the household necessities. Loans, whether large or small, are not easily available from *Muthalalis,* commission agents and others without any security.

Thus, in general, the fisherfolk of Kerala, irrespective of religious affiliations, pawn their valuables to moneylenders for loans. Professional moneylenders in coastal fishing villages of Kerala give them loans either against gold ornaments or brass utensils at exorbitant interest rates. Sometimes the fishermen take loans against security from non-professional moneylenders in the fishing villages or outside at more than 100 per cent interest.

During our study, it was observed that whenever the fisherfolk sustain loss in their business or fail to get adequate catch, they are forced to sell their fishing boats, nets, ornaments and utensils. In this manner they become paupers and the economic power is gradually shifting to the hands of a few rich fishermen and the non-fishing communities. They are controlling the means of production which is represented by the catch while the majority of fishermen are converting into fishing labourers. This tendency is also affecting the social structure of the fishing communities.

Until recently, the fisherfolk of Kerala somehow survived by means of fishing. But matters have come to such a pass that their traditional occupation no longer seems to be capable of ensuring their survival. Many of them have lost their fishing boats and nets and hire themselves out to the *Muthalalis*. But their earnings for the last few years have been too meagre to maintain their families. In consequence they are heavily indebted to the owners, money-lenders, relatives and local shopkeepers. Most of them are now left with nothing except their labour power for survival. They are, therefore, forced to work as fishing labourers in the fishing units owned by the owners of big canoes and fibre boats. This is mainly due to the fact that the income from fishing was utterly inadequate to meet their minimum domestic requirements. So unless they are assured of their daily bread, some fishermen may be compelled to give up their traditional caste occupation.

As the economic situation of the fishermen is characterised by chronic insecurity and scarcity, they are compelled to take loans for a living. When they are in dire necessity, they are ready to stake anything for their survival. They then depend to a great extent on the mercy of the money-lenders, nationalised banks, private banks and *Matsyafed*.

Dwindling Supply of Fish

During our field work, frequent complaints were heard from the fishermen that the catch from the Arabian Sea was gradually dwindling. Due to this, their economic condition was deteriorating and they were landing into more debts. Based on their knowledge of the habitat and habits of fish that was gained from the experience of past generations the fishermen gave several reasons for the scarcity of fish in the ocean.

In addition to the fall in catch from the sea, the number of fishermen has increased due to the increase in population in the same marine environment. Besides, the constant plying of a large number of mechanised boats and trawlers and canoes scatters the shoals of fish in the area where the nets are set. The traditional wooden crafts have also become obsolete due to the receding of fishing grounds further in the sea. For the past two decades, prawn's eggs have been extensively captured in the sea. Along with them, small fry of other species get destroyed by the fine-meshed bag nets. The fishermen think that this has had an adverse effect on the stock of species in the ocean.

Scientifically, it has been found that clean drinking water is a very essential humane requisite for sustenance of fish life. Clean water is also *sina qua non* for the development of fishery resources (Jhingran, 1982: 121). Polluted water is one of the main reasons for the decrease of fish in the sea. There are a number of factories located on the banks of the Periyar river and the Arabian sea and industrial effluents are being discharged into them without any pre-treatment. The effluents reduce the acidity of the water required for the survival of the fish and increase the ammonia instead, which is harmful to them. The caustic soda in the effluents also adversely affects them. The pollutants in the river destroy fish food such as phytoplankton, zooplankton and benthic algae, and fish spawns and larvae. Fish life is also adversely affected by the highly localised pollution near Cochin and Beypore.

Pesticides and insecticides, used extensively in modern agriculture, aggravate the problem by spreading and diffusing the hazards of pollution through drainage to ponds, lakes and streams all over the country, putting the interests of agriculture and aquaculture at cross purposes.

"Whatever be the reasons for the dwindling of catch, it is clear that preservation and destruction should go side by side. There must be a balance between the two. An otherwise aquatic ecosystem may bring severe biological imbalance which, in turn, may cause ecological disaster" (Jhingran, 1982: 122).

Change in the Traditional Credit System

Traditional money lenders are being slowly replaced by service cooperative banks, private banks, nationalised banks and Matsyafed. The interest charged on the loans by the banks is from 12 to 14 per cent per annum. The Matsyafed charges from 14 to 22 per cent against the loans taken by the fishermen. It does not pay the loan in cash but adjusts the amount against the fishing crafts and gears purchased by the fishermen.

Thus, it has been noticed that the credit system plays a very important role in the development of the fishing economy. Credit is fundamental to the fishing economy as it facilitates not only capital investments but also helps in the marketing of the catch. The kinsfolk, friends, neighbours and even villagers extend their cooperation in raising capital for a particular fisherman to help him to invest in fishing equipment. Traditional money lenders are being slowly replaced by banks and Matsyafeds. The scriptural injunction that Muslims should not take interest on money lent is also not strictly followed by the Mappila.

In this section we have tried to analyse the economic activities connected with capital, saving and credit among the fisherfolk of Kerala. The emphasis has been on describing the concrete economic activities and conventions rather than on elaborate quantification. The important reasons for avoiding elaborate quantification are that no actual statistics exist. Even when data like the accounts kept by the cooperative societies were available the figures were not found to be authentic. The fisherfolk of Kerala generally do not keep accounts and do not reveal them; even if they do, the money lenders and creditors are reticent to provide authentic information. Therefore, data had to be collected selectively from a few informants just to illustrate the points.

The capital of the fishermen are mainly represented by traditional boats, mechanised fibre boats with outboard engines, *Valiya Vallams* with inboard engines, trawlers and trawl nets, boat seine nets, shore seine nets, nylon nets, rudder, oars, coir net, ropes and mast for sailing. Poor fishermen own one or two nets and traditional non-mechanised boats and hire casual fishing labourers. But this helps them to become a shareholder in a fishing unit. Among the few neo-rich fishermen the mechanised boat and trawler are significant pieces of equipment. The capital value of a mechanised fishing unit is higher than that of a traditional unit.

Land is considered by the fishermen as one of the best forms of investment. The fisherfolk of Kerala believe that investment in land is secure as it cannot be lost, stolen or destroyed. Ownership of land itself confers prestige on the fishermen. The fisherfolk also invest their funds in jewellery which comes handy for pawning.

There are a number of priorities over whatever little surplus is available, than investing it. This is particularly true of the Mappilas and Hindu fishermen living in a well-integrated village community. There are demands for gifts and loans from kinsmen and neighbours.

The Mappila fishermen who wish to climb up the social ladder spend lavishly on ritual feasts, funerals, ceremonies such as wedding and special feasts like *Cayakalyanam* and *Kutikalyanam*. Apart from acquiring social prestige, the organisation of such feasts helps the fishermen to mobilise capital.

Affluence is the way to political leadership in fishing villages of Kerala, certainly for membership to the Village Panchayat, Block Panchayat, Zilla Panchayat and the local Muslim League. Most Panchayat members, particularly the Mappilas and the Latin Christians are wealthy individuals whose source of income is not confined to fishing. Very few Hindu fishermen are members of the three-tier panchayat system due to their poor economic conditions and lack of political influence.

The affluent fishermen tend to invest their income in lands. As the demand for rice lands by the fishermen during the last few years has increased, the land price in the region has gone up sky-high. Therefore a study of the rising land values in different fishing villages and their effects on the social organisation of the fisherfolk would be of a great sociological value at the moment. Our general impression is that spurt in land values as well as the great demand for fish, particularly, the prawns in markets abroad, will have a considerable impact on the traditional social organisation of the Mappilas in general and the Latin Christians in particular. A new class of entrepreneurs has emerged among the Mappila fishermen of Tanur due to their migration to the Arabian Gulf countries in very recent years. They advance loans to fishermen for buying nylon yarn and nylon nets and other equipment. Many of the Gulf-returned fishermen, irrespective of their religious affiliation, have started constructing houses and investing their earnings in fishing crafts and gears.

Trade and Market

Marketing is the most complex of all aspects of the fishing economy. The fishing industry on the Kerala coast has been developed remarkably well during the last two decades. The producing units are widely distributed over the entire coastal belt and, therefore, direct supply to consumers is extremely difficult without the intervention of middlemen and dealers. Earlier the marketing of fish in Kerala used to be concentrated mainly in the hands of non-fishing communities. These influential people still exercise control over the marketing of fresh, cured and dried fish. However, a few of the fisherfolk of Kerala have become traders in recent years.

Writing on the market system among the Mappila fisherfolk Mathur (1978) observed: "The striking characteristic of the marketing system in Tanur is the adherence to accepted business ethics. A trader who has promised to deliver certain species of fish on a particular date tries his level best to fulfill his commitment even if it means a financial loss, for it is essential that he keeps up his word. If the trader is found to be dishonest or undependable, the fishing community sometimes bring strong informal sanction, social and economic, to bear on him. Another interesting feature of the marketing system is the grant of credit for the mere asking, and a trader who gives it knows he is assured of future catches. The advent of modern technology and innovations has made the rich fishermen wise and they sometimes send some of their members to Cochin, Trivandrum and other wholesale markets to negotiate large purchase direct from individual fishermen so that they have a large supply of fish of uniform quality, which they can dispose in one transaction at a reasonable price. In other words, if the wealthy fishermen are not satisfied with the price offered by the wholesalers in the village and are under no obligation to sell the catch to them, they may get market intelligence by sending a member to wholesale markets outside the village to find out the possibilities of sale there. If he succeeds in striking a deal, the group hires a truck to transport the fish and sends one or two of its members to receive the payment."

Traders

The traders are people who act as intermediaries between the producers (fishermen) and the consumers. Their most important concern is to provide immediate market for the fish as soon as the catch is landed. They make available the catch to the consumers who live in far off places. They also act as fish curers. The fish dealers in Kerala perform duties of wholesalers as well as retailers and most of them depend on fishing for their livelihood. The wholesalers may be divided into three groups: (1) iced fresh fish merchants, (2) hawkers and pedlars, and (3) middlemen.

An examination of the relationship that subsists among the dealers on the one hand and the businessmen and the producers (the actual fishermen) on the other gives us the modus operandi of the wholesale organisation, which is of considerable importance to the fishing economy.

The most significant feature of the marketing organisation among the fisherfolk of Kerala is the existence of a number of fishmongers who are engaged in the disposal of the catch. There are fish merchants who sell uncured fish, iced fish curers, pedlars or basket-men who dispose of fresh fish carrying them on head loads, and hawkers who use bicycles to distribute fish with little loss of time.

The number of fish traders fluctuates very often due to the following two important reasons. Firstly, there are fishermen who take up temporary dealership just because their fishing units or implements are damaged. These men return to their calling as soon as their tools are repaired or replaced. Secondly, there are a few fishermen who have a regular employment but engage themselves in trade as a subsidiary occupation.

Another interesting feature of the fish dealership in Kerala is that some of the traders take up distribution in addition to their normal occupation. Some of them have a large business in cured fish, while a few others are the owners of traditional fishing units in addition to bulk business in fresh fish.

Many owners of traditional fishing units are also fish curers. A few of them have three to six fishing units. The fish curers generally advance loans to fishermen against the security of their catches. They are, therefore, the main source of finance for fishermen. A few commission agents go out to sea from time to time to supervise the catch, while others engage themselves in procuring nylon yarn and nets from Bombay, Madras, Cochin and Cannanore. In fact the commission agents are either individuals or employees of the wholesalers.

Fresh Fish Merchants

The fresh fish merchants are the most important link between the producers and consumers. When the boats land with their catch, which is usually, just after midday, a large number of dealers, including the fresh fish merchants, assemble at the landing stations. Some of the dealers are experts in telling the size of the catch just by noticing the speed of the returning boats. When the boats are landed the dealers assist the crew to drag them up. They generally gather around the boats with great eagerness to see the catch, for which the nets are hurriedly removed.

The bargaining about the price between the dealers and the owner of the units then commences. The wholesale fresh fish merchants generally wait and see the outcome of the bargains, but sometimes they join in the bid so that they are not out-bid by dealers. It appears from this that there is competition among the dealers and wholesalers to a limited extent. Besides competition, higgling and haggling is the other feature of the bargaining.

Sometimes a single dealer cannot afford to buy the entire catch of a fishing unit. In such a case, as soon as the price is settled, the whole catch is divided among two or three dealers. In this case the buyers do not bid against each other. On the contrary, they join hands to bring down the price.

As soon as the price is settled, the fish is scooped out from the boat into baskets for retailing in the market by carrying it in motor trucks, rail or other means of transport. Ice is added before transporting the fish in order to preserve it. The wholesale disposal is generally made on credit. The fishermen have standing agreements for supplying the superior quality fish to traders from whom loans have been taken for buying the fishing gear. However, no written agreements are made between the fishermen (sellers) and the buyers/traders in this respect.

The wholesale disposal of fresh fish suffers from a number of difficulties. The main reason is that dealers are not interested in wholesale, as they want to dispose of the fish immediately and release their investments daily. Circulation of money is also affected by retail trade. The wholesalers thus knock down the catch at a cheap price to reap their profit for their investments in capital assets such as trucks and ice plants.

Retailers

The *Cillarakaccavadakkar* (retailers) may be divided into three categories, viz., bicyclemen, basketmen and middlemen. The retailers buy fish from the wholesalers or from the fishermen themselves and supply it to markets, which are not far off from the fishing village. Bicycle men and basket men are found in a large number at Blangad and Puthiyappa. The retailers are debarred from having direct contact with the fishermen. However, they are also allowed to bid for the catch if it is small. The *Cyclekar* (bicylemen) and *Kottakkar* (basketmen) mainly belong to the non-fishing Muslim community. At times, the basketmen work as head-load workers. They buy fish paying cash on the spot. On the contrary, the wholesalers are the local Mappila and rich Hindu fishermen and they buy fish on credit. This compels the retailers to buy the fish mostly from the local wholesale merchants giving the latter a quick return and a reasonable profit.

When the fish is in short supply the retailers are the worst sufferers as the price quoted to them is generally exorbitant. They often grumble that though they want to buy direct from the fishermen the latter won't supply them first as they are "outsiders". Another important impediment in their business is the fall in the price when there is a bumper catch.

A number of buyers participate in the retail market. There are iced fresh merchants and fish-curers who buy the major portion of the catch of the day. There are traders who make profits by acting as intermediaries in transactions between the fishermen and the wholesalers. There are also womenfolk at Saktikulangara and Kasargod who sell a small quantity of fish in the neighbouring markets which they have received direct from a number of units. There are numerous fish sellers at the fishing villages of Kerala coast who move from one fishing unit to another and run from one landing station to another. Everywhere there is haggling and bargaining. This is the picture of the fish market during the fishing season.

The price fluctuates hour-by-hour, day-by-day, and even minute-by-minute depending upon the arrival of the catch, the quality of fish and the bargaining capacity of the individual retailers. Down-cash payment in retail transactions induces fishermen to flock to the market, whether they are adults or children. The children prove themselves to be as shrewd as their elders in the fish trade. In this manner they are initiated into the art of bargaining, which is the striking feature of the fish market. Retail prices vary very widely at the market. The retail buyers underquote the rates prevailing in the neighbouring markets to make the fish cheap and this induces the fishermen part with their catch. An interesting feature of the retail transactions is that a concession in retail price is often given to the kinsfolk and friends. The fishermen and wholesalers reduce the price to retailers and their regular customers.

Factors of Price Determination

The basic factors that determine the price in any transaction, particularly the wholesale price, observed by Mathur (1978) are given below:

1. During the wholesale transaction of the catch the buyers and even bystanders are free to make any comment on the quality of the fish and the bidding. This helps in giving information not only about the nature of the catch landed earlier in the day but also about the fishing units that arrived and the price at which the lot was sold.
2. The middleman (*Tarakan*), or in his absence the *Chukkankaran* (steersman), is responsible for quoting the price when the boat first lands. Other members of the crew generally do not question the adequacy or otherwise of the price quoted either by the middleman or the *Chukkankaran*. The prices quoted in the first instance are generally above the market rate.
3. The buyers (fresh fish merchants and fish-curers) are generally of one mind, though occasionally out-bidding is resorted to. This concerted action of the buyers helps to reduce the price which enables even "a small dealer" to earn his bread by staking a little amount.
4. Bargaining is "the sine qua non" (Firth, 1946) in the fixation of price. This leaves freedom both to the buyers and sellers to quote the price within their means. The fishermen (sellers) have no intelligence of the ruling market price as they have been at sea. They do not even know the price fetched at Cochin on the previous day. On the other hand, wholesalers (buyers) have no idea of the quantity of the catch. There is no means to predict whether the arrival of earlier boats is to be followed by other boats with bumper catches. They are, therefore, unable to know

whether there is likely to be supply in excess of demand or vice versa on a particular day. These factors, which are beyond the control of sellers and buyers, help reaching a price satisfactory to both parties. However, a clue is obtained from the initial quotation. Higgling and haggling are, therefore, essential in fixing the price of each lot of catch.

5. Another significant factor, which settles the price, is the minimum price below which the seller is unable to dispose of his catch. This is considered the dead end in bargaining. In practice this is often revised as the fishermen have no means to preserve the catch and are compelled to dispose it on the same day. The fishermen are often found resigning to their fate in this respect.
6. Freight, cost of ice, curing, etc., are also calculated before settling the price.
7. Heads of some of the fishing units act as buyers from time to time in order to prevent the wholesale merchants from cornering the fish at a throw away price.
8. Another important factor, which determines the price, is the time of the arrival of the catch at the shore. "The Mappila traders used to say that 'the train (62 down passenger to Shornur) bound for Shornur had already left' and the 'Cochin passenger had already arrived at Tanur railway station', 'it is already dark and there is no transport', etc. What they really mean is that if they miss the trains, the loss is greater. Two choices are left to the fishermen when the trains have left. One is to sell the catch to the wholesale fresh fish merchants at a low price whose responsibility it is to transport the fish to Cochin. The other way is to sell the catch to fish curers at a very low price. It is evident from this that there is a distinction between the fresh fish market and the cured fish market. The buyers are found to be extremely impatient towards the afternoon. As soon as the sun sets the traders start complaining of 'darkness'. This is not without its effect on the price. The catch that is landed after 4 p.m. goes dead cheap as there is little demand for old fish. It is usually bought by the curers. But if the catch happens to be prawns the price is unlikely to come down to a low level as is the case with other lots of fish. The price offered in the evening is admittedly lower than the amount paid in the morning. When there is no

bumper catch and fish like prawn, shark, catfish, skate and ray, etc., are in short supply, the advantage lies with the fishermen (sellers). When such a situation arises the fishermen have the upper hand in fixing the price of their catch without much higgling and haggling" (Mathur, 1978).

9. The variety of fish, their size and other factors also have their bearing on the price of the fish. Similarly the price of larger species of fish like shark, skates and rays, cat-fish, etc., are settled on the basis of their size apart from the forces affecting supply and demand.
10. The sellers and buyers follow "the concept of just price" (Firth, 1946) in their business transactions. This has prevented the exploitation of the fishermen by the traders to a certain extent. The hardest bargainers are unpopular and generally looked down upon by the fisherfolk. There are a few fishermen who make no concessions to the traders and there are, in turn, some wholesale merchants who do not pay the price agreed upon with the sellers. These persons are often nicknamed as "backdated fishermen". The point underlined here is that a fisherman, whether he is a buyer or seller, must make out a living somehow or the other. They are, after all, members of the same community and some of them are friends and kinsfolk. They are also economically interdependent. There exists a fellow-feeling among the sellers and buyers.

Marine fishing is limited to the coastal belt and the catch is brought to the shore immediately to avoid loss. Fish is a highly perishable commodity, and the tropical conditions of the Kerala coast hasten their decomposition. Once on land the fish has to be transported at once to the market, factories for processing, or to a cold storage. Decay of the fish starts immediately after they are removed from the nets. Fishing centres are deficient in the essential requirement of an abundant supply of clean water, good roads and vehicles to transport the catch to the interior retailing centres and storage accommodations during the peak fishing season. Nearly half the total fish caught in Kerala is marketed afresh. Unfortunately, there is no direct contact between the fishermen and the consumers. The middlemen are the only links between these two factions.

General Observations

The monsoon season accounts for 70 per cent of the total fish production in Kerala. The average annual monsoon production from Kerala is 1.4 lakh tonnes, which forms 30 per cent of the total annual production from the state. Ninety per cent of the production comes from the motorised country crafts and the entire fishing industry looks forward to the monsoon season.

There are various ancillaries associated with fishing, namely, peeling sheds, ice plants, processing units, manufacture and sale of fishing equipment, etc., and their respective work force.

Fishing is thus a specialised economic activity in the social and cultural context of Kerala. Both traditional and modern fishing units continue to exist side by side. The Mappila fisherfolk, Latin Christians and a dozen Hindu fishing communities are actively engaged in this sector. During 1970–72, it was observed that the *Beppukar* (the hook-and-line fishermen) have stuck to their primitive technology but the *Valakkars* (users of nets) have adapted several innovations. In 2006, however, both the *Beppukar* and the *Valakkar* of Tanur were found to have adapted modern fishing nets and mechanised boats. The *Beppukar* had discarded their traditional *Chunda* (hook-and-line) and had adapted modern hook-and-line fishing and nylon nets and fibre boats, particularly mackerel nylon nets and *Ozhukkuvala* (gill-net). Similarly, they had discarded their traditional dug-outs and adapted plank-built small fibre boats. In the sociocultural domain, in the pasts there was no intermarriage between *Beppukar* and *Valakkar*, but these are now taking place to some extent. In both the traditional and modern fishing units there is a distinct division of labour according to the age, sex and skill. The organisation of the fishing activity depends upon cooperation and coordination. Entrepreneurship and sound financial judgment ensure maximum reward, but exploitation by the wealthy continues in traditional and modern units, although kinship ties bring about a social equilibrium. Acceptance of modern technological innovations has strengthened the solidarity of the kins. The cooperative movement and the functioning of the Matsyafed have not exerted any appreciable influence on the welfare of the fisherfolk of Kerala. Recent studies have brought to the limelight that a few fishing labourers have become the owners of fishing boats and nets. No less than 500 Mappila fisherfolk from

Malabar work in the Arabian Gulf countries as fishermen and fish traders. While some of them have erected palatial bungalows spending as much as 25 lakhs rupees, many others have returned to Tanur without making a mark and have again started working as fishing labourers.

Fishing is not only a subsistence occupation but also a commercial avocation. The economy of the fishermen depends upon the market condition for which the freshness of fish is crucial. This market is governed greatly by the availability of transport as fish will not keep fresh for long in hot climates, particularly in the state of Kerala. Alternatively fish are frozen or packed in ice, cured in a number of ways, or processed or canned. The lives of fishermen revolve around fishing and fishermen spend most of their leisure with talks of boats and nets, and wind and weather. They rarely have any secondary occupation. Their womenfolk, however, do not take any active part. Exceptions to this are noted among the Latin Christians of the southern districts and the Mogavirar and the Bovis-Mogeyar of Cannanore and Kasargod. They either sell fish caught by their menfolk or buy them directly from the fishermen. The women also cure fish in the districts of Cannanore and Kasargod.

The financial condition of most of the fisherfolk is pathetic with no capital of their own. They have to often borrow to meet both their ends. The reason is that the major share of their profits are taken by middlemen who advance money to the fishermen and exploit them. The relationship of fishermen with the buyers of fish and the rules that govern the social and economic ties between them are a universally occurring phenomenon. They are aimed to sustain the social organisation of this domain and explain it in terms of attempts at rational behaviour, rather than in terms of psychology, custom or some other non-rational motivation. It is best explained from the perspective of the economic sphere of small-scale commercial fishing. As the fishermen must sell the fish they catch, they are inextricably linked to the market mechanism. The connection is further characterised by the nature of the commodity in that it is produced in unpredictable quantities and must be marketed quickly to preserve quality.

Since the supply is unpredictable, the fishermen are usually forced to continue to fish even when prices drop. The possibility of holding out for a higher price is never considered by the traditional fishermen although trawler and purse-seine operators do so at times, especially when ice factories and freezers are readily available. Missing the seasonal run cuts into a fishermen's income to a fraction. In the absence of a back-up fishery this would prove to be disastrous.

Since fresh fish must be sold immediately, production cannot be withheld until market conditions improve, prices rise, or a satisfactory market is found. The urgency to sell their catch is further compounded by the desire to return to the sea as time spent on land is viewed as time and money lost. Refusal to fish until a satisfactory price prevails is not the norm here.

It has been argued most powerfully by Dumont (1980: 201*ff.*) that "alternative identities do not last long in caste society without striking some compromise with the principle of hierarchy". He refers particularly to the identities supplied by minority religions such as Islam and Christianity. But his argument is too general; it must be qualified by a close attention to the particularities of local configurations of power. In the coastal villages, the Jesuits were quite willing to compromise with the principle of hierarchy and fit into the caste structure, but the people they converted had a powerful basis for keeping themselves aloof from caste society in the fishing economy. Among the Mukkuva Christians of Trivandrum, Catholicism has found its place within a structure of difference rather than of hierarchy.

The members of the Bovis-Mogeyar community in the northern part of Kerala held that they were palanquin-bearers and body-guards to royal families before they took up fishing, whereas in the south, Mukkuvas and Valans claimed to have been always fishermen having no other occupation whatsoever. The Bovis-Mogeyars claimed that although they had abandoned their traditional occupations, their caste name associated with it had persisted.

Fishing, though held and viewed as an unclean profession, continued to be pursued by the Hindu fishermen for different reasons. However, many of them suffered social oppressions and untouchability. Traditional fishing needed group cooperation, and kinsmen or fishermen of the same village usually teamed up together, but the lack of infrastructure in their villages, long and total involvement in fishing, and the financial instabilities, forced the fishermen everywhere in Kerala to depend on middlemen.

The middlemen, in turn, exploited these helpless fishermen in many ways. This phenomenon is a universally occurring one. Thus, the economy of fishing is an intricate and complex one involving not just the fishermen alone. Although the fishermen are aware of the exploitation by the middlemen, they are helpless to resist this, and their attitudes and behaviour towards the latter reflect their dependency on them. In some fishing villages of the districts of Cannanore, Thrissur, Calicut and Malappuram several members of the fishing and non-fishing communities have ventured into fish trade as middlemen.

Mechanisation was introduced in the fishing sector for various reasons and the response to it by the different fishing groups in Kerala has varied. The impact of modernisation, therefore, had different consequences and results here. Fishing cooperatives set up all over the world to help traditional fishermen have gone through varying experiences as is understood from different studies. The cooperatives that were set up in Kerala failed in their objectives, particularly among the Hindu fisherfolk. The only exceptions were those of the Bovis-Mogeyar of the districts of Kasargod and Cannanore. The temple organisation of the Bovis-Mogeyar helped them to adapt modernisation like the Muslim and Christian fisherfolk.

Throughout the ages, different technologies and innovations have been introduced to perfect fishing as an avocation. Though modern crafts and gears have appeared on the scene, the fisherfolk of Kerala, for the major part, are still dependent on their traditional techniques. Despite the fact that their technology has undergone a great deal of change for combating the environmental factors, their livelihood is still absolutely dependent on ecological factors.

"The technology of fishing is related to the ecological conditions. These conditions govern, to a great extent, the distribution of fishing communities, the type of settlement pattern, the nature of fishing economy, rituals and cosmological ideas" (Mathur, 1978). These people who have been fishing for all their lives, over generations, have expert knowledge of the marine life and its behaviour, and of the sea itself, and, as mentioned, have modified their techniques to meet the changing circumstances. The traditional methods of fishing are confined to the coastal belt.

There are different kinds of crafts in use all along the coast of Kerala. Their variations depend upon the nature of the shore-line, climatic conditions and the financial position of the fishermen. Canoes of the same kind have different names at different places. There are mainly three kinds of non-mechanised boats, viz., (1) *Cheriya Toni* (2) *Odam*, and (3) *Repputoni*. In Tanur, the traditional *Repputoni* has been discarded. The *Cheriya Toni* (plank-built boat) is about 20 feet long and takes in two to four persons. These are now replaced by fibre-covered plank-built boats at Tanur and Blangad. However, dugouts are still used in rivers, river mouths, estuaries, and for shore-fishing along with shore-seine net by the Hindu fisherfolk of Kerala. In the south the term *Vallam* is used instead of *Cheriya Toni*. The *Odam*, which is also called *Vanjii* or *Valiya Toni*, is about 30 feet long and takes in about 6 to 10 people in Kasargod and Cannanore. It must be pointed that fibre covered bigger boats (*Valiya Vallam*) with inboard engines are operated at Blangad and its neighbouring villages and the Mappila Bay of Cannanore District. The *Repputoni*, which can accommodate 2–5 persons, is about 30 feet long and is the craft used for deep-sea fishing by the Muslim fishermen of Cannanore and Kasargod. Small *Kettuvallam* covered with fibre are used at Blangad for fishing. Similarly *Valiya Kettuvallam* (big fibre-covered boats with inboard engines) and a crew of 30 to 40 are also used at Blangad and its surroundings.

There may be slight variations in the construction of these crafts from place to place but their functional aspects are the same everywhere. *Thanguvallam* is another craft used in the southern districts, besides *Catamarans* in the south of Varkkala (Trivandrum district). The distribution of the craft is dependent on the kind of the fish caught and the type of nets that are launched, giving allowance for local preferences. A detailed account of crafts, types, construction, gears, etc., is to be found in Mathur's 'Mappila Fisherfolk of Kerala' (1978: 120–179).

The limits of sea assigned to the trawlers and purse-seine boats are always violated and there are constant frictions and quarrels between the traditional fishermen and owners of the mechanised boats. At Blangad and many other villages, the motor boats come right upto the shores and cast their trawler nets, preventing the traditional fishermen from launching their boats and dugouts. These fishermen protest against the 'poaching' by mechanised boats in the 15 km. limit assigned to them by the Fisheries Department. Though the owners of trawler and purse-seine nets have sophisticated gears, they prefer to fish in the same grounds where the indigenous fishermen operate, as beyond the specified limits the water is deep, in consequence of which fish do not thrive. Conflicts between the traditional fishermen and the owners of mechanised crafts are regular occurrences, especially during the season of prawn fishing from August to October. At times, even District Collectors have to intervene to settle the disputes and ultimately the matter has to be taken to the court. The High Court of Kerala has recently held that the state government has no jurisdiction over the coastal area beyond 5 kms and hence has no right to prescribe the 15 kms limit. The ruling of the Court allows mechanised units to fish near the coast and this has affected the catch of the indigenous fishermen.

The traditional fishermen contend that the purse seine nets, by sweeping across the sea-beds, destroy fish rose and fries before they can attain maturity and reproduce themselves. Moreover, the noise of the motor engines scare away the fish. Mechanised boats anchor along at sea and keep casting their nets again and again, while smaller crafts shuttle to and from with the catch. This creates a glut and brings down market prices, besides reducing the fishermen's catch drastically.

The districts of Ernakulam and Kollam have the maximum number of trawlers and mechanised boats; but hardly one per cent of them is owned by the Hindu fishermen. They are mostly owned by the Muslims and Latin Catholics.

The Mappila, Hindu and Christian fisherfolk spend about 25 per cent of their income for the performance of rituals connected with economic activities and religious beliefs. Thus, among the fisherfolk of Kerala where the fishing economy is still rural, ritual occupies an important place in the economic activities. Although there has been some ritual change after the Hindu fisherfolk embraced Islam and Christianity, they have retained numerous Hindu elements in the spheres of rituals and magic. There is also a tendency among the neo-rich fishermen to gain ritual status through political influences. Sometime they show more fanaticism in celebrating Islamic and Christian festivals than the non-fishing communities.

According to Zachariah (1996), "There is a fishing community found in mainly Madapally and Edakkad areas in the north of Kozhikode District, which claim to be the same as Mogeyar and whose Association had applied for admission to the Bovis-Mogeyar Association in 1980. This community is the Mukaya (also known as Mokaya and Mogeyan, and having four *kulam*), which had traditionally both marine and backwater fishing. They used both hook and line, and nets for fishing and dived to collect shellfish and cockles. Anthropological evaluations indicate that this community is not the same as the Bovis-Mogeyar community under study" (Mathur and Zachariah, 1983). The Mukaya community has also, for many years now, assumed the name Moger, a caste included in the Scheduled Caste list, and many of their members have thus been able to obtain high education and are mostly well positioned, all obtained through the official caste reservation."

It was observed that almost all these religious organisations have a political and cultural role to play, besides the predominantly religious one. The social fabric of these fishing communities hinges on these variants. The assumption of an economic role was possible only by some communities. Among the Bovis-Mogeyar in an already existing circumstance, i.e., the presence of the temple, has traditional economic dealings also with its members, with the whole functioning as a social group. The generosity of their affluent members inside and outside the state played significant roles to provide practical financial assistance to the temple committees.

Economic viability and scopes have given these temple organisations the ability to forge ahead into the new fields of modern technological innovations. The community had an added advantage of exposure to new technologies in the neighbouring state of Karnataka in general and the fisherfolk of Mangalore in particular. They were convinced by their kin groups and elders, and encouraged by their kinsmen's initiatives also. They had therefore fewer difficulties than many other Hindu fishing communities in the region to change from long entrenched traditional technologies.

Some of the activities of the religious organisations of the Latin Christian and Muslim communities for the development of the fishermen have shown how their timely and practical aid to their fisherfolk helped. The problems faced by the other fishing communities that hindered their development have also been observed and the example of the deviant group from the Mogavirar community outlines that the consequence of defiance to the traditional fishing organisation, may lead to the supports being withdrawn from them.

Technological Change & Economic Development

In describing and analysing the impact of technological innovations, we must first of all make clear that we are not discussing the first exposure of some isolated ethnic communities to economic development and modern technology. We have already stated that the fisherfolk of Kerala, particularly the Mappila and Latin Christians, are very much proud of responding to modern technological innovations and readily adapt foreign ideas to their own conditions. The fishermen are doubly exposed to wider economic development and foreign influence of maritime traders and to the agrarian economic systems. Historical records on the fisherfolk of Kerala are very scarce but we can usefully rely on the methods adapted by Kosambi for material culture. In this connection, Ram (1991: 129) observes that "the *Catamaran*, for example, is conjecturally of Polynesian origin, while the *Vallam* bears distinct traces of Arabic influences (Kurian and Mathew, 1982). Types of fishing techniques also bear the mark of overseas conquerors and maritime traders. Kurian and Mathew 1982 hold that the boat seine technique is unmistakably of Spanish origin, while the shore seine or *Karamadi* was introduced by the Portuguese. Possibly the only indigenous techniques in Kerala and Kanyakumari are the use of hook and line and cast nets."

Migration

The migration of the fisherfolk themselves are reflected in the gradual regional shifts that have taken place in the various types of fishing gear and crafts. To take just two regions that influence the technology used in Tamil Nadu and Kerala, it may be noted that the *Catamaran* is the characteristic of the eastern coast of Tamil Nadu where the seas are rougher. The craft typical of Kerala is the *Vanchi*, a kind of dugout canoe, predominant today in the northern districts of Kozhikode, Cannanore and Kasargod.

The *Catamaran* begins to make its appearance in Virinjam, Trivandrum district. It is said that it was the fishermen of Tamil Nadu who brought the *Catamaran* to Kerala. According to Mathur (1978: 130), when the logs are tied together with coir they become curved and shaped like a canoe. The logs are cut square at the sterns and tapered at the bows. Thus the *Catamaran* is shaped like a canoe rising above the surface of water, forming the stern from where the steersman uses the rudder. The craft is rowed with the help of a bamboo pole. Neither trade nor monetisation nor even technological innovations have brought about dramatic changes among the Hindu fisherfolk of Kerala that have swept over the Latin Christians and the Mappilas. This requires to be stressed because development is often equated with a change from a simple subsistence economy to a market economy based on monetisation (Platteau, 1982).

Planners and administrators as well as national policy makers, referring chiefly to the technology generally agree with the biases of social theory in this respect that conceives the fisherfolk to be a primitive and crude and changeless rural community. They also hold that existing methods of fishing are largely of a rudimentary primitive character carried on by ignorant, unorganised and ill-equipped fishermen.

An anthropological research on the impact of planning and development has interpreted social change primarily in terms of technology. We are left to assume that the great majority of the Hindu fisherfolk, who have not been direct recipients of the modern technological innovations, remains traditional, unchanging and perhaps 'primitive'. The reason is that studies of the areas of the Kerala coast which have received large inputs of funds and technological aid, both foreign and national government funding, have revealed that these communities have also progressed proportionately (cf. Platteau, 1982; Gulati, 1983; Kurien, 1985).

Nevertheless the use of non-mechanised technology is itself virtually no guide to the nature of the social relations of production within which the technology is embedded and utilised. Thus, while traditional technology appears to be seemingly unchanged in its technical aspects, it is in fact now employed in producing for a world capitalist market. Fishing for the commodity at the heart of the new thrust towards foreign markets and foreign exchange – prawns and crustaceans – has been largely caught, not by the modern, mechanised fishing boats, but by men working with artisanal technology. Over 70 per cent of the prawns caught in India is by these traditional fishermen.

The natural migration dictated by the location of the fish and the evolution of craft and gear, and the more recent type of migration in search of jobs rather than fish, coexist but mutually interact. The account given by the fishermen of Blangad, Tanur and Puthiyappa indicated a great deal of flexibility in moving from one type of work to the other. Those who found life as a fishing labourer on mechanised boats and canoes difficult would revert to taking their *Catamaran* and other equipment to various parts of Kerala and working as hook-and-line fishermen in the lean months. A more serious problem in regarding the latter kind of migration as natural is indicated by the case histories we collected of previous generations of fishermen in these fishing villages. There was no one at Tanur who had picked up patterns of seasonal migration from his father or grandfather. The earliest instance of seasonal migration dated back to forty years, followed by another which began twenty-five years back. The bulk of migration undertaken by those working with *Catamaran* (hook and line), falls within the period of the last ten years, which is also the period in which the migration to centres of mechanised fishing has become established.

Male Migration

Development, if by this term one means the introduction of new technology and its widespread utilisation, has not taken root among the Hindu fisherfolk of Aleppey and Trivandrum, nor has it generated the promised benefits of local employment. The annual out-migration of the men makes more sense when we further consider the paucity of locally available alternative forms of employment. The only employment for men in Virinjam and even in Puthiyappa other than fishing is the ferrying of goods to the big canoes that occasionally call in at the natural harbour of Puthiyappa, Beypore and Mappila Bay. The irony of this is that the men are, for the most part, ferrying out sackfuls of sand that has been mined from their own beaches, places on which their houses are erected. According to Ram (1991), "the ships, calling from Japan, UK, France, Yugoslavia, Taiwan and Korea, take on as cargo sand mined by the government-owned Indian Rare Earth Company. The sand contains oxides of uranium and rutile, and the mining is conducted along open-cut lines, affecting all beaches between Colachel and Manavala Kuruchi, a distance of some sixteen kilometres of heavily settled coastline. The erosion of the beachfront has already had an impact on the poorly constructed dwellings of the fisher people, which are now more vulnerable than ever to the hazards of the wind and rain during the monsoon season."

There is no compensatory employment available for fishermen in the mining company itself. The only work it has generated for them is loading sand and ferrying it to the ships in the large *Vallam* crafts. The work is arduous and dangerous, with a high rate of accidents. The work is disliked intensely by most, but poorer households have no choice but to send their men. With this the range of options for local employment is virtually exhausted. Work in the agricultural sector of the coast line of Kerala is impossible to find in an already overstrained labour market, where the class of agricultural labourers grew in far greater numbers than the class of owner-cultivators.

The relative underdevelopment of Kasargod, Aleppey and Trivandrum fisheries is directly related to the development of other parts of the western and eastern coastline of South India. They are the by-products of the logic of profitability. Although the districts of Kasargod, Aleppey and Trivandrum have not been the target of technological revolution, the work performed by the men as seasonal immigrant labourers in other districts and even the work they perform with artisanal technology, are all part of a supra-regional, and potentially a worldwide, set of market relations.

Traditional Fishermen's Opposition to Trawlers

The traditional fishermen's yearly journey to work as seasonal labourers for mechanised trawlers does not signify a complete acceptance of new modes of fishing. An increasingly vocal and organised fishermen's lobby is placing considerable pressure on the State to exercise control over the operations of mechanised vessels. Organised action is a new and foreign experience for the men of the coastal villages. Far more to their taste is quick and direct action – sabotage, the scuttling or burning of a trawler in reprisal for a slashed net or the loss of daily catch, a physical confrontation with the crew of mechanised boats (in which stab wounds and death are possibilities). It may be noted here that such methods have not been left behind. These modes of dealing out justice have, however, proven inadequate in resisting the systematic transgression of the moral economy of artisanal fishing by the activities of large-scale producers.

Due to the traditional neglect of the fishing community by organised political parties and unions, the unionism within the fishing community fiercely proclaims its independence and its non-political nature. In the Indian context, this can only mean autonomy from direct affiliation with any of the major political parties. The agenda of the new organisations is, if anything, more potent and wide-ranging than that of the by-now bureaucratised parties of the Right or Left. Their demands go all the way from arguing for the limitation of the fishing operations of mechanised boats, to an analysis of the social causes of poverty in the fishing community. The tactics adopted in recent years to press for their demands have been increasingly agitations staging demonstrations, resorting to hunger strikes, picketing, and organising road-blocks and 'fill the jail' campaigns.

The organisations have repeatedly drawn attention to the impact of mechanised fishing on the ecology and livelihood of the thousands of small-scale producers using non-mechanised technology, who provide, an estimated over forty per cent of the total catch. Specific criticisms have been made of each of the three new techniques gill-netting, shrimp trawling and purse-seining. Gill-netting is criticised for the wall-like deployment of the nets, which prevents the movement of fish from offshore to inshore waters, and for scaring away fish that would have otherwise found their way into the traditional inshore nets. Shrimp trawlers are opposed for their operation through June–August, which is the breeding season for most species of fish, and for causing damage to the eggs and young fish. The ploughing and sweeping of the bottom sea by the trawling nets is felt to lead to the depletion and ultimate destruction of marine resources. The purse-seiner is opposed because it competes with the traditional fishermen for their traditional catch: oil sardine, whitebait and mackerel.

Scientific opinion remains equivocal about the technical accuracy of these claims, but recent economic trends have been alarming enough to move the state government of Kerala to appoint a committee of government officers, scientists and trade union representatives to examine the dramatic downturn in the total catch of several important species of fish in the state. According to the report of this committee (1981), sea-fish production rose steadily until the mid-seventies, both in Kerala and at the national level. However, after 1976, while the national all-India figures continued to show an increase, in Kerala the figures fell from four hundred thousand to three hundred thousand tonnes between 1976 and 1980. Among the most severely affected species were oil sardines and mackerel, both targets of the purse-seining technique.

The agitation of the small-scale traditional fishermen runs directly counter to the immediate inclination of both regional and national governments, which is to take advantage of the quick and easy profits from exporting frozen prawns to an international market. Any commitment on the part of the government to treating fisheries as a source of cheap protein for low-income rural and urban consumers, a commitment much in evidence through the fifties, has been long abandoned in favour of capital intensive technology, which effectively prices the product out of the reach of traditional consumers. Since the 1970s, the Central Government has been encouraging international capital with Indian subsidiary interests (companies such as Tata, Kelvinator, Union Carbide, Indian Tobacco Company, Parry, Britannia, etc.), offering them loans at low rates of interest to invest in export-oriented marine fishing. Trawlers acquired in this way are eligible for rebate on excise duty.

Inevitably this policy has encouraged the growth of a class of purse seine operators and merchants enjoying links with the international export market whose political muscle far outweighs the fishermen's lobby. In 1982, the purse-seine operators in Cochin went on strike against legislative restrictions on trawler fishing, backed by the Mechanised Boat Owners Association, the Fish Merchants Association and the Ice Manufacturers Association (Kovakanandy, 1984). Such pressure has ensured that concessions extracted from the government by the traditional fisherfolk (conclusions such as the 1977 Ordinance designed to restrict the operation of trawlers to specific depth zones, areas and seasons) remain largely empty pieces of legislation, neither enforced nor policed.

The fisherfolk, however, although aware of the destructive tendencies inherent in the unchecked operation of mechanised fishing, and offered at best only seasonal employment as fishing labourers on the craft of the more wealthy and powerful, show little inclination to abandon their traditional methods of fishing. In the last decade, far from giving way gracefully to the technological harbingers of modernity and progress, the fishermen have showed a stubborn tendency to persist and even proliferate their artisanal craft and gear.

The work experience of the men of Tanur, Puthiyappa and Beypore is not directly explainable, in the terms we have inherited from Marxism, as a transition to capitalism or an irresistible movement forward in history. Nor do the complications introduced into the model in the 1970s in these fishing villages such as notions of 'quasi' or 'partial' proletarianisation, adequately escape the underlying teleology of this narrative form. In a style not dissimilar to that experienced by other (urban) labouring classes in India, the exposure of the Mappila, Latin Christians and Bovis-Mogeyar to global economic relations in the post-Independence era has only encouraged an exaggeration of 'old' forms of work, which now are given a fundamentally new significance. The division of labour between the sexes among the Latin Christians, Bovis-Mogeyar and Mappila follows a similar logic of exaggeration and reinterpretation, as the female half of the potential labour force remains behind in fishing villages during the annual exodus of men.

7

Fishing Rituals, Culture and Development

The Mappila, Hindu and Latin Christian fisherfolk of Kerala perform a number of rituals and observe numerous magico-religious practices in the course of fishing. The fisherfolk of Kerala, irrespective of religious affiliations, believe that sorcerers can cause not only illness but also death by performing black magic. They engage a host of mercenaries such as fortune tellers, magicians, exorcists and medicine men in order to ward off the effects of sorcery, witchcraft and black magic. There is no denying the fact that the fisherfolk of Kerala seek the assistance of supernatural powers by the medium of rituals and magical practices whenever the natural phenomenon is beyond their comprehension or whenever they face calamities.

Fishing Rituals among the Mappilas

According to Mathur (1978), "the Mappilas have to face great dangers during their fishing expeditions in the deep and violent seas. Anxiety and uncertainty are writ large in their economic pursuits. Their beliefs and rituals and magical practices infuse in them a certain amount of confidence in order to face such dangers which would otherwise be overwhelming. These beliefs and magico-religious practices help fill gaps in their technology of fishing and thus narrow the areas of doubt and uncertainty. Thus, ritual activity co-ordinates and regulates the corporate behaviour of the Mappila fishermen in a large number of social situations. Rituals, forces and powers like Allah, angels, spirits and ghosts are very important in the life of the Mappila fishermen, who do not disregard them at their peril."

While describing and analysing the rituals and superstitious beliefs, sorcery and magic among the Mappilas, the important questions that came up uppermost in the researcher were:

1. To what extent are the rituals and beliefs of the Mappila fisherfolk related to their techno-economy of fishing?
2. What are the motives and notions that govern the behaviour of the Mappilas and how do they find expression in their rituals and beliefs? and
3. To what extent do the Mappila and Hindu fisherfolk share the common fishing rituals and magico-religious practices?

The above questions can be answered by reproducing almost all the data from the *Mappila Fisherfolk of Kerala* (Mathur, 1978).

Rituals

Maulud

Maulud is the most important and commonest among the rituals performed by the Mappila fisherfolk, in which the birth and life history of the Prophet and his predecessors and descendants or other saints, are recited.

Mathur (1978) had classified the Mauluds into five major kinds, viz., (1) those associated with the technology of fishing (manufacturing and launching of new nets and boats); (2) those concerned with the birth of the Prophet and ending of the fast: (3) those concerned with the social integration of the village – the Nattu Maulud (ceremony for propitiating the village spirits); (4) those associated with the direct descendants of the Prophet, and (5) those concerned with the predecessors of the Prophet.

The Maulud performed exclusively in honour of the Prophet is called Mongoos Maulud (big Maulud). The Maulud performed on the birthday of the Prophet exclusively in his honour is known as Rub-ul-awal Maulud. The Mangoos Maulud is usually celebrated in honour of the Prophet by the Mappila fishermen either on a Thursday night or Sunday night in their homes to mark the launching of a new net or a new boat. This Maulud is sometimes performed for getting a bumper catch, protection of fishing implements and crew, curing of diseases, warding off of evil spirits, releasing the kin from imprisonment, begetting children, and as an insurance against hazards and turbulence of the sea.

Pattern of Maulud

Maulud sessions are usually held at the residence of the fishermen on Thursdays, just after the sunset and the evening prayer. Maulavis (religious instructors) are called upon to conduct the ceremony upon their arrival. They take their seat on mats spread on the floor. The bedroom where the Maulud is being held is filled with the fragrance of burning incense and perfumes. When all guests have arrived, a sumptuous feast is given to them before the Maulud ceremony begins. The Maulavis take their seats facing Quibla whereas the guests crouch on mats spread on the floor. When the fisherman finishes his explanation for holding the ceremony the Maulavi begins to recite the Fateha, a prayer that precedes the reading of the Holy Quran. The Maulavi then proceeds to recite the passages in toto from the *suras* (chapters) of Lkhlas (sura No. CXXX). Mizzamil (sura no. LXXIII) and Muddaththir (sura no. LXXIV). When the recitation of these three *suras* are over a *Doarkkal* (special prayer) is chanted in honour of Allah. The Maulavi apprises Allah the reason for celebrating the Maulud in honour of the Prophet, or any of his descendants or predecessors or other saints. He invokes the blessing of the Prophet on behalf of the host for getting bumper catches. His blessing is sought for the protection of the host from all dangers at sea and his net from the attack of whales (*Timinnalam* or *Ana*). Blessings are also sought to cure illness caused by sorcery, barrenness and to get ownership of fishing units and other more mundane objects. The assembled fishermen utter the chorus of *Amin Amin* whenever the Maulavi pauses. When the *Doarkkal* is over they rub their palms on their face as if they are getting up from slumber. When the prayer is over, the Maulavi begins to read loudly stories connected with the birth, life, miracles and sayings of the Prophet or his predecessor or his descendants or the saint in whose honour the ceremony is performed, as the case may be. The reading is done in candlelight and takes about two hours. When the reading is over, the Maulavis are invited to a *cirani* (feast). The Maulud session comes to an end with the *Doarkkal* (prayer) in which the Maulavi invokes the blessings of Allah and the Prophet in favour of the host, the owner of the fishing unit, by uttering 'Oh Pataccone' (Oh! Creator), 'Enre Rab' (my lord). The sum and substance of the prayer is that the host be blessed with bumper catches, be protected from hazardous situations, and his net and other fishing equipments be safe. The host and the fishermen assembled for the Maulud say "Amin" in unison.

The Mappilas perform MaulMauluds in honour of the Prophet, his first four companions Abu Baker, Ummar, Usman and Ali, Refai Shaik (21st grandson of the Prophet), Shaik Mohiuddin Abdu Qaadir Jeelani (Mohiuddin Shaik in short), Khadeeja and Fatima (wife and daughter of the Prophet respectively), Hamsa (Prophet's father's eldest brother), Hussain and Hassan, Nagore Sahul Hammeed, Khawja Moinuddin of Ajmir and Mambram Tannal, who are the most important worthies in whose honour Mauluds are held on their birthdays every year. The predecessors of the Prophet including Hilar Nabi, Iyyas Nabi, Sulaiman Nabi and Issa Nabi are also honoured by holding Maulud sessions.

Mongoos Maulud

"The Mappila fishermen perform the Mangoos Maulud in honour of the Prophet to mark the launching of new nets and boats. This may also be performed for getting bumper catches. A new net is launched (*Valananakkal*) by the Mappila fishermen either on a Friday afternoon or on a Sunday after sunset. The night before the launching of either a new net or a boat, is ear-marked for performing a Mongoos Maulud. The Prophet alone is propitiated to mark the launching of new nets and boats. The content of the Mongoos Maulud is the same as the one described earlier. The Mappila fishermen firmly believe that failure to perform Mongoos Maulud, especially before launching new nets and new boats, will cause their fishing equipment to sink in the sea. They are thus deprived of protection from the violence of the waves and tides. The performance of Mangoos Maulud is an insurance against uncertainties and anxieties" (Mathur, 1978). Apart from the Mongoos Maulud performed before the launching of new nets and new boats, the Mappila fisherfolk observe a number of other ceremonies in this connection. When the building of a boat is complete the Mappila fishermen perform a *kuttipuja* (ceremony performed by the carpenter on finishing a house or a boat).

Khattam Otikkal

According to Mathur (1978), "Pattu and Khattam Otikkal are the reading and recitation of the full text of the Holy Quran by one or more Maulavis in honour of (1) the Prophet (2) Mohiuddin Shaik (3) Rafai Shaik (4) Sahul Hameed, (5) Khawaja Moinuddin Chist, (6) Fatima and Khadeeja, (7) Badriyangal (8) Hamsa, (9) Companions of the Prophet Abu Baker, Ummar, Usman and Ali (10) Mambaram Tannal, and (11) Predecessors of the Prophet Iyyas Nabi, Sulaiman Nabi, Issa Nabi, Yunnis Nabi, etc., for getting bumper catches, prevention of capsizing of boats, curing of illness caused by sorcery and spirits, getting ownership in fishing units, restoration of the irregular menstrual periods of fisherwomen, easy and natural delivery, etc. The contemporaries, associates, relatives, descendants and predecessors of the Prophet in whose honour Mauluds are performed are also propitiated with Khattam Otikkal. One to ten Maulavis are generally engaged for performing this ceremony. Arrangements depend upon the nature and the intensity of the vow that the Mappila fisherfolk had taken."

Pattu and Khattam Otikkal are performed on several occasions in honour of the Prophet for a number of objects such as getting bumper catches, curing illness caused by sorcery and spirits, becoming owners of fishing units, etc. The Mappila also conduct the Pattu and Khattam Otikkal in commemoration of the Prophet's ascension to meet God on the 27th day of the lunar month of Radjeeb and receive the first collection of words of Quran and commandments for observing the Salat and Fast.

Some of our informants averred that the Pattu was not generally conducted in honour of Iyyas Nabi, the guardian of the sea. But the ritual of the Khattam Otikkal in his honour is performed by every Mappila family in the month of July, when there is no fishing. The minimum number of Maulavis required for the purpose is ten. They start the Khattam Otikkal on the first Thursday night of the month of July and complete the reading on the night of the last Thursday of the month. The reading of the Quran is concluded with a prayer and feast. In the prayer the Maulavi would inform God that the host is holding the ceremony in honour of Iyyas Nabi for getting protection from violent waves and the hazardous sea.

Bait

The Bait refers to the chanting of poems in honour of the great Muslim saints by singing elegies. The Mappila fishermen of Tanur conduct even today the Badari Bait for getting bumper catches and for curing diseases.

Badari Bait

Writing on Badari *Bait* Mathur (1978) observed: "the Badari Bait is performed in honour of 313 Badariyangal (the participants in the battle of Badar) for forwarding of the attack of goblins and spirits, curing illness and for getting bumper catches. The minimum number of Maulavis employed for the performance of the Badari Bait is four. One of them chants the names of Badariyangal whereas his colleagues repeat the same. There is a strict code of precedence. The chief Maulavi first chants the names of the Prophet and his four companions, namely, Abu Baker, Usman, Ummar and Ali. This is followed by the recitation of the names of the next five close associates. In other words, the names of Rasul and his associates are chanted first for three times, but the rest of the names of Badariyangal are chanted only once. When the names of Rasul and his four associates are chanted the chief Maulavi informs them and Allah the reason for celebrating the *Bait*. The whole chanting ceremony lasts for about three hours. The Badari *Bait* is generally conducted between 7 and 10 p.m. on a Thursday night at the fisherman's house. Incense is burnt profusely for the ceremony. The neighbours, kin and friends are given a feast the menu of which includes ghee- rice, beef, dal and pappad."

Ratheeb

Ratheeb refers to a special ritual exclusively performed in honour of Mohiudddin Shaik and Rafai Shaik for favours received by the Mappila fishermen.

Mohiuddin Ratheeb

The Mohiuddin *Ratheeb* is generally conducted at the village mosque on the 11th day of every lunar month at midnight for getting bumper catches, getting rid of incurable diseases, obtaining the release of kins from prisons and for getting ownership of fishing units. Eleven Maulavis are employed for the performance of the ceremony, of whom one must be the village Qazi. Before commencing the recitation of *Baits, Dikkar, Salats*, the *Fateha* and a prayer are said. The Qazi informs God the intention of the fishermen behind the celebration of the *Ratheeb*. After offering prayers to Allah, the Qazi proclaims that the Ratheeb is conducted in honour of Mohiuddin Shaik, who has helped in fulfilling certain wishes. The rest of the Maulavis and fishermen assembled for the Ratheeb would say, Amin, Amin. The actual Ratheeb commences with the initial invocation, followed by the recitation of prayers and *Dikkars* (Confession of faith) which Mohiuddin Shaik himself recited in honour of Allah. He also recites some poems which deal with the miraculous deeds of the Shaik. It is interesting to point out that the seven *Dikkars* recited by Mohiuddin in honour of Allah are repeated by each of the 10 Maulavis 10 times. When the recitation of the *Dikkars* is concluded, rose-water is sprinkled over the assembled people. *Fateha* and three *Suras* are thereafter recited. Again the prayer is said.

Rafai Ratheeb

"The Rafai Ratheeb is performed generally by very rich fishermen once a year in order to prevent capsizing of fishing boats, curing incurable diseases like leprosy, insanity and cancer, besides getting rid of illness caused by sorcery and witchcraft. A large number of Maulavis, not less than 50 in number, are employed for the performance of the *Rafai Ratheeb*. These Maulavis are headed by a Chief, called *Talayanikkaran*. The striking feature of the ceremony is the use of four kinds of lethal weapons by the Maulavis. The weapons used on the occasion are (1) Kadhari (dagger), (2) Kavilotti (weapons used for piercing the cheek), (3) Katti (knife), and (4) Damboos (a dagger to stab on the head). The drum locally called Tambur (a five stringed guitar) is also used. The weapons and drum are used during the recitation of *Baits* in honour of Rafai Shaik" (Mathur, 1978).

Healing, Sorcery and Magic

We shall describe here the practices of healers, sorcerers and ritual specialists.

The Mantravadi: Healer, Sorcerer and Ritual Specialist

"The Qadiyar (the chief *Imam*) plays a pivotal role in the religious life of the Mappila fisherfolk of Tanur; but the *Mantravadi* (magician) takes his place in respect of extra religious practices dealing with spirits and magical practices. The Mappila fisherfolk often proclaim that there is no God other than Allah worthy of being worshipped. Allah is considered omnipotent and is the ultimate cause of any event. The demons whom the *Mantravadi* controls are of great socio-logical interest to us. There are *Mantravadis* belonging to the Mappila fishing community and the Hindus at Tanur. Religion is immaterial as far as a *Mantravadi* is concerned. *Mantravadis*, irrespective of their religious and caste affiliations, are invited to diagnose and cure diseases caused by supernatural agencies. *Mantravadis* of Tanur are not full-time practitioners of sorcery and magic" (Mathur, 1978).

The Qadiyar and the two Qazis are the accredited sorcerers and magicians of the village. They are called Seharu Ceyyunnavan (sorcerers who perform the magical rites according to the Holy Quran). There are *Jinn Hadaraths* (who possess Islamic Jinns and whose healing powers are not permanent as spirits do not have permanent abode in their bodies), besides Maulavis and Tannals, both from the fishing and non-fishing Muslim communities who are reputed curers by employing spells and herbal medicines. Besides this host of the doctors of the Black Art among the Muslims, there are Hindu *Mantravadis* in the neighbouring villages.

Curing Techniques

"The fame of a *Mantravadi* is proportionate to his curative powers. The *Mantravadi's* treatment has three important stages, the first stage being the *Olivunokkal* (suitability of a particular *Mantravadi* for a patient). A Hindu astrologer or a Muslim diviner is often consulted to ascertain which *Mantravadi* is able to cure a specific disease of a particular patient. The *Mantravadi* selected in a particular case may be famed in his power of divination and astrology and might have cured more serious diseases than the one referred to him but there is no guarantee that he can cure the illness of a particular person. Sometimes the *Mantravadi* selected through *Olivunokkal* fails to cure a specific disease. In that case the patient goes on trying one *Mantravadi* after another until he is completely cured, without bothering about the cause for the failure of a particular occultist. The second stage in the *Mantravadam* treatment is the diagnosis and selection of the proper technique of treatment. The third stage witnesses the real application of the treatment. Diagnosis is generally based on any one of the three important techniques or a combination of them: viz., (1) *Olivunokkal* of *Kanakkunokkal* (predictions based on astrological charts) (2) trance/intuition through meditation and (3) analysis of the symptoms of the disease. The *Mantravadi* in his trance calls the name of the patient and his/her mother's name. In his trance he orders the spirits that caused the illness to vacate the body of the patient and commissions more powerful spirits to overpower the lesser spirit. Thus, he may call upon spirits including those of his departed ancestors (who are *Mantravadis*) to direct him further in his search for the cause of the illness and its cure. He sometimes enquires of the spirit concerned the reason for its possession. If the spirit is identified, the *Mantravadi* enters into a dialogue with it. Spirits, in most cases speak through the medium of the patient; sometimes it does so through the mouth of the *Mantravadi*. The *Mantravadi*, after ascertaining the reason for the possession of the spirit, tries to exorcise it by reading mystic formulas or passages from the Holy Quran, placing a jar of water or a string as the media. The water is given to the patient to drink. The string is tied around the patient's wrist. The *Mantravadi* allows sufficient time for the spirit to abdicate the patient. The spirit is generally asked to leave the patient on or before the 7th, 14th, 21st or 41st day after the trance and the tying of the string or drinking of the water. The fixation of days for the spirit to leave the patient is called *Avutivekkal*. On the expiry of the *Avuti* (days fixed) a magical rite is performed to propitiate the spirit and also to ward it off. Thus the curing of illness require both contact between the *Mantravadi* and the spirit that caused the illness in addition to the use of herbal medicines according to the magical rites prescribed. It must be, however, pointed out that the technique of trance is a very difficult one, for it requires years of practice and concentration of mind" (Mathur, 1978).

"The diagnosis of magical afflictions is done in the same manner as the illness caused by a spirit. The magician is called in to diagnose the complaint of a person when he suffers from mysterious fevers, stomach aches, etc. When black magic is suspected, the magician puts the questions to the patient to find out the exact nature of the illness and the media by which it has been caused and the person behind it. Counter magic is also often employed in this case, Sometimes, the spirit causing the illness is sought out and destroyed. The person responsible for unleashing the spirit will be asked to bridle it up or if the disease is caused by a spirit under the direction of a *Mantravadi* it will be dealt with in the same manner as other spirits. If the magician refuses to give up his magical practices when he is detected in his nefarious trade or if he denies having knowledge of it, he faces loss of his life" (Mathur, 1978).

Possessed Curer

There are two kinds of persons who cure diseases caused by spirits: (1) *Murttisevayullavar* (people who possess malevolent spirits by inheritance) and *Jinn Hadaraths* (owners of Islamic spirits). The *Murttisevayullavar* are often consulted not only for curing illness but also for causing illness by applying sorcery and witchcraft. The *Jinn Hadaraths* are usually invited to cure illness caused by spirits and sorcery but not vise versa. The powers of the *Murttisevayullavar* are based on mastery and inheritance whereas the *Jinn Hadaraths* derive their power when they enter into their persons. There are certain families with whom malevolent spirits, Jinns, and demons are associated from time immemorial. These spirits are at the beck and call of *Mantravadis* of such families. These *Mantravadis* are therefore called *Murttisevayullavar* or *Sevamurttiyullavar*. Most of the *Sevayullavar* are Hindu males, but *Jinn Hadarths* are invariably Tannals, rarely Muslim women. The former are numerous in the area whereas the number of the latter can be counted on fingers. The power of the former is greater in efficacy and last for longer periods but the powers of the latter disappears in no time. *Jinn Hadaraths* appear to be psychologically imbalanced but the *Murttisevayullavar* or *Sevamurttiyullavar* are psychologically and economically sound.

The techniques employed for curing the illness caused by malevolent spirits like (1) Karinkutty, (2) Orramulakkaratti, (3) Akasa Gandarvan, (4) Kutticcattan, (5) Cerra, (6) Caittan (7) Latcca, Ulsa, Bhutam and Gulikan (8) Ruhan Jinn, (9) Kalladimuttam, (10) Patinettu Apasmara Murtikal, etc., are (a) Homam, (b) Ulinnuvanal (c) Takiteluttu, (d) amulets and other appropriate mantric remedies. Homam and Ulinnuvannal techniques are described below.

Writing on the curing techniques of illness Mathur (1978) has observed: "Homam is performed by making a metre deep pit on the southern corner of a patient's house lighting fire in it with the help of fire wood. This pit is called Homakundam. The patient is seated on a wooden stool facing the western side of the pit. Facing the east of the pit the *Mantravadi* inscribes *Mantrams* (incantations) on 21 *erikkinre ela* (Calotropis gigantia – Siva's leaves). The Mantram ran as follows. "*Innakulukkum Adduvum Mobeen*" (surely there is an enemy in you). When the writing of the Mantram on all 21 leaves is complete they are taken round the patient reciting the *Sura Yasin*. These leaves are waved over the head of the patient after reciting the *Sura Yasin* in full three times. After the conclusion of the recitation of the *Yasin* for each time, seven leaves are thrown separately into the pit. In other words, in the course of the recitation of the *Yasin* in full, the term '*Mobeen*' occurs seven times. When the term *Mobeen* (self evident truth or clear) occurs each time a leaf is thrown into the pit. Thus, when the *Yasin* is once fully recited seven leaves are put into the Homa pit. For throwing the 21 leaves into the pit after reciting the necessary incantations, the *Mantravadi* recites the *Yasin* thrice. The leaves are waved over the patient's head for 21 days and are consigned to the Homa pit. The pit is kept ablaze for 21 days. Apart from this, the Yasin is recited for 21 days during the period of sacrifice, both in the morning and evening. When the term "Mobeen" occurs in the course of the recitation the Mantravadi utters some mystic formulae over the patient. He drops a few ounces of water into an earthen pot, chanting the *Yasin*, and uttering incantations over it for 21 times. The patient is seated in a clean place and the water is poured over his head. The patient is bathed in this manner in the evening. This continues for 21 days. On the 21st day of the sacrifice there were elaborate rites. Three coconuts, half a litre of coconut oil, three eggs, 21 *tiris* (cotton wicks) and *Kuruthi* (water mixed with turmeric and lime) are kept ready for the performance of the rite. Each of the 21 cotton wicks is dipped in oil, lighted and taken around the patient's head three times accompanied by the recitation of the *Yasin Sura*. Each wick is waved over the head of the patient when the term *Mobeen* occurs during the recitation of the *Yasin*. When these wicks are thus waved, they were thrown into the *Kuruti*. The magician then writes the following *mantram* on each of the three eggs. *Kullohiya Ilaiha Annahastamaha Nabarum Minnal Jinni*, the meaning of which is: "You must tell which *Jinn* you are. For the sake of Allah's Holy Quran, you must leave the patient. You must take care of yourself". The eggs are then waved round the patient's head thrice. The eyes of three coconuts are then pierced and the *Yasin* chanted. When the word *Mobeen* occurs in the course of the recitation, the *Mantravadi* blows some charms into the eyes of the coconut. In fact, the *Yasin Sura* is recited in full for each of the three coconuts. When the recitation of the *Yasin Sura* and blowing over the coconuts are finished, the eyes of the coconuts are sealed and the recitation of the *Fateha* begins. When this is finished each coconut is taken around the patient's head in a clockwise direction thrice. Thus, for each coconut, the *Fateha* is recited for seven times and for each time the coconut is taken around the patient's head thrice. In other words, each coconut is taken round the patient's head for 21 times by reciting the *Fateha* for seven times. Then the coconuts are rolled forward and backward for three times in the direction of *Quibla*. All articles, except the coconuts used for the *Homam*, are thrown into the *Homakundam* and reduced to ashes. The coconuts are taken home by the *Mantravadi*. The ashes are then collected from the pit and thrown into the sea facing the west. The patient undergoes ritual bath twice daily as long as the sacrifice lasts."

"The term *Ulinnuvannal* refers to the magical rites for warding off the wrath of evil spirits and ghosts (including *Gulikan*). Articles used for the *Ulinnuvannal* are a fowl, *koti* (a five-feet long brand new cloth), one ripe coconut, one tender coconut, 12 Cadjan leaves (*takitus*), 12 eggs, 5 bananas, 101 nails (*Anis*), a buffalo head with its horn (buffalo killed on the same day), *Kuruttola Kuruti*, 101 *tiris*, one litre coconut oil, a big earthen pot, puffed rice, two measures of raw rice powder, turmeric powder, burnt husk and green powder prepared from Neem leaves (these five coloured powders are called

Pancavarnam), 12 pieces of molasses, black thread and plantain leaves" (Mathur, 1978).

"The *Mantravadi* prepares a *Kappal* or *Bali* with *Valapola* (plantain stem) in the shape of right angles. Eighty nine lighted *tiris* are kept in this *Kappal* and it is waved over the heads of the patient/members of the crew and placed on the *Kalam*. When the *Ulinnuvannal* of the *Kappal* is finished, a cock is waved over the patient/each member of the crew separately three times. The cock is also taken around the *Kalam* thrice. The *Mantravadi* then draws 101 figures on the skull of the buffalo and hammers a nail on each of them. In other words, 101 nails are pierced on 101 figures drawn on the skull. The skull is then waved over the heads of the patient/members of the crew. This skull is buried in the *Khabar* by the *Karanavan* of the fishing unit, facing its head to the north. When the skull is taken to the burial ground, the *Karanavan* is instructed by the *Mantravadi* that he should recite a particular spell so that he will not be attacked by the spirit. The *Karanavan* and party return home without looking back to the grave after the burial. The burial of the skull is called *Khabar Homam*. Sometimes, instead of a buffalo skull, a wooden image of a man is substituted for the *Khabar Homam*. The number of figures drawn on the wooden image is the same and nails in each number are pierced. This is taken over the heads of the members of the crew. The burial is done in the same manner as the buffalo skull. The burial is accompanied with the funeral ceremony. The image is covered by a *Koti* after tying knots in three places. This elaborate ritual is confined to the appeasement of *Gulikan*" (Mathur, 1978).

"*Ulinnuvannal* and *Homam* are performed by both Hindu and Mappila fishermen to ward off the wrath of *Gulikan* and *Kutticattan* or as a measure of counter-magic. Once they perform the above magical rites they believe that they would get bumper catches" (Mathur, 1978).

Sorcery, Illness and Curing Methods

Sorcery is largely in the hands of *Mantravadis*. When a Mappila fisherman wants to commit sorcery on some of his enemies, he hires a Hindu/Muslim *Mantravadi* to do the same. Most of the *Mantravadis* in Tanur denied that they ever practised sorcery, but our field experience rebuts this denial. The remedy against sorcery is sorcery by the employment of a more powerful *Mantravadi*. If one's *Mantravadi* is more capable than his enemy's he can throw the latter's magic back upon the perpetrator, and the adversary will fall a prey to the sickness which he wants to unleash. There are several kinds of sorcery, all of which have a family similarity. Some of them may cause death of the enemy. The general symptoms of the sorcery are vomiting of blood, disorders in stomach, mysterious fevers, etc.

Importance of Spirits, Magic and Sorcery in Economic Pursuits

"As fishing is the most uncertain field in economic pursuits, the Mappilas have placed utmost importance to the influence of spirits and their propitiation. They are aware of a large number of spirits, particularly, nautical spirits and water spirits, that affect the catch and therefore rely on the efficacy of magic and sorcery. Water spirits are generally dealt with directly by the Mappila fisherfolk. According to them *Ifireeth* is one of the most important water spirits that runs on the beach with a lighted torch in its mouth. It can assume the form of a tree or a bamboo pole. It drinks the blood of a fisherman if it enters into his person. A fisherman, possessed by *Ifireeth*, is too weak to go out for fishing. *Ifireeth* is rarely seen in the village since it has been electrified a decade ago. In order to prevent it from causing illness to the fishermen or causing fish to escape from the nets, sea spirits like *Ifireeth* is propitiated" (Mathur, 1978).

Magic and Bumper Catches

"The Mappila fisherfolk employ magic and sorcery in their economic pursuits. For instance, they seek the assistance of *Mantravadis* in getting bumper catches. Whenever they fail to get good catches, they approach either Hindu Mantravadis or the *Qadiyar* of the village. The *Mantravadis* chant some spells over a string and it is given to the fishermen to tie it to one end of the net. If a fishing unit fails to get a bumper catch even after tying such a charmed string, the steersman reports the matter to the *Mantravadi*. Then he prepares an amulet and gives it for fixing under the stem or prow of the boat. When they launch their boats and nets, an egg and a tender coconut over which incantations are written are thrown into the sea. In such an event, they usually get a good catch" (Mathur, 1978).

Whenever the Mappila fishermen fail to get large species like cat-fish and shark during the first week of the New Moon and Full Moon, they consult the Hindu astrologer and ascertain the reason for their failure to get any catch. If the failure to get a good catch is due to the displeasure of *Siharu*, astrologers advise the fishermen to contact the priest of the local Ganapathy temple. They are also given generally strings over which spells are chanted to tie them to their nets. Besides, *takitus*, on which *asmavu* is written, and other spells are fixed below the stern of their boat. The Mappilas say that these things cannot be done by a Mappila *Mantravadi* and they have to consult Hindu wizards.

Kutticathan and Bumper Catches

The spirit *Kutticattan* (poltergeist) is propitiated by both Hindu and Mappila fishermen of Kerala for getting bumper catches. There are *Kutticattan Sevayullavar* (the possessors of *Kutticattan*) among the Hindu and Muslim community in the area and their help is sought by the fishermen in getting bumper catches and curing illness.

Whenever there is no catch the Hindu and Mappila fishermen consult the astrologer for divining the cause. If the diviner thinks that the failure is due to the wrath of *Kutticattan*, the fishermen approach *Mantravadis* who are well known as *Kutticattan Sevayullavar*. A *Kutticattan Sevayullavan* generally chants some spells over a string and the same is given to the fishermen to tie it to their net. A special *takitu* with a spell written on it, is given to the fishermen, which they fix below the prow or stern of their boat. The Mappilas say that once the string is tied to the net and *Takitu* is fixed on the boat, the *Kutticattan* appears in the sea and drive the fish into the net from distant areas. This enables the fishermen to get bumper catches.

Mutterikkal

"*Mutterikkal* refers to a magical rite performed by the Mappila fisherfolk to ward off the evil eye. In the event of failure to get good catches, the Mappila fisherfolk consult their diviners or Hindu astrologers for finding out the cause therefore. If the failure of the catch is due to *Karinkannu* (evil eye) and *Karinakku* (black tongue) the diviner advises the fishermen to conduct the *Mutterikkal* rite. The *Mutterikkal* ceremony is generally conducted by the *Maulavis* belonging to the Mappila fishermen or the *Musliar* of the village who belong to the non-fishing section of Muslims. Coconuts are the main ingredients in the *Mutterikkal*. The number of coconuts required depends upon the number of members of the crew. If the fishermen are very poor only one coconut is used. Sometimes 21 *tiris* (cotton wicks) and *Kuruti* are used, provided the evil eye is very serious. First a coconut is cleaned and its eye is pierced with a knife. The *Maulavi* chants passages from the *Sura Yasin* and spells into the eye of the coconut, keeping them on the head of each member of the crew. The coconut is waved over the head of the members of the crew for seven times in the clockwise direction. (Each coconut is taken over each member of the crew and an equal number of coconuts is used for a ten man crew). Before waving the coconut over the heads of the members of the crew, eyes of the coconuts are closed. When the waving of the coconuts is over, the members of the crew are taken around the nets and boats of the afflicted fishing unit. The *Kuruti* is also waved over the head of the members of the crew for 21 times. Similarly, each of the 21 *tiris* is also waved over the head of the members of the crew and thrown into the *Kuruti* with the chanting of the *Mantram*. "Salamam Qulam Mirabi Rahim" (May there be peace from God the Benevolent)" (Mathur, 1978).

"When the waving of the coconuts is over, they are placed on a *vatilpadi* (door step) and the afflicted members of the crew are told to leap thrice over the coconut forward and backward. The coconut is then cut into two halves with a hatchet. The falling of the two halves of the coconut upside down is considered an ill omen. The Mappila fishermen in that case anticipate some untoward incident in their families. The *Maulavi* chants some magic formulae over dammer and it is smeared over the members of the crew as well as members of their families to prevent such misfortunes. In case both the halves of the coconut fall upside down (*malannuvilal*), it is considered auspicious. On the other hand, if one of the two halves of the coconut falls upside and the other upside down, it is believed to be 'not bad'. The *Kuruti*, coconut and *tiris* used in the ritual are thrown into the sea. The *Mutterikkal* is concluded with a small *Cirani* (feast). Sometimes the coconut used in the rite is used for the feast also. Kunhi Mohammad Maulavi told me that if a Mappila can chant the spell "Salamam Qulam Mirabi Rahim" for 1479 times, he will not be affected by Black tongue and evil eye; he will also get bumper catches" (Mathur, 1978).

Gulikans and Failure of the Catch

The Hindu as well as the Mappila fishermen believe that the wrath of the Gulikan (Gulikan, son of Saturn) causes illness to the members of crew of the fishing unit, besides infight and quarrels. Above all, if the Gulikan is not taken care of it may cause failure of good catches for the fishermen. Members of the Mappila and Hindu fishing communities believe that if the Brahma Gulikan is not propitiated, he causes failure of good catches. Similarly, if they incur the wrath of the Asura Gulikan the members of the crew would fight among themselves and behave like lunatics and drunkards. Sometimes the members of the crew fall sick. In such an event the steersman of the unit consults either a Hindu *Mantravadi* or his Muslim counterpart to ascertain the nature of the displeasure of the Gulikan and appease the enraged spirit. This will generally ward off the wrath of the Gulikan.

"The Mappilas say that they can detect themselves the wrath of the Gulikans by examining the cooked tapioca in their house. If the tubers of the tapioca in cooking are joined together (*Pulunniya Kappa*) or sticks (*Ottuka*) they take it for the wrath of the Gulikan. This portends a bad catch and some untoward incident in the family. If the failure of the catch is due to sorcery practised by an enemy, the Mappilas believe that the steamed tapioca may disappear from their kitchen for two days. The appeasement of Gulikan by Hindu and Muslim fishermen by conducting a magical rite is called *Ulinnuvannal* of which the basic ingredients have already been described earlier" (Mathur, 1978).

Katavukettal and Kutotram

"Sometimes an owner of a fishing unit goes on getting bumper catches for many days while his neighbouring fishing unit fails to get any catch at all. The latter becomes jealous and feel frustrated. He approaches a *Mantravadi* and tells him. 'My unit is not getting any catch while my neighbour's unit is getting bumper catches daily. The shoal that appears in the sea must come to my nets. His boat must capsize and the members of his crew should fight among themselves. Thus, sorcery is performed against the enemy for capsizing his boats, failure of the catch and infight among the crew. This is called *Katavukettal*" (Mathur, 1978).

"An earthen pot, *Kuruti* and *takitu* are used for the *Katavukettal*. Human figures are drawn on the *takitu* and *mantrams* written on it. The *takitu* is propitiated. The *takitu* and the *Kuruti* water are put into the pot. The mouth of the pot is covered and buried in the enemy's compound without his knowledge or in the place where his boat is anchored. A fight ensues between the members of the crew and the owner of the fishing unit on the one hand and between the members of the crew themselves on the other once this is done. The members of the crew do not go out for fishing. Sometimes they fall sick. Since the *takitu* and pot are used, the sorcery is also called *Kutotram*. Probably the term *Kutotram* is a contraction of *Kutu* (amulet) and *patram* (pot). In order to cure the illness and ward off the effects of this sorcery, the *Mantravadi* first of all finds out the pot and throws it into the sea. *Mantrams* are written on four *takitus* and buried in four corners of the compound of the household to prevent the recurrence of sorcery. The house and fishing equipment are purified by sprinkling *Kuruti*. This prevents any type of sorcery being sent against the fishermen with the help of *Kutotram* to cause the failure of catches and fight among the members of the crew" (Mathur, 1978).

Curing Techniques and Related Issues

We may note from the foregoing description that there are three main elements that constitute the curing technique: (1) the herbal medicine, (2) charm and (3) what Malinowski called "the *condition of the performer*". Just as in Trobriand, in the Mappila society too, spells and rites are essential practices, rather than medicines. The medicine plays a secondary role whereas spells and magic form the basis of the Mappila healing technique. The condition of the *Mantravadi*, the spiritual strength and miraculous powers of the magician or the performer, are the vital elements in the curing techniques. When I asked Enikutty Maulavi, 'what qualifications should a good

Mantravadi possess to become a powerful magician' he answered that "a good *Mantravadi* should have the capacity to go into trance and diagnose the malaise so that proper treatment is prescribed. He must possess superior mantric powers than that of the enemy's *mantravadis*. This power must be inherited from his ancestors or *gurus*". Sometimes, each of the three elements, i.e., spell, medicine and mantric power of the *Mantravadi*, may be employed separately. Medicines may, at times, be used with or without the knowledge of the *Mantravadi*. The patient may be cured sometimes without taking any medicine, but through magical powers by meditating his problems. There are occasions when diseases are cured without uttering spells and taking medicines. Sometimes, the Mappila fisherfolk get rid of their diseases by chanting spells by *Maulavis* who do not give any medicine or have no power of spells. "All three elements are present in most cases as interrelated and interdependent parts of a unitary curing technique in which spell and medicine are employed by the powers and wizardry of the *Mantravadis*. The Mappila fisherfolk believe that the spell of the *Mantravadi* reaches God, or the spirit concerned which is under his control because his intuition through meditation drives the spirit into its consciousness. The spirit obeys the commands of the *Mantravadi* in response to the spells when he chants them" (Mathur, 1978).

Supernatural Beliefs and Fishing Rituals among the Hindu Fisherfolk

The Hindu fishing communities are deeply religious and are steeped in rituals; they spend a large percentage of their income in the performance of rituals and festivals. Their means of livelihood is highly risky and erratic. This uncertainty has led the Hindu fishing communities to evolve an elaborate system of rituals and beliefs to offset their fear and feeling of total dependence on the elements of nature. This aspect is vividly described by Mathur in 'The Mappila Fisherfolk of Kerala' (1978) and 'The Marine Hindu Fisherfolk' (1979). In his Mappila Fisherfolk, he says: "The Mappilas have developed elaborate rituals and supernaturalism to insure the risk of uncertainty" (p. 382) "... through deep and intimate association with rituals, mythology, social structure and social sentiment, fishing is essentially a moral activity … . The superstructure of supernaturalism and rituals do not distort the inductive rational core of their ethnoscience" (p. 387).

To overcome their fear and dependence on Nature's elements, the fishermen set apart a major portion of their earnings. Religious ceremonies are not only performed in relation to their occupation of fishing, but also with regard to their health. Almost the entire activity of a fishing village is centred around its religious edifice. They spend lavishly for the celebration of their festivals which incidentally coincide with the fishing season. They also go on pilgrimage to famous shrines during the same time.

The fishing communities, in the past, were not allowed to enter the temples of caste Hindus and had to keep a certain distance. They had to deposit their offerings at a particular place, move back and call the attention of the templekeepers, who accepted them. The remnants of the divine worship (*Prasadam*) were wrapped in leaves and thrown to these devotees. Some among the fishing communities like the Bovis-Mogeyars and the Mogavirars have their own temples. A few more Hindu fishing sub-castes later on built their own temples called *Palliara.* Brahmin priests officiated in many of these temples in the past, and religious rites. These days the Hindu fishing communities participate actively in the temple festivals and celebrations, performing *thullal, urullu, kavadiattam* etc. at Sabarimala, Ponnani, Guruvayur, and other places. The Kodungallur temple opened its gates for the Hindu fishing communities during the *Bharani* festival. This was because of their historical association with this temple.

With the exception of the Mogavirars and the Nullayans, all other Hindu fishing communities have Bagavathy and Badrakali as their patron goddesses whom they claim to be their *Kuladevis.* The Mogavirars worship the goddess Malaraya Sharda Devi. It is rather possible that this is yet another name for Bagavathy since the Bovis-Mogeyars have one of their eleven temples dedicated to her at *Uchillam*, where is also located the important shrine of the Mogavirars.

The Nullayans do not attach must importance to Bagavathy but worship the deities of the Tamilians in general like Krishna, Siva,

Subramania, Saraswati etc. The Hindu fishing communities also worship the other deities of the Hindu pantheon and as such celebrate most of their festivals. Except the Bovis-Mogeyars, the Mogavirars and the Nullayans, other sub-castes celebrate *Onam* on a grand scale. Festivals like *Vishu*, and *Kumbha Bharani* which coincide with the Purolsawam of the Bovis-Mogeyars and Mogavirars, are celebrated on a lesser scale. The first of every month (Malayalam calendar) is celebrated and is called *Shankaranthi dinam*. The New Year day is *Shankaranthi*.

Supernatural Beliefs

World View

The Hindu fishermen, particularly the Mukkuvas and the Arayans, say that the world has been created by Daivam or Iswaran (God). According to our informants, "God is everywhere and in all beings. God is eternal, omniscient, shapeless and subsist everywhere". Krishna Kunhu, an informant, held that "God could be seen in human and non-human beings. God can see everything without eye and listen without ears. He is a power permeating the whole of the universe. When God created the universe, he also created all the necessary things for the existence of man". But Raghavan, another informant, said that there was no caste and community distinctions when God created the first man and woman. He held that the Hindus were the first human beings in the world.

Creation of Sun, Moon and Sea

According to Raghavan, the Sun was created by God. But he could not say as to when and how the Sun was created. But almost all fishermen whom the author interviewed held that God created the Sun, but none was able to say the circumstances leading to the creation of the Sun. When an explosion took place in the Sun, it was broken and a part of it fell on the another earth where the water sprang up and thus sea and rivers were formed. The Hindu fishermen believe that the Moon was created by God to give light to the mankind during nights. They propitiate the Moon, particularly after the New Moon. They believe that it is very auspicious to see the Moon on Fridays, Thursdays and Mondays, immediately after the

Karuthavavu (New Moon). But seeing the Moon on other days is considered inauspicious. They believe that it brings chaos and misfortune if one sees Moon on Saturdays, Wednesdays and Thursdays just after the *Karuthavavu*. To ward off such misfortunes, resulting from seeing the Moon on inauspicious days, the Arayan pluck the plants or steal the crops of others. The Arayans also predict the fluctuations in the price of commodities in South and North India by observing the position of the Moon. For instance, they hold that if the northern side of the Moon is higher than its southern part, the price of commodities is likely to shoot up. On the other hand, if the southern part of the Moon is higher than its northern side, the price of commodities in South India is likely to go up including that of fish. Thus, the Hindu fishing communities predict the fluctuations in the price of fish, particularly in South India, where they have their markets by observing the relative positions of the Moon just after the *Karuthavavu*. The Hindu fishermen believe that the Moon can be conquered by man. But this has been refuted by the Mappila fisherfolk of Tanur. They say man cannot land on the Moon and the Americans are fooling the people. On the other hand, the Hindu fishermen, particularly the Arayans and the Mukkuvas, held that the American astronauts have really landed on the Moon.

Belief in Soul

The Hindu fishing communities believe that the *Atmavu* (Soul) is eternal and immortal. The *Atmavu* or *Jivan* is part of God. Sometimes they refer *Atmavu* as ones's spirit and God as *Paramatmavu* (the Supreme Spirit). When the *Atmavu* leaves the body of a person, he dies. The soul after its release from the body of human being goes to the Upper World (*Paralokam*), the home of God. Thus, the fishermen hold that there is transmigration of souls as the *Atmavu* or *Jivan* leaves the body of one person at death only to enter into another body. To the author's question, when the soul actually enter into its abode, Raghavan said: "It enters the body while it is in the process of making; it animates the foetus in the womb and makes it grow". He philosophised the soul and compared it to a worn cloth. He quoted, in support of his views, the following passage from the Bagavad Gita. "Just as a man discards old clothes and puts on new ones, the soul discards worn-out bodies and enters into new ones". The Hindu fishermen believe that Yama Dharman, the Lord of Death, carried the soul to God when a person died. Chitra Gupta, the Accountant of Yama, submits an account of the good and bad deeds done by a man when he dies. A man takes nothing with him when he dies and leaves this world. For instance, when a child is born it folds its hands signifying its future but when a man dies, he unfolds his hand indicating that he carries nothing from this world.

Law of Karma and Dharma

The Hindu marine fisherfolk believe that it is the law of *Karma*, which determines births and deaths. They attribute to the law of *Karma* the reason for some souls being born as animals, birds and insects while others as human beings. Chitra Gupta keeps the account of good and bad deeds of every person. For each good deed the person gets a 'merit entry' and for each wicked thought or deed, he is given a 'demerit entry'. When a person dies he is taken to the Court of the Yama, Chitra Gupta reads out the records of his thoughts and deeds in his worldly life in order to present a true account. If his good deeds outweigh his sins, he is sent to heaven or reborn in the world in a better form of life. However, should the evil deeds outweigh the good ones, the soul is sent to hell or reborn in a base form or in a polluting caste like the Arayans. Krishna Kunhu, our informant from Kanhiranchera village, said that a person who led his life according to the law of God would attain *Moksha* (Salvation) and his soul enters into *Brahmavu* (Creator); he will not be reborn. Another informant said that every deed in the present life was controlled by the past action (*Karma*). The sufferings of a person and his birth in a low caste are all due to the sins committed in his previous birth. Krishna Kunhu and other informants held that good deeds were firm belief in God (Maha Vishnu, Siva) and His supernatural powers. A pious man would always remember the powers of God and would offer prayers to Him and perform good deeds in all situations. He would love all beings irrespective of their caste and creed. When such a man dies he attains immortality. For instance, they believe that Gandhi and Nehru attained immortality after their death on account of their good deeds. They loved all people of India irrespective of their caste and religious affiliations and they sacrificed everything for the sake of the Nation. This is one of the reasons why statesmen from all over the world came to India to pay their respects when they expired. From this exposition it appears that the Hindu fishing communities have definite ideas of good and bad deeds. Adultery, stealing, drinking, murder, cheating, betrayal, atheism and disbelief in His powers, disrespect to parents etc. are some of the bad deeds counted by Hindu fishermen. Thus, a man's *Atmavu* is influenced by his conduct and behaviour. The Hindu fishermen held that those who have done evil deeds would become evils. By holy deeds people become holy, by sinful ones they become sinful, an Arayan fisherman said.

The Hindu fishing communities believe that righteousness is *Dharma* and conversely all bad, wicked and wrong doing are *Adharma*. *Dharma* refers to the righteous conduct. It is also a synonym for righteousness, goodness and virtuousness. It stands for an austere, pure and spiritual life. Living, according to *Dharma*, will be rewarded, while the violation of the *Dharma* will be punished both here and hereafter. If one observes the rules of *Dharma* he will be born in his next birth in a high caste and would be rewarded with fortunes. If he does not observe the rules of *Dharma* he will be born in a low caste like the Arayan, Mukkuva and other Hindu fishermen or animals like the donkey and dog.

Heaven and Hell

The Hindu fishing communities of Kerala, particularly the Arayans and the Mukkuvas, held that heaven and hell exist in the *Paraloks* (Upper World). But this view has been contested by some young fishermen who held that heaven and hell existed here, in this world. But both our old and young informants agree on one point that good deeds lead to spiritual merit and bad deeds in the accumulation of sins (*papam*). In other words, good deeds will entitle a person to go to heaven and the soul of a bad person is sent to hell. In the case of the former, he attains *Moksha* (Salvation) and enters *Brahma* (Creator). The marine Hindu fisherfolk believe that hell is a place of fire and torture where the sinners are condemned to serve their terms. Some of the sinners are put in the burning fire in hell. On the other hand, heaven is a place in the Upper World where one gets all kinds of comforts to lead a spiritual life. Heaven is full of several kinds of flowers and scented plants. One can satisfy all his desires in heaven. Speaking on the existence of hell one young Araya fisherman held that those who did evil deeds went to hell during their lifetime itself. For instance, such sinners died generally of cancer, leprosy and mysterious kinds of sores all over the body and from which sometimes germs would issue causing bad odour. In other words, the sinners will *Puzhuthu Chakum*! The Hindu fishermen hold that if a person dies from a protracted illness that in itself is a hell. Thus, one can predict whether a person is entitled for heaven or hell by observing his life, particularly at the end of his life. There is no place in the world for a sinner. Krishna Kunhu believes that those who do not give the share of the catch to the fishing labourers are 'sinners and thieves'; they are fit to be sent to hell. Similarly, those who do not look after their parents are regarded as sinners.

In olden times marriage outside the caste was treated a sin. Nowadays, although inter-caste marriages are looked down upon by the conservative fishermen, they are not considered a sin. For instance, at Kanhiranchera village of Alleppey district, a number of inter-caste and inter-community marriages have taken place in the recent past. But to our question, "Whether a marriage between an Araya man and a Pulaya woman (Scheduled Caste) is a sinful deed or not". Raghavan confided to us that mere birth in a Brahmin family does not necessarily mean that a person is pious. One's good deeds alone make him holy. In this respect there is no difference between a fisherman and a Brahmin. Sometimes innocent persons are found suffering from incurable diseases on account of sins committed in their past life. Raghavan told this investigator that a wealthy Christian in his village of Elamkunnapuzha 'saw the hell on the earth itself before his death'. He amassed wealth worth several lakhs by cheating and exploiting the poor workers. He persuaded police to beat the innocent. He was popularly called *Janadrohi* (the enemy of the people). He died after a long protracted disease, worms coming out of his gaping wounds. Thus, he died of gangrene (*Puzhuthu Chattu*). Suffering of this nature signifies that a person has gone to hell.

Rebirth

The very fact that Hindu fishermen believe that good deeds entitle a person a free passage to heaven or rebirth in a better form of life and bad deeds relegation to the hell or rebirth as an untouchable or as an animal indicate that they have a firm faith in the doctrine of rebirth. There are several stories current among them to support their belief in rebirth. We shall make an attempt to analyse two such incidents.

About 80 years ago one Shenoi died. He belonged to a very rich family. His sons consulted the village astrologer whether their deceased father was displeased with them. The astrologer predicted that a few years after the death of Shenoi he would be reborn as an ox at Kaladi, the birth-place of Adi Sankaracharya. The astrologers directed the sons of the deceased Shenoi to Kaladi and see the oxen for themselves. When they reached Kaladi an oxen ran towards them after breaking its rope. Shenoi's sons immediately brought the oxen to their home. The animal was given all comforts and treatment as if it were their own father. The animal was provided with a cosy bed. It was bathed every day and ashes and sandal paste were applied to its forehead. Its horn and legs were always adorned with golden bangles. This treatment continued for 20 years. When the oxen died a funeral ceremony was conducted. Our informant said that his grand-father used to visit the oxen for several years.

Another interesting story is told that very long ago a king and a minister prayed to Bhadrakali for her presence before them and for granting them two boons. When the goddess appeared before them she demanded a human sacrifice as a reward for the two boons for which they failed to get any person on the spot. The king's minister told him "You can sacrifice me in honour of goddess Bhadrakali". But the king did not agree to the proposal. He suggested that the priest of the temple should be sacrificed and his son would succeed him and land assigned to him. The priest was greatly distressed at the decision of the king. The priest then struck at the image of Bhadrakali and said, "I served you with my right hand for the last 70 years, yet why do you want my blood". Bhadrakali appeared before the priest and told him that the king and his minister were goats in their previous birth and the priest sacrificed them in her honour with his right hand and gave their blood to her. It was now his turn. "I want your blood. If you do not agree, you will be killed by me. The king and his minister are in front of you". The priest obeyed the order of Bhadrakali. Accordingly, the priest was sacrificed to the goddess. Then the goddess asked the king and his minister what boon they would like to have from her. Their reply was that the life of the priest might be restored. The boon was granted immediately.

Ancestor Worship

When a death occurs in their family the marine Hindu fishermen do not go out for fishing for 10 days. They observe death pollution for 16 days. On the 16th day of death, friends, relatives and neighbours are fed and a *Pindam* (a rice-ball) is offered to the departed soul. The mortuary rites are conducted by the caste priest. He is paid for his services. This offering is popularly called *Bali*, and obsequies are performed by heirs by feeding crows. After feeding the friends, neighbours and relatives on the 16th day, they are offered *Pan* (betel leaves and nut to chew). Failure to offer a *Pan* to the invited on this day is almost tantamount to disrespect. Betel leaves and arecanuts are a must on the occasion in the performance of the mortuary rites among the Hindu fishing communities. On the night of the 16th day after death, the spirit of the dead is propitiated by the next of kin by offering meat, fried fish, which the departed liked best, betel leaves, and arecanuts. This is one of the rites to symbolise the end of the period of mourning. The mourners also grow beard for 16 days and on this day have their head and beard shaved clean.

On the *Karkkidavavu* day (the New Moon day in the month of Karmitakam – July-August) and on the New Moon day in the month of *Tulam* (September-October) *Pindas* are offered to the spirits of the departed. The marine Hindu fishermen do not go out for fishing on these two days. A day prior to the New Moon, the mourners, especially the eldest son of the departed, keeps fast and celibacy. On the New Moon days in the months of *Karkkidakam* and *Tulam*, they take their bath early in the morning, preferably in the sea, and offer prayers to the spirits of the dead. Some Hindu fishermen visit the local Siva temples on these days. In some parts of the former Travancore-Cochin state, the *Pindas* are prepared out of cooked rice (*Vellari*). The cooking is done in a new earthen pot by the man who keeps fast. The *Pinda* is carried on his head to waist deep water and dipped. While taking the dip in the sea, the mourner must wear a new cloth (*Koti*). After his ritual bath, he throws a few coins and the new cloth into the sea. At noon, mutton, fish (*Ayikkora, Avoli* and *Pala Meen*), toddy, arrack and various kinds of vegetarian dishes are offered to the departed. A room is reserved for the ancestors in every Hindu fisherman's house. The dishes are offered on a plantain leaf and the departed is propitiated in the ancestor's room (*Karanvante*

Muri). Incense, *agarbathy*, (joss-sticks) benzoin etc., are burnt. A *Puja* is conducted by the eldest member of the household. The dishes are taken out in the same plantain leaf after the propitiation and kept on the southern corner of the courtyard of the house. Crows are invited to partake of the offerings by clapping hands. This is called *Danamvekkal*. *Payaru Payasam* (pudding prepared out of jaggery and pulses) and *Chakara Payasam* (pudding made of rice and molasses) are offered to the departed on this night. On the New Moon days of *Karkkidakam* and *Tulam* a feast is the highlight of the rituals. Apart from this, the spirit of the dead is propitiated on the New Moon of every month by offering *Pindas* and *Payasam*.

Peedam Vekkal

We have already mentioned that a room, preferably the one facing the south, is reserved for the departed in every home of Hindu fishermen. Sometimes shrines are also built in honour of the dead. A *Mukkali* (tripod) is kept in the shrine or in the ancestor's room. It was found that four such tripods were placed inside a shrine at Quilandy representing four ancestors. Each of the tripod is called a *Peedam*. The picture of Lord Ayyappan and Lord Subramania adorn the walls of the shrine since 1969. When a *Peedam* is installed in honour of a person who died of smallpox, it is known as *Vannalamurthi*. The *Peedam* of Vannalamurthi is kept in the ancestor's room or in a shrine. The installation of a *Peedam* is called *Peedam Vekkal*. A wick-lamp is lighted every day. When *Peedams* are installed in honour of ancestors who had natural deaths, they are called *Gurukaranavar*.

The *Peedam Vekkal* is widely practised by the Mukkuvas of Malabar. A caste priest or a magician belonging to the Kaniyan (astrologer) or Irava conducts the ceremony. A *Madam* or shrine is generally located on the south-east of the homestead. The Mukkuvas believe that the east is sacred. The *Peedam Vekkal* ceremony is conducted with the recitation of certain *Mantrams* (incantations) and other formulae accompanied with offering of rice and flowers. The members of the fishing crew, neighbours, friends and kin are fed with a sumptuous meal to mark the installation of a *Peedam*.

At Uttaman's house in Quilandy village a *Peedam* has been installed in honour of his kin who died several years ago of smallpox.

There is a feminine oracle of *Vannalamurthi* who nowadays lives at Badagara. She visits the village on the New Moon day in the month of *Karkkidakam*. She becomes possessed by the spirit of the dead, and after her communion with the spirit, gives instructions as to how to propitiate it and other ancestors. The oracle in her frenzy tells the fortunes and causes of illness or epidemic in the family and the village. Whenever there is an attack of smallpox and cholera or an epidemic in the village, the Vannalamurthi's oracle is invited. She takes her stand behind the patient and the relatives are free to put questions to her, such as (1) whether the patient will recover from smallpox, and (2) why he is suffering from such a disease, so on and so forth. The oracle decides the day when the patient, suffering from smallpox, should take bath. She usually chooses the 14th or 21st day; if the attack is virulent, she prescribes a later day.

Kudiyiruthal

Among the Arayans of Kanhiranchera village, it was found that unhewn stones representing the ancestors were housed in the precincts of the local Bhadrakali temples belonging to them and other castes like the Iravas and Nairs. *Kudiyiruthal* generally refers to the consecration of stones representing the ancestors in the village temples. This is similar to that of *Peedam Vekkal* of the Mukkuvas of Malabar. In the *Peedam Vekkal*, the *Peedam* is used as the altar of the ancestor. The Arayans of Alleppey, after performing the mortuary rites buys a spot in the precincts of the local Bhadrakali temple for the installation of unhewn stones representing the departed. At Kanhiranchera, it is generally installed either in the Alissery Bhadrakali temple of the Iravas or in the Arayan's own temple dedicated to the same goddess. A large number of such stones representing the ancestors have been installed at these two temples spending huge amounts. Rituals are conducted in these temples in honour of the departed before the *Kudiyiruthal* (consecration ceremony). Sometimes, miniature silver representations of the dead are given to the village priest for worship from 21 to 41 days before the actual consecration ceremony takes place. Then they are offered to the deity. Apart from the rites conducted at the Bhadrakali temple and propitiation of the dead, a sacrifice called *Chakrahuti Homam* is conducted at the house of the deceased. The spirit of the departed enters the body of the eldest son or the body of the father, if the dead is his son, while the *Homam* is being performed. The oracle of the dead issues instructions to the living members in his frenzied state. The Arayans commonly talk about their *Karanavar* who ordered to do this and that; it takes time for an outsider to understand that they are referring to the oracle's orders. The oracle in his communion with the departed tells his kin at which temple the spirit should be installed and as to how it should be done. These instructions only confirm and support what has already been decided by the living members of the family. The oracle sometimes dances before the *Homam* pit and tells what species of fish the dead liked best and gives instructions to offer only those. The dance of the oracle is called *Pithru Thullal.*

When the performance of the *Homam* is over, the unhewn stones or wooden images representing the departed are consecrated on the New Moon day. Formerly a goat was sacrificed on the occasion of the *Kudiyiruthal* ceremony. It is now replaced by a grey gourd. The spirit of the dead is asked to enter the unhewn stone or wooden image by the priest or magician before the installation ceremony. This is called *Avahichu Edukkal*. It is estimated that a *Kudiviruthal* ceremony costs a fisherman a sum of at least Rs 150–200, which includes the fee of the priest of the temple and the magician who performs the *Homam*. The departed ancestor is propitiated at the temple once a month on the New Moon day by offering pudding, plantains and molasses. All their ancestors are propitiated at home by the fishermen once in a year, on which day they died, by offering their choice fish, meat and liquor. This sometimes involves borrowal of funds.

Besides the ancestor-rooms in the household (*Mandapams*) platforms (*Karanavar Taras*) are built around the Pagoda tree. The *Mandapam* is generally a one-roomed shrine facing the east or south-east. Apart from the practice of housing the spirits of the ancestors, we have found that it is the practice of the Hindu fishing communities to consecrate the spirits in the precincts of the local temples dedicated to Bhadrakali. Wooden figures representing the departed ancestors, crude human figures embossed on sheets of copper, silver, gold or even slabs of stones with figures sculptured are kept in the room reserved for ancestors (*Karanavante Muri*) or in the shrine or sometimes on the platform. It is very common among the Mukkuvas of Malabar to install *Mukkalis* in the shrines representing ancestors. Many such shrines are found in Quilandy. The Arayans have built a number of such shrines at Elamkunnapuzha. One Araya family has two such shrines in honour of two such ancestors, namely Chemban, Muthappan and Unnikka Muthappan. Rough unhewn stones have been installed at these shrines to represent the two ancestors. The ancestors are propitiated periodically and on all important festivals. When an ancestor is propitiated with offerings like liquor, meat and fish, a wick-lamp is lighted and kept for sometimes.

A shrine dedicated to the ancestors contains sculptured images, unhewn stones and sometimes tripods. Where there are no shrines, households have ancestral rooms (*Karanavante Muri*). Richer fishermen seem to build ancestral shrines. Earthen platforms around milky trees dedicated to an ancestor yields place to a shrine. Ancestor platforms are the rule among the Mukkuvas of Malabar. Unhewn stones embedded in earthen platforms around trees seem to be an evolution of a shrine. The shrines represent the pinnacile of worship. An elaborate system of propitiation of ancestors is prevalent among the Hindu fishing communities of Kerala. There is no uniform rule to the propitiation.

Belief in Spirits and Demons

We shall presently enumerate a number of spirits who cast their spells on the behaviour of both Hindu and Mappila fishermen. However, we confine ourselves here to describing the beliefs of the Hindu fishing communities connected with them.

Pretham (Ghost)

The *Pretham* (ghost) of a person who commits suicide hovers around the place where he took his life or by the side of his house. The Hindu marine fishing communities believe that when a *Pretham* enters the body of a person he begins to behave like a lunatic. When such symptoms are shown they lose no time in consulting the Kaniyan who is the professional village astrologer, to identify the *Pretham* and locate its abode. The astrologer, after identifying the ghost, directs the next of kin to contact a particular *Mantravadi* (sorcerer) for conducting a *Homam* (propitiatory rite) in order to get rid of the *Pretham*. Accordingly, the *Mantravadi* is invited to perform the

Homam. The *Mantravadi* draws a *Kalam* (cabbalistic figure) in the courtyard of the subject with coloured powders. The figures of the supposed *Pretham*, *Gandharvan* and his wife *Yakshi*, are drawn in the *Kalam*. When this is over the *Homam* is performed. The belief is that once the propitiatory rites are conducted the *Pretham* is consigned to flames. Sometimes, an amulet is prepared for wearing around the waist of the possessed.

Akasa Gandharvan

Akasa Gandharvan, an ethereal goblin, is believed to possess supernatural powers by the Mappila fishing community of Tanur. The Hindu fishermen believe that *Akasa Gandharvan* was responsible for causing dissension between Lord Krishna and Arjuna. Krishna Kunhu, our Araya informant, said that once Lord Krishna was offering prayers taking water in his right hand. The *Gandharvan* was taking his aerial ride in his chariot, which caused disturbance to the Lord's mind. The Lord's prayer was interrupted and froth from the *Gandharvan's* horse's mouth fell on his palm, thus polluting the water. When Lord Krishna opened his eyes, he found that the *Gandharvan* was riding in the air. He became infuriated and took a vow to teach the *Gandharavan* a lesson for his misbehaviour. He decided to kill the *Gandharvan* before the sunset. The *Gandharvan* was very much distressed at his impending death. He wanted to save his life. He approached Brahma, Vishnu and Siva, but they did not oblige him. Meanwhile, sage Narada appeared on the scene and advised the *Gandharvan* to seek the help of Arjuna who was spending his life incognito in the mountains together with his brothers. Narada then informed Lord Krishna that the *Gandharvan* had gone to Arjuna. The sage decided to bring about a solution. He appeared in disguise before Arjuna and told him to affect sleep when Lord Krishna killed the *Gandharvan*. He promised that he would restore the life of the *Gandharvan*. The Lord killed the *Gandharvan* before the sunset as he had decided, but his life was restored by Brahma at once. Thus, the Hindu fishermen believe that the *Gandharvan* is still at large.

Kadutha

The Arayans of Cochin have a belief in a large number of spirits like Kuttichathan, Karumkutty, Pookutty and Kalladimuthan. The spirits of the dead wizards belonging to different castes like the Nairs, Iravas and Cherumans or Pulayans are also capable of making mischief, it is believed. For instance, the Arayans of Elamkunnapuzha propitiated on all important occasions, *Kadutha*, the spirit of a dead Nair sorcerer of the village. The spirit is housed in a shrine. This Nair sorcerer is believed to be a great devotee of Lord Ayyappa, whose shrine is at Sabarimala in the High Ranges of Kerala. Thus, *Kadutha* is worshipped by the Arayans for getting bumper catches and warding off diseases. Special propitiatory rites are performed in *Kadutha*'s honour. The oracle of *Kadutha* dances before the shrine and predicts whether there will be good catches or not and about recovery from illness.

Marutha

The Arayans of Elamkunnapuzha believe that *Marutha*, the spirit of a Pulaya woman sorcerer, has supernatural powers in curing incurable diseases and preventing capsizing of boats at sea and also in getting bumper catches. Sometimes the Arayans worship her making special offers of liquor and meat ostensibly for getting ownership in fishing units. During the *Mandalam* (fast days from 1st of *Vrischikam* to the 10th of *Dhanu* in Malayalam Calendar) *Marutha* is offered beaten rice, puffed rice, molasses, plantains and camphor. It is interesting to note here that although *Marutha* is the spirit of a Pulaya woman (Scheduled Caste), she is housed in a special shrine by the Arayans and propitiated on all important occasions, besides offering special worship.

Kotha

The Arayans of Elamkunnapuzha say that *Kotha* possesses miraculous powers in curing illnesses and saving the fishermen from the hazards of the sea. *Kotha* is the spirit of a departed woman sorcerer who belonged to the Pulaya caste (agrestic serf). She is believed to be the sister of *Marutha*. The spirit is not infrequently worshipped by the Arayans by offering liquor and meat. A shrine has been built in her honour in the village for offering worship to her. Mention may be made here that during the *Mandalam* period only vegetarian dishes are offered to gods and goblins.

Chemban Muthappan, Unnikka Muthappan and Suranat Kaimal

We have elsewhere stated that the spirits of the ancestors are propitiated by the Hindu fishermen on all important festive occasions. The Arayans of Elamkunnapuzha village have installed three unhewn stones in a shrine representing three ancestors, namely, Chemban Muthappan, Unnikka Muthappan and Suranat Kaimal. Chemban Muthappan and Unnikka Muthappan belonged to the Araya caste whereas Suranat Kaimal was a Nair. These three spirits are propitiated for 41 days during the *Mandalam* days by making offer of vegetarian dishes.

Thottam Kazhikkal

The *Thottam Kazhikkal* ceremony is performed on the last three days of the *Mandalam*. Pulluvans and Velans are commissioned in the performance of the *Thottam Kazhikkal* ceremony. They draw cabbalistic figures using coloured powder, five in number. They also draw human figurines of deities like Vigneswara and Yessaswari. Pulluvans and Valans beat their drums and sing songs in praise of the deceased sorcerer. The female oracles, possessed of the spirits of Chemban Muthappan, Unnikka Muthappan and Suranat Kaimal, dance before the *Kalam* (the figures drawn on the floor) and circumambulate the shrines. The ceremony lasts for about 6 to 8 hours. The oracle proclaims whether or not the ancestors are pleased with the propitiation. They predict the future welfare of the fishermen. The Pulluvans and Velans are paid between Rs 10 and Rs 15 for their services.

Munnodi Appan

Munnodi Appan, the spirit of the dead ancestor of a family belonging to the Village of Tarayakadappuram in the Alleppey district, is propitiated by the Arayan fishermen. The spirit has been housed in a shrine and it is worshipped as a Dewan (male deity). The local tradition is that long ago the fishermen offered their prayers to different gods for getting bumper catches to meet their livelihood. Munnodi Appan, an Arayan, employed his magical powers and got bumper catches to the entire villagers. Thus, he saved the villagers from starvation. A shrine was built in his honour by the villagers when he died. Whenever the Arayans of Quilon, Alleppey and Trivandrum do not get good catches, they take a vow to offer special *Pujas* in honour of Munnodi Appan either at home or in the shrine. They are seen usually blessed with bumper catches. They make votive offerings consisting of meat, liquor, fried fish etc. when their prayers are heard. Now vegetarian dishes only are offered in the shrine of Munnodi Appan. When Munnodi Appan is propitiated at home, the species of fish, which the fishermen want to haul, are offered to him. *Avoli* and *Ayila* (mackerel) were relished by Munnodi Appan. Apart from fish and meat, liquor and *Pappad* (wafers) are offered to this spirit.

The Hindu fishermen of erstwhile Travancore State in general and those of Alleppey district in particular make periodical offerings and perform special rituals in honour of Munnodi Appan at his shrine, if they get bumper catches, or if any serious illness is cured within a given time. Once their wishes are fulfilled, they conduct the special ritual at his shrine spending as much as Rs 500. The mode of worship at the Munnodi Appan's shrine is very interesting. The office of the priest is hereditary in the family of Munnodi Appan. The womenfolk are forbidden from entering the shrine and are obliged to keep a distance of a furlong during the propitiation of Munnodi Appan. The priest is not allowed to wear any cotton clothes when he performs the worship. He has to observe vegetarianism and celibacy. The priest's dress on this occasion is made of banana leaves. This apparel is called Ada. The Arayans of Travancore generally set apart 3.5 per cent of their day's catch for the Munnodi Appan's worship. When the amount swells to Rs 500 the special ritual at the Munnodi Appan's shrine is conducted. Sometimes they fail to save as much as Rs 500 for his worship in the course of one full year. In that case, they do not hesitate to borrow funds from moneylenders to propitiate Munnodi Appan.

Parassinikadavu Muthappan

Both the Hindu and Muslim fishermen of Malabar send their votive offerings to the Parassinikadavu Muthappan. His shrine is at Parassinikadavu, a village in the district of Cannanore, and belongs to a Thiyya family. The local tradition says that once Lord Siva, under the spell of *Sani* (Saturn), went to the forest of Parassini for penance.

He attained *Moksha* during his penance. A structure was raised by the Thiyya family in His honour at Parassinikadavu, then in the forest, and later on it came to be known as Parassinikadavu Muthappan's shrine. This Muthappan is a non-Sanskritic deity. Meat, fish and liquor are offered to him. It is noteworthy that the members of almost all castes in this region propitiate the Muthappan. His devotees are generally given free lodging and boarding. Even Namboodiri Brahmins and Nairs are sending their offerings to the Muthappan in fulfilment of their wishes and vows. The Hindu fisherfolk of Malabar, especially the Mukkuvas and the Mogayans (Mukkayans), propitiate the Muthappan at home as well as in his shrine for getting bumper catches, relief from illness, protection from the hazards of the sea and for becoming owners of fishing units. His favourite dishes are a mixture of fried mullets, *Pappad* (wafers) and *Katala* (Bengal gram). The Hindu fishermen not infrequently vow to set apart 5 per cent of their daily haul for the Muthappan before launching their nets, if they are blessed with good catches. They exhort thus: "Oh Muthappan! Let our net be filled with bumper catches; we shall send our offering to you". Sometimes a fishing unit is staffed by both Hindu and Mappila fishermen. In that case the expenses in connection with the propitiation of the Muthappan is borne by the entire crew.

The Muthappan is worshipped daily; but special offerings are made once a week, once a month and once a year, when his festival is celebrated. The propitiation of the Muthappan is called the *Pazhamkuttivekkal*. Another important offering to the Muthappan is called the *Vellattam* or *Tiruveppu*. Two or three Mannans (washermen) impersonate the Muthappan for *Vellattam* and perform a dance in front of his shrine. This performance of the dance is conducted as the main *Vazhivadu* (votive offering) for getting bumper catches and rid of incurable diseases. It costs Rs 13 for conducting this ceremony. On the other hand, only one Mannan is required for *Tiruveppu*. He assumes the guise of the deity and dances before his shrine. The Mannan's fee for performing this rite is Rs 1.25. The *Vellattam* is performed at night times.

Gulikan

Gulikan or *Mandi* is propitiated by the Hindu fishermen all over Kerala. The *Gulikan* is connected with diseases and death.

Horoscopes are cast by most Hindus of Kerala in which the position of the *Gulikan* is clearly indicated. The *Gulikan's* position indicates when the person is going to die. When death takes place in a fisherman's house, it is customary to plant a stone outside the house representing the *Gulikan*. An informant confessed us that a stone representing the *Gulikan* will never be installed in the room reserved for ancestors or in any part of the house. However, the fishermen believe that once the *Gulikan* is propitiated by installing a stone on an earthen platform it will prevent deaths. The *Gulikan* is generally worshipped to get big fishes like *Etta* and *Avoli*. They vow to conduct a special *Puja* in honour of the *Gulikan* by offering *Etta* and *Avoli*. They fulfil their vow by making the votive offerings when they get such species of fish. The *Gulikan* is periodically propitiated by the offer of *Kuruthi* (water mixed with turmeric and lime), toddy, bananas and flowers. A wick-lamp is lighted when the *Gulikan* is worshipped. A red fowl and a ram are sacrificed in the *Gulikan's* honour on every important festive occasion like Onam and Vishu. Sometimes, the oracle of the *Gulikan*, in his trance, directs the fishermen what to do at any crisis of life. The oracle reveals the causes of illness and discloses the remedy.

The Mukkuvas of Malabar believe that there are four kinds of *Gulikans*: (1) Brahma Gulikan, (2) Vishnu Gulikan, (3) Abhimanyu Gulikan, and (4) Asura Gulikan. They further believe that if the Asura Gulikan is displeased tensions in the family are bound to arise, besides infight among the members of the crew. In such a case, the Mukkuvas consult the Kaniyans (village astrologers) and seek their advice to ward off the wrath of the enraged Asura Gulikan. The Kaniyans prescribe the remedy in the form of a magical rite called the *Puramneekkal*, which is similar to the *Uzhinhu Vangal*. A magician is invited to conduct this rite. The articles required for this purpose, prescribed by the magician and the astrologer, are: a cock, fried fish (*Etta*, *Ayikkora*, and *Narimeen*) coconut oil, potato, ladies' finger, Bengal gram, beaten rice, and puffed rice etc. The thigh of the cock is roasted in the fire and given to the Asura Gulikan. The total expenditure for conducting the rite comes to Rs 100.

Brahma Gulikan

The Brahma Gulikan is worshipped for getting bumper catches and also for warding off the misfortunes and illnesses caused by him.

Sometimes the Mukkuvas call the Brahma Gulikan as Brahma Yakshassu. This has been contested by a Namboodiri Brahman who held that the Brahma Yakshassu is the spirit of a dead Brahman. The Mukkuvas believe that if the Brahma Gulikan enters the body of a person he becomes a lunatic. This *Gulikan* is capable of causing separation of a husband and wife. An amulet is prepared by the magician and given to the patient to wear or a *Puramneekkal* ceremony is conducted to appease him. Sometimes the *Mantram*, 'Oh Gulika, Brahma Gulika, Vishnu Gulika, Asmara Gulika and Abhimanyu Gulika' is chanted for 21 times over a string for putting on the waist of the patient. The same *Mantram* is written 21 times on a copper plate and an amulet is prepared for wearing. Sometimes rice powder mixed with molasses is given to the Brahma Gulikan in order to appease his wrath. The Mukkuvas say that the Vishnu Gulikan and the Abhimanyu Gulikan are not wicked. The Abhimanyu Gulikan is the spirit of Abhimanyu, son of Arjuna, who died in the Kurukshetra war.

Pottanmar

The Mukkuvas of Malabar believe that *Pottanmar* are the bodyguards of the goddess Bhadrakali. They are propitiated on festive occasions like *Mandalam – Karkitika Shankaranthi* and *Shivarathri*. The *Pottanmar* are offered vegetarian dishes consisting of balls of rice, jaggery and coconut. The ordained oracle of the *Pottanmar* speaks a *Mukabhasha* (gesture language) in their frenzy and declare what offerings would please the goddess. They predict the future events that affect the fishermen and divine the causes of illness in their families.

Kandakarnan

The Mukkuvas of Malabar hold that *Kandakarnan* is the defence minister of Bhadrakali. They say the deity does not like meat and liquor. It seems he has been elevated into a Sanskritic deity. The main offerings of *Kandakarnan* consist of boiled rice, jaggery and coconut. He is worshipped for getting bumper catches, rid of illnesses particularly smallpox, and for protection from the hazards of the sea. Mention may be made here that the Iravas of Palghat propitiate this deity by sacrificing a fowl.

Serpent Worship

There was no Hindu family in Kerala in olden times without having its own serpent-grove. A platform was raised where unhewn stones were kept to represent the serpent gods. Vegetation was permitted to grow around the platform. Women in their periods were prohibited to approach the grove. When they violated this code, expiatory rites were performed to appease the serpent-god.

Hindus all over Kerala, and the Mukkuvas of Malabar in particular, propitiate the serpent-god on certain occasions. The Hindu fishermen like other Hindus are reluctant to kill cobras found either in shrines or serpent-groves. Serpent-groves are found even today in Malabar in almost all households of Nairs and Namboodiris. Among the Mukkuvas, it has been observed that two to four serpent-groves are clustered together. These groves are located in the south-western corner of the compound of their houses. There is a Nagayakshi temple at the Blangad (Manathala) village of Trissur district. Propitiation of the cobra-deity in this temple is done on festive occasions. The Mukkuvas of the village have appointed a Namboodiri priest to conduct *Pujas* in this temple. He worships the serpent-god by reciting certain Sanskrit *mantras*. This shows that the serpent-worship among the marine fisherfolk has the same orientation as among the high castes like Namboodiris and Nairs in Malabar. It may be mentioned here that the cobra, in Kerala, is not connected with the deity Subramania, "whereas everywhere in South India cobras are identified with Subramania, the son of Siva" (Srinivas, 1952: 168).

This does not mean that the fishing community does not worship Subramania. They worship this god when the constellation of *Makara* appears. They do not go out for fishing on the *Makara Puyam* day. Most Mukkuvas visit Palani, where the presiding deity is Subramania. They do not have their own temple dedicated to Subramania at the village level. However, they do not forget to send their offerings to temples dedicated to Subramania, which are owned by Brahmins, Nairs and Iravas.

The Hindu fishing communities all over Kerala send their periodical offerings to the temple at Mannarasala at Haripad (Alleppey district) dedicated to the serpent-god. This temple is the most famous serpent-worship centre in Kerala. The Mannarasala temple is located two furlongs to the north of the Subramania temple at Haripad. It is surrounded by very beautiful groves. The temple is managed by a local Namboodiri *illam* (Namboodiri house). The owner of the temple is the eldest female member of that *illam*. She lives in the *Nalakathu* (inner circle) of the *illam* and utmost care is taken not to defile her. According to the local tradition a Namboodiri lady was given in marriage from Vettikottu, another important centre of serpent-worship, to Mannarsala. She gave birth to a snake-boy. The serpent grew up and gave birth to a number of adders. They were all removed to a place, on the south of the Namboodiri *illam* where a beautiful grove sprang up in due course. On the star of *Ayilyam*, in the month of *Tulam* (October-November), a large number of Hindu fishermen offer their worship to the cobra deity of this temple. The Hindu fishermen often take vows to send offerings to Mannarsala temple, if they are cured of incurable diseases like leprosy, leukaemia etc.

We have already mentioned that the Mukkuvas of Blangad (Manathala) village have their own temple dedicated to the serpent-goddess, which is locally called Nagayakshi *Kshetram*. They conduct the Sivarathri festival at this temple in the month of *Kumbam* (February-March). Every household contributes a sum of Rs 20 towards the festival. Figures of cobras are drawn by the Pulluvans on this day. They beat their drums and sing songs in praise of the serpent-goddess on the occasion. The oracles of the cobra-gods are generally women. They give oracular audition to the fishermen about the remedies for incurable diseases and their welfare. The ceremony is called *Kalam Kazhikkal*. This is performed for the whole night. Until 1969 a Mukkuva priest conducted the *Puja* in this temple, but since 1970 a Brahmin is officiating. Only vegetarian dishes like milk, fruits, coconuts etc., are offered to the serpent-god. Special votive offerings are made during festival days. *Pujas* are performed in honour of the Nagayakshi for curing illness and getting protection from the hazards of the sea.

The Mappila fisherfolk of Tanur do not lag behind their Hindu cousins in cobra-worship. Four serpent-groves are found among the Mappilas. The worship is conducted by a Namboodiri priest in all these serpent-groves. One of the fishermen (Edakapuram Kunhimon) has set apart a room for the worship of the serpent-god regularly. Whenever the Mappila fishermen see cobras in their homesteads or compounds they vow to send offerings of coconut oil for lamps, milk and bananas to a temple dedicated to the serpent-god. They believe that leprosy and leukaemia are caused by the wrath of serpent-god. Kunhimon's relatives consulted the local Hindu astrologer and he told them that it was due to the negligence of the serpent-god. He advised them to take the patient to Mannarsala. Kunhimon told us that his mother was taken thrice to Mannarasala but to no purpose. A separate room now has been set apart for the god and propitiation is being conducted weekly. Sometimes, Kunhimon's family members pray the serpent-god that their enemies are punished for their misdeeds and betrayal. He prayed the serpent-god to cause illness to the cheats. The fisherman became mad and his relatives apologised to Kunhimon's father. They also sent offerings to the serpent-god. The patient became all right in course of time.

It will not be out of place to mention here that the Pazhayakath Muslim family (non-fishing community) has its own serpent-grove. The author's informants asserted this family was originally Namboodiris. All members of this house were converted during the invasion of Kerala by Tipu, the Sultan of Mysore. The family has retained its serpent-worship. A Namboodiri priest is invited once a year to propitiate the serpent-god by offering milk, plantains and flowers.

Rituals Connected with Temple Festivals

There is a considerable body of rituals directly connected with the festivals of village deities. Of all village deities, the goddess Bhadrakali is considered the caste deity of the Hindu fishermen of Kerala. Besides their village deities, the Hindu fishermen do not omit to send votive offerings to the most reputed temple dedicated to her in Kerala at Cranganore. Bhadrakali is considered to be the patron of all the Hindu fishermen of Kerala. The festivals of the goddess Bhadrakali both at the village and State level are the chief source of their ritual idiom. Most of the villages inhabited by the Hindu fishermen have their temples dedicated to Bhadrakali. All other deities in a fishing village are subordinate to goddess Bhadrakali. The goddess Bhadrakali is represented by either a stone image or unhewn stones throughout Kerala. These are either housed in shrines or temples. Daily worship is performed either by a Hindu fisherman priest or a Kaniyan (astrologer) in these temples. There is a tendency among the Hindu fishermen to appoint a Brahmin priest in their Bhadrakali temples. Such a priest has been recently appointed to perform the ritual at the Bhadrakali temple of Kanhiranchera belonging to the Arayans. There is daily worship at this temple. The daily worship is offered only in some Bhadrakali temples, but in others once a week or once a year, when the festival of the goddess comes. Such festivals in the past lasted from 3 to 11 days with the sacrifice of fowls and goats. Goats and fowls have been done away with these days in temples and hence the festival is concluded with *Arat* (ceremonial bathing of the idol on the last day of the festival). The idol is taken in a ceremonial procession to the nearest tank, river or sea, with the accompaniment of instrumental music. The image is washed after offering oblations in the presence of a large number of devotees.

Every fishing village individually or two or three villages conjointly celebrate the annual festival of the goddess Bhadrakali in the month of *Kumbam* (February-March) and *Meenam* (March-April). These festivals have a similar structure all over Kerala and a short description of any one of them is sufficient to get an idea of the ritual idiom of the Hindu fishing communities. This, however, does not mean that there is no difference at all between the festivals observed in one village and another. A comprehensive view about the rituals connected with the festival may be obtained by examining the following two festivals, one at the village level, and the other at the State level.

Festival at Kanhiranchera

The festival of the Jagriam Bhadrakali temple at Kanhiranchera in the district of Alleppey, described below, may be considered typical of the Arayans. The festival at the village lasts for 11 days, involving a very elaborate and complicated ritual and imply the cooperation of several castes and not infrequently of two or three villages. There are three Bhadrakali temples in this village, the festivals of which are held on different dates. The Nairs celebrate the festival of Bhadrakali, Subramania and other deities. Thus, there is complete cooperation among the various castes in the celebration of the festivals of the different deities. This is evident from the celebration of the festival of Bhadrakali temple of the Hindu fisherman. The festival commences in the month of *Kumbham* (February–March) and lasts for 11 days. The *Kodiyettam* (hoisting of the temple banner) is the first act of the festival. The members of the Araya community of the village and nearby places gather before the temple on this day and the Namboodiri priest offers worship to the deity. The oracle of the goddess becomes possessed on the occasion and in his frenzy gives instructions to the fishermen as to the manner in which the festival is to be conducted. In large numbers Hindu fishermen flock to the temple from far and wide to make votive offerings to the goddess from the day the flag is hoisted. Special *Pujas* are conducted on the second day. On the third, fourth, fifth, sixth, seventh, eighth, ninth and tenth day *Kathakali, Harikatha, Ottamthullal, Panchavadyam, Nadaswara Cutchery* and religious discources are conducted. The performance of the *Kathakali, Harikatha, Ottamthullal,* etc., require the whole-hearted cooperation of several castes, not only of the village but also those of the neighbouring ones. The important castes that participate in these performances are the Namboodiris and *Ambalavasis* (castes attached to the temple) like the Nambiar in Travancore-Cochin, Pisharodi, Nambisan, Marar and Kurup. They perform specific duties in the temple. On the last day, the deity is taken on a caparisoned elephant, accompanied by *Panchavadyam* and *Nadaswaram* for its ceremonial ablution in a neighbouring river, sea or tank. The deity is taken back in procession to the temple. The oracle of the goddess, ordained or unordained as the case may be, follows the procession when the deity is taken for the ceremonial bath. The *Ezhunellippu* (carrying the deity in procession) is on a caparisoned elephant. The number of elephants may be always in odds: 1, 3, 5, 7, 9, 11. Their number depends upon the funds at the command of the temple or the number owned by it. *Panchavadyam* is conducted during the whole night. The festival comes to an end with the *Thalapoli* (the offering made to the goddess Bhadrakali, a ceremony in which unmarried girls circumambulate the temple with *Thalams* or plates in their hand which contain lighted wick-lamps, rice, flowers etc.), and fireworks. The 11th day of the festival is the most important one. Visitors, particularly fishermen, come to the temple from far off villages with their votive offerings to participate in the festival. Sometimes the Bhadrakali *Pattu* (songs in praise of Bhadrakali) is conducted by the Mannans (washermen) during the festival days from three to seven days depending upon the nature of the *Vazhivadu* (vow). The ordained oracle of the goddess plays the most significant role in the festival on the concluding day. It is he who gives directions to conduct the festival at every stage and these are obeyed to the minutest details as he is supposed to be possessed by the goddess. It is interesting to point out here that while the Namboodiri Brahmin is engaged in the worship of the Bhadrakali inside, reciting Sanskritic *Mantrams*, the oracle of the goddess stand in front of the temple and answers the questions put to him by the fishermen. Sometimes the fishermen ask questions about their catches and family affairs. A fisherman may ask whether his fishing unit will get a catch or his wife will recover from her illness. Another may ask why his fishing unit's net was lost in the sea or his boat was capsized. A third man may ask whether he should become a fishtrader or start some other business. The oracle is identified with the deity when he is possessed. Even the Brahmin priest forgets for a moment that the oracle is a fisherman. At the end of the festival, when the oracle regains his normalcy (that is in the early morning of the 12th day) the priest shuts the door of the temple and goes home. The temple remains closed for the next six days. On the seventh day, when the doors are again opened, a large number of devotees visit the temple with their votive offerings.

The expenses connected with the festivals are met by raising contributions from the fishermen of the village. Each fishing unit sets apart 2 per cent of its catch daily. This amount is accumulated. Besides this they make contributions of Rs 5 per head for the festival. The Araya Jana Sabha (a caste association) collects subscriptions from other castes of the village as well. The total expenditure for the festival comes to about Rs 5,000.

Bharani at Cranganore

Before describing the rituals connected with the *Bharani* festival, we shall examine the importance of Cranganore (Kodungallur) and its origin. Cranganore, the ancient port capital of Kerala, lies 25 kms south-west of Thrissur. It was here that Hinduism, Judaism, Christianity and Islam flourished side by side till A.D. 1500, under royal patronage. It will be of interest to mention here that the first missionaries of Christianity and Islam set their foot in India at Cranganore.

The Hindu fishermen of Kerala, like other castes, believe that the goddess Bhadrakali was born from the third eye of Lord Siva. They also believe that the temple dedicated to goddess Bhadrakali at Cranganore is the most famous all over Kerala and was consecrated by Parasu Rama, the mythical sage who created Kerala. There is a myth associated with the rituals followed in the celebration of the festival in the month of *Meenam* (March-April). This is to the effect that the festival is celebrated in commemoration of the victory of the goddess Bhadrakali in killing the demon Darika. The story goes that when Parasu Rama, after taking revenge upon the Kshatriyas for the murder of his parents, redeemed Kerala by throwing his battle-axe to the Arabian sea and settled Brahmins in sixty four villages. Unfortunately, smallpox broke out and many died of it. This caused sorrow to the sage. He then took penance and prayed Parvati, consort of Lord Siva. The latter appeared before him in the form she had assumed when she killed the demon Darika. She told him: "If you consecrate me here and make offerings of toddy, fowl, and goat, I shall banish smallpox from this place". When Parasu Rama expressed his inability to propitiate the goddess with liquor, she told him that meat could be offered instead to her for seven days prior to the asterism of *Bharani*. She also said that if she were given these offerings the same would fetch as much benefit as one would get from her eternal worship. She then disappeared. Thus, the popular belief is that it is Parasu Rama who introduced the celebration of the *Bharani* festival.

The festival begins with the hoisting of the banner of the temple on the *Bharani* day (the second constellation) in the month of *Kumbam* the day on which the Bhadrakali is believed to have killed the demon Darika. Members of various castes including Hindu fishing communities flock to the temple from all parts of Kerala, ten days prior to the festival, accompanied by a number of oracles, either men or women, of the goddess. The fishermen observe fast and avoid sexual communion during the period of the festival and come to the temple in a group of 25–50 with other devotees who are members of the armed force of the deity. Fowls and goats were sacrificed in large numbers prior to 10 days of the festival. Such killings are no longer practised. The very day the banner is hoisted on the southern entrance of the temple signified the actual commencement of the fight between the goddess and demon. On the Aswati day (1st constellation day) the ornaments of the goddess are removed by the chief priest (*Adikals*) and sandal paste is smeared in order to enhance the beauty of the image. A special *Puja* is conducted. *Darshan* (ceremonial look) of the deity after the performance of the special *Puja* is regarded as rewarding. The *Pulaya* chieftain climbs upon the *Althara* (platform around the Banyan tree) and opens his silken umbrella to mark the *Kavutheendal* (polluting the temple) by the agrestic serfs. The right of Kavutheendal is vested in a Pulaya family known locally as Palakkal Velan's family. When the local ruler unfurls his umbrella the Palakkal Velan accompanied by his party and oracles enter the temple through the northern gate in procession and circumambulate the sanctum. Formerly they used to carry deadly weapons in order to scare away the demons. Nowadays they carry only sticks, symbolic weapons.

In the early morning of the *Bharani* day the chief priest worships the goddess by offering rice-pudding. Then the goddess is removed ceremoniously to a raised platform for public worship. At the same time, the temple-flag is hoisted at the eastern gate on the flag-staff, signifying the victory of Bhadrakali over the demon Darika. The temple remains closed to the public for six days following the *Bharani* day. However, the deity is worshipped once a day. On the seventh day when the temple is reopened, countless devotees throng to have *darshan* of the deity, for they believe that a glimpse stolen will be sufficient to get the blessings of the goddess. On the same day all the flags hoisted are lowered to mark the end of the festival.

Rituals and Technology

Nets

The traditional cotton-nets are manufactured by the Hindu fisherman. The Kaniyans (astrologers) are consulted for finding out the auspicious date and time for starting the work. The Hindu fishermen, particularly the Arayans and the Mukkuvas, choose Monday and Friday for commencing the making of nets. They avoid *Rahukalam* (inauspicious time) for starting this work. Women in their monthly periods are not allowed to make the nets. They are not even allowed to touch the yarn meant for making the nets. Nylon nets brought from the market do not command such taboos as these are manufactured by machines and therefore there is every possibility of their being polluted by women who work in such factories. The making of nets among the Hindu fishermen is called the *Valakettal* (entwining nets).

Rituals and Valakorkkal

When piece-nets are ready they are joined together by the members of the fishing crew on an auspicious day, preferably either on Friday or Monday, avoiding the *Rahukalam*. The joining of piece-nets together is called the *Valakorkkal*. The Karanavan of the fishing unit or the most elderly fisherman of the village is invited to preside over the *Valakorkkal* ceremony. He takes a handful of yarn and asks the crew, "Shall I start the joining of the piece-nets"? "You may do so by blessing us", the members of the crew would say. After the formal inauguration, the Karanavan hands over the yarn to a member of the crew with his blessings for getting bumper catches and protection from all hazards at sea. The Karanavan receives a sum of Rs 5 for his services. The sunrise is considered the proper time for *Valakorkkal*.

Launching of a New net

The launching of a new net (*Valananakkal*) is performed either on Friday or Monday. The Hindu fishermen conduct certain ceremonies before launching the new nets. They perform special *Pujas* in the local village temple in honour of the goddess Bhadrakali and Lord Subramania. On the launching of a new net, two flower garlands made of *techipuvu* are sent to the temple of Subramania at Elamkunnapuzha, near Fort Cochin. They also make an offer of *Palpayasam* (three sweet dishes). Similarly, special *Pujas* are offered on this day in honour of their village deity, Bhadrakali, by offering flowers and rice-pudding. On the day of the launching of a new net, it is taken in a boat just opposite the Subramania temple of the Elamkunnapuzha village, and launched from there facing east. The flower garlands that adorned the deity are tied to the prow and stern of the boat before the ceremonial launching. A coconut (which has been offered to Lord Subramania) is kept in the cod-end of the net before launching it. Above all, a bunch of plantain is tied to the prow of the boat. After the ceremonial launching, the fishermen paddle their boat to their fishing grounds where the plantains are eaten by the members of the crew. When the net is hauled sometime in the afternoon the coconut is removed from its cod-end and cut into two halves without draining its water. The water is kept in the two halves of the coconut for the purpose of the divination. A flower is plucked from one of the garlands and put into one of the halves of the coconut. If the flower happens to touch the coconut kernel, the Hindu fisherfolk believe that it is auspicious and they are sure to get bumper catches. On the other hand, if the flower sticks to the centre of the coconut without touching the kernel, it is considered most inauspicious. They attribute this to the wrath of their ancestors and village deities. Vows are immediately taken to make appropriate offerings to them. The blood of a fish (from the day's catch) is squeezed out and it is smeared on the net. The Hindu fishermen believe that the net drinks the blood of the fish. At night, the members of the crew, their friends, neighbours and kins are fed with a feast and liquor. The relatives and friends make contributions varying from Rs 5 to Rs 25 to the owner of the net.

The departed ancestors are usually propitiated on the night of the day of launching of a new net by making offerings of toddy, fried fish, mutton and arrack. It is compulsory that the fish caught on this day must be offered to the departed. Again, the fish the departed liked best must be there.

The Hindu fishermen send images of fish made of silver and gold, particularly those of mackerel and sardines, to important Sanskritic deities like Guruvayur Krishna, Sabarimala Ayyappan, Chottanikkara Bhagavathi, Kodugallur Bhagavathi and Palani Subramania on the day of launching a new net for getting protection from all hazards of the sea and also for bumper catches. It is also important to note here that the Arayans of Cochin and Travancore arrange for saying a special mass in honour of Virgin Mary at Vallarpadam Church in Ernakulam, on the day new nets are launched. The Mukkuvas in Malabar send, while launching new nets, white or silk clothes for covering the tombs of Mambarm Tangal and Quilandy Seethi Tangal, the descendants of Fatima.

Building of Boats

The Hindu fishing communities of Kerala believe that Thursday and Monday are the most auspicious days for the commencement of building fishing boats.

Ulivekkal

The carpenter first starts his work on a trunk of a tree meant for building the boat either on a Monday or a Thursday. He generally avoids the *Rahukalam* for commencing the work. The god Ganapathy is propitiated by the head carpenter before the work is commenced by making an offering of betel leaves, beaten rice, puffed rice, flattened rice, molasses, plantains, coconut and flowers. After propitiating the deity the carpenter would ask the members of the crew, "Shall I start the work?" "Yes, you may", the owner of the boat and other assembled fishermen reply. Then the carpenter begins the work. But he works only for an hour for the day. He receives from Rs 5 to 10 as *Dakshina* (fees) on this day for this service.

Vanchinanakkal

It takes 20 days at least for five carpenters to finish a boat. Each carpenter is paid wages varying from Rs 7 to 12 besides lunch and tea. The actual cost of a boat used to vary from 2,000 to Rs 3,000 including the wages of carpenters. When the boat is ready for launching (*Vanchinanakkal*) an auspicious time and day is fixed by an astrologer. The new boats are launched either on Sundays or Mondays.

Kuttipuja

Just before the launching of a new boat or on the night previous to it, a ritual called *Kuttipuja* (ceremony performed by the carpenter on finishing a house or boat) is performed in the boat. The head carpenter officiates as the priest for the ceremony. The articles required for the ritual are toddy, cock, white clothes, four coconuts, rice, puffed rice and rice-powder. The clothes offered to the deity are presented to the carpenter. The expenses are met by the owner of the boat regardless of his religious affiliation. The fowl, after its decapitation in honour of goddess Bhadrakali, is given to the carpenter. The assistants of the head carpenter are presented with a pair of new cloth and shirt. Besides, they get the *Koomecha paisa* (coins thrown on the *Koomecham*, part of the boat where the sail is attached by the kins and friends) which varies from Rs 100 to Rs 200. The head carpenter gets Rs 25 as an honorarium for his services. Formerly the head carpenter received a silk cloth (Pattu) and a golden bangle (*Vala*) as his reward. The kins, friends and well-wishers make liberal contributions, from Rs 5 to Rs 25, to the owner, the quantum being determined by their economic status and kinship ties. Sometimes the owner of a new boat gets from Rs 500 to 1000 by way of donations. He makes use of this amount for arranging a feast as well as for repaying his debt. The head carpenter goes to the waterfront and hands over the rudder to the steersman on the day of launching. In return, he receives a payment Rs 5 to 10 from the owner of the boat.

The Hindu fisherfolk, particularly the Arayans of Elamkunnapuzha village, Ernakulam district, and Kanhiranchera village of the Alleppey district, propitiate deities like Bhadrakali and Lord Subramania on the day of launching of a new boat. They also send their offerings on this day to the Subramania temple belonging to the Nairs. On the other hand, special *Pujas* are conducted in their own Bhadrakali temples. Kaniyans were the priests in Bhadrakali temples belonging to the Hindu fishermen. They were discarded sometime ago and the fishermen appointed their own castemen (Arayans or Mukkuvas) as priests. Brahmin priests in recent times have replaced the Arayan or Mukkuva priests in their Bhadrakali temples.

When a new boat is ready for launching, it is first of all taken to the church of Vallarpadam for getting the blessings of Virgin Mary. This practice is confined to the Arayans of Cochin. The *Achan* (Vicar) of the church sprinkles a few drops of holy water on the boat and recites some passages from the Bible. Just like Latin Christian fishermen, the Hindu fishing communities of this region, dedicate themselves to Virgin Mary for life or a fixed number of years. This custom is called *Atimayakal*. The priest blesses the fishermen for better catches and also for protection from the hazards of the sea. The fishermen pay from Rs 3 to 6 for the service of the priest. As soon as the performance of the rites in the church is over, the boat is taken to the Subramania temple of the Elamkunnapuzha village, just opposite the church, for propitiation. Garlands are generally offered to the deity and the same are tied to the prow and stern of the boat after consecration when the boat is launched. The plantains are distributed among the members of the crew. The Mukkuvas of Quilandy, Calicut district, take their boats to the tomb of Seethi Tangal for ceremonious launching. Before launching a new boat they send coconut oil for lighting the tomb and new clothes for covering it.

The catch on the day of launching is generally distributed among the members of the crew but a portion is set apart for the worship of ancestors. The members of the crew, carpenters who built the boat, friends and kins are fed with toddy, arrack and sumptuous dishes in the evening of the launching. At night the departed ancestors are propitiated by offering liquor and fried fish, which they relished. As has already been mentioned elsewhere, the Hindu fishermen send their offerings to the Muthappan's temple at Parassinikadavu (North Malabar) on the day of launching a new boat.

The Hindu marine fisherfolk of Kerala observe all calendar festivals like other castes. Their caste deity is goddess Bhadrakali. However, worship of *Kadalmma* (goddess of sea), is prevalent, particularly among the Arayans of former Travancore State. It is relevant to note here that the worship of Lord Subramania is widely prevalent among them, which is not more than 80 years old. The author's informants could not recollect a pilgrimage to Palani Subramania temple before 1900. Sanskritic deities like Guruvayur Krishnan and Sabarimala Ayyappan are also visited for the past 25 years or so. They send their offerings to these deities for getting bumper catches.

The Hindu fishermen believe in the law of *Karma* and *Dharma* as any other caste Hindu of Kerala. They also have firm belief in the existence of hell and heaven and life hereafter. Sometimes, they blame for their poor economic status to their birth in an untouchable caste. The dead ancestors are carefully looked after by offering the fish they liked best and by conducting memorial ceremonies. The propitiation of the serpent-god is prevalent among them as any Hindu caste of the region. They also observe the rituals connected with life crises. It is interesting to note that the Hindu fishermen in general, and the Mukkuvas and the Arayans in particular, have adopted Sanskritic mode of worship in their Bhadrakali temples by appointing Brahmin priests.

Calendar Festivals

There are some important common calendar festivals observed by the Hindu fishermen and other members of the Hindu community. We shall give a short description of some of the important festivals in order to understand the environmental, social and economic background of the Hindu fishermen and the rituals connected with them.

Ponkala

The Arayans of Kanhiranchera village, Alleppey district, perform an annual festival called *Ponkala* in honour of *Kadalamma* (mother-goddess of sea). It is reported that this ceremony is also conducted by the fishermen of Trivandrum, Quilon and southern parts of Ernakulam. *Ponkala* refers to the cooking of rice or pudding in the open air by women as an offering to *Kadalamma* or in honour of a certain Bhagavati (goddess). The Arayans propitiate the *Kadalamma* for 41 days during the *Mandalam*. During this period the *Kadalamma* is worshipped daily for which a *Mandapam* (open shed) is erected. The offerings consist of flattened rice, puffed rice, jaggery and *Navadhanyam* (nine pulses), *Ghee*, camphor, benzoin, sugar cane and coconuts. The *Mandapam* is decorated with tender coconut and mango leaves before the actual ceremony takes place. Fisher-women who have reached their menopause from each household assemble on the morning of the 41st day at the sea coast with *Ponkala* pots containing the necessary rice, jaggery, coconut and fire-wood. The ladies who cooked the *Ponkala* should abstain from sexual intercourse and observe vegetarianism for the day. They cook two kinds of *Ponkala*, one with rice, jaggery, coconut shavings and plantains and the other without jaggery. Fortyone women participated in the festivals of 1971 at Kanhiranchera village in Alleppey. Each lady cooked the *Ponkala* in a new earthen pot and propitiated the *Kadalamma* jointly. Formerly the *Ponkala Payasam* (pudding) along with the pots was thrown into the sea after their mouths were sealed. This practice has been discontinued in recent times. However, it is reported that this custom is still in vogue in the Azhiakkal village, Alleppey district. Charms and sacred formulae are uttered while throwing the pudding into the sea. Krishna Kunju, the author's informant, told us that in olden days, the *Ponkala* pots thrown into the sea used to return the following day to the ovens on which the pudding was prepared. Many of his ancestors, he claimed, had seen such miraculous deeds of the *Kadalamma*. A few coins and a little rice pudding only are thrown into the sea today instead of the *Ponkala* pots. A night-long *Nadaswaram_Cutchery* (a music concert played with Nadaswaram – a snake pipe with 12 holes – with its accompaniment) is held on the occasion.

Onam

Onam, the national festival of Kerala, is celebrated by all communities, particularly the Hindus, in the month of *Chingam* (August–September) in commemoration of the golden age of Mahabali, a legendary ruler of Kerala, who is supposed to revisit the country to see for himself the welfare of his subjects. The festival is celebrated for 10 days from the lunar asterism *Atham* in the month of *Chingam*. The most important days are *Uttaradam*, *Thiruvonam* and *Chathayam*. The whole season is marked by mirth and merriment among the Hindus, particularly among the fishermen. The Hindu fishermen do not go out for fishing for 4 to 7 days. Houses are cleansed and whitewashed. Loans are taken from the owners of the fishing units and even moneylenders, to celebrate the festival on the grandest possible scale. There is a proverb, "*Kanam Vittum Onam Unnanam*" (One may sell even his landed property to celebrate Onam). It is an occasion for cementing the social and economic relations between the agricultural labourers and their masters. Formerly the fishing as well as agricultural labourers used to present gifts to their patrons on the eve of Onam and in return they received new clothes, vegetables and rice and loans for celebrating the festival. Nowadays, this practice is dying out in Kerala not only among the fishermen but also among the peasants.

As Onam has been declared the national festival of Kerala all communities celebrate it. Provision has been made by the State Government for celebrating the festival at the State level. One of the highlights of the celebration is boat races in different parts of Kerala, particularly a grand one at Aranmula. A number of Hindu, Muslim and Christian fishermen participate in it. In 1970, one of the Mappila fishermen got a trophy in the boat race.

Vishu

Vishu is another important festival observed by the Hindu fishing communities on the first day of *Medam* (April-May) which is the astronomical new year day. The Hindu fishermen, like other castes, believe that one's fortune for the whole year depends upon the nature of the articles one see in the morning of the *Vishu*. A *Kani* (an auspicious sight) is prepared by each family on the previous night for seeing it in the morning of *Vishu*. A few grains of raw husked rice is kept in a circular bell metal tray over which a newly washed cloth, neatly folded, is spread. A golden gourd, betel leaves, betel nuts, a metallic mirror, yellow golden flowers of *Konna* (Cassia fistula), golden coins, and ornaments are aesthetically placed over the cloth in the tray. Two coconut halves, filled with oil and lighted with cotton wicks, are also placed inside the tray. The traditional bell-metal lamp filled with coconut oil and lighted with cotton wicks is kept burning near the tray. Sometimes, a picture or a clay figurine of Lord Krishna is kept inside the tray or near it. Early in the morning of the *Vishu* day between 3.30 and 5 a.m. the eldest married woman (whose husband is alive) of the family gets up and lights the lamps and looks at the *Kani*. Utmost care is taken to prevent anyone from looking, by chance, at something else. The *Kani* is either taken to the bedside of the members of the family or it is kept at one place and the members are led there blindfolded. It is interesting to point out here that the cattle must be shown the *Kani*. There is an interesting practice of seeing the *Kani* among the Hindu fishermen of Kanhiranchera village. The Kaniyans (astrologers) prepare the *Kani* and it is taken to every household of the fishermen. While showing the *Kani*, they sing songs in praise of Lord Krishna. The Kaniyans are paid a sum ranging from Rs 2 to Rs 5 by each family depending upon their economic status for this service. It is reported that the Kaniyans have been replaced recently by Iravas and Nairs in showing the *Kani* to the Hindu fishermen in this part of the country.

Kaineetam (present) in the form of few silver coins is given to the younger members by the eldest member of the family together with betel leaves and betel nuts or with *Konna* flowers and some rice after seeing the *Kani*. This is repeated by other members who in turn give such presents to their younger members, relatives and friends. After the *Kaineetam*, the children explode crackers to herald the *Vishu*.

There is a tendency today among the Hindu fishermen and other castes to pay visits to Sanskritic temples at Sabarimala and Guruvayur on the previous day of the Vishu. They sleep in the precincts of the temple and get up early morning to take their bath and have *darshan* of the deities. They believe that seeing the deities on the morning of *Vishu* is the most auspicious sight and would bring fortune to them.

On the morning of the *Vishu* all members of the fishermen's household take ritual bath as other Hindus and put the caste marks with sandal paste or ashes and visit the local Bhadrakali temple to offer their prayers. After the worship, members of certain other castes are fed. The fishermen feed themselves with *Kanji* (rice gruel). Coconut shavings are added to the rice gruel. Leafy spoons are used for taking the Kanji, made from the leaves of jack fruit trees. It is called Kumbil.

The fishing labourers are given cash and vegetables for the celebration of the *Vishu* by their patrons or the owners of fishing units on the previous day. Similarly, landlords used to give rewards in kind to their agricultural labourers on the previous day of the *Vishu*. Acceptance of this reward in kind is still considered a bond that strengthens the economic ties between the patron and the labourers among the fishing and peasant communities. The acceptance of such rewards is on the clear understanding that they shall work under their masters till the next *Vishu* festival.

Navarathri

The Hindu fishermen observe the Navarathri festival (a festival lasting for nine days observed by Sakti worshippers in the month of *Kanni*, September-October). In fact, the Navarathri festival refers to the propitiation of the Divine Mother or the Universal Mother. This festival is celebrated all over India. In eastern India it is called *Durga Puja* but in Mysore, Maharashtra and Madhya Pradesh it is called

Dasara. In North India it is called Saraswati *Puja* or Ram Leela. Thus, during the Navarathri days, the Divine Mother is propitiated in her different incarnations such as Durga, Saraswati, Kali etc.

The Navarathri festival is celebrated for nine days by all Hindus of Kerala including the Hindu fishermen. The last three days, Durga Ashtami, Mahanavami and Vijaya Dasami, are considered more sacred than other days for the propitiation of the Divine Mother. The Hindu fishing communities believe that offering of prayers on the last three days is capable of securing favours of the goddess and is conducive to get good catches.

Ayudha Puja (worship of implements) is performed by the Hindu fishing communities like other Hindu castes. They perform this *Puja* on the Durga Ashtami. All fishing implements and tackles are worshipped on that day. They propitiate the Divine Mother for getting bumper catches on this day. On the Durga Ashtami day a ritual called *Puja Veppu* is conducted in the evening. It is performed by rich families of higher castes only in a village. Local temples and schools are also used for the same purpose by the poor. Books – religious and educational – are carefully arranged in a room, which has been sanctified previously and worshipped with a picture or image of the goddess Saraswathi in front. The Hindu fishermen keep their fishing equipment in a room or a clean place and propitiate them. No fishing is done during the *Puja Veppu* day. The goddess is worshipped by offering fruits, beaten rice, puffed rice and molasses. When the propitiation is over the articles offered to the deity are distributed among the devotees present. All economic pursuits including fishing are kept in abeyance just before the *Puja Veppu*. On the Mahanavami day, goddess Saraswati is worshipped both in the morning and evening by offering pudding, molasses, puffed rice and beaten rice. Devout fishermen observe fast on this day.

On the Vijaya Dasami day, after the *Puja* in the morning, the books and implements are removed from the room and this ritual is called the *Puja Eduppu* (the breaking of the *Puja*). This marks the end of the Navarathri festival. This is followed by *Vidyarambha* (initiation to letters) and work. An auspicious moment is chosen by the Hindu fishermen to do some work using their fishing equipment that have been sanctified. Again, at this auspicious moment, children are initiated to the mysteries of reading and writing for the first time.

This is called *Ezhuthinnu Irruthal*. It is customary to use sand or raw rice for writing. A piece of gold or a golden stylus is used to write 'Hari Sri Ganapathaya Nama' on the tongue of the child. The child is entitled to write or read only after this ceremony. This ceremony is mostly observed by the Sanskritised fishermen of the Vellayil village near Calicut.

Maha Sivarathri

Maha Sivarathri refers to the 'great night of Lord Siva' or the 'great holy night'. In Kerala, the Hindus including Hindu fishermen observe the Sivarathri festival in the month of *Kumbam* (February–March). According to Raghavan, the author's informant, the Sivarathri is observed in commemoration of the day on which Lord Siva is believed to have saved the world by drinking the *Kalakuta*, the deadly poison obtained by churning the milk of ocean. The marine Hindu fishermen in Kerala do not go out for fishing on this day. Both the women and men observe fast and keep awake the whole night. The fishermen throng to temples dedicated to Lord Siva in their locality along with other Hindus to offer tender coconuts, *Bilva* leaves, flowers, oil, ghee, camphor, etc. Pious Hindu fishermen spend the whole night in the precincts of the Siva temple and attending the *Puja*. They break their fast on the next day morning by taking the water of the tender coconut, which has already been offered to Lord Siva. Women and children outnumber the menfolk in observing the ritual fast on the Sivarathri day.

Sometimes special *Pujas* and rites are performed by Brahmins in their homes. This special *Puja* is called *Chadurdasi Puja*. There will be a *Puja* at every quarter of the night. There are sometimes temples where 24 *Pujas* are performed on the Sivarathri night. *Abhisekham* (pouring of some liquid or smearing of ashes or sandal paste) and *Dhara* (continuous pouring of liquids like milk, oil, curd, water etc., for a fixed time) are the important rites of the *Chadurdasi Puja*.

The Sivarathri festival at the Siva temple of Alwaye, on the banks of the Periyar river, is attended by thousands of devotees from all over South India including the Hindu fishermen. It is regarded as a miniature *Kumbamela* held at Triveni, the confluence of Ganga, Yamuna and the invisible Saraswathi.

The Mukkuvas of Malabar prepare a *Kappal* (boat) using plantain stems. This boat is decorated with *Kuruthola* (tender coconut leaves) and lighted with wicks. There are two kinds of *Kappals*: small and big. In the case of the small *Kappal*, 7 coconuts, and 7 tender plantain leaves are used for the performance of magical rites. On the other hand, for the big *Kappal* 101 plantain rinds are used for the magical rite. The *Kuruthi* (mixture of turmeric and lime) is poured over the *Kappal*. The tender coconuts are broken on the *Kappal*. The *Kappal* is thrown into the sea after the propitiation on the night of the *Sivarathri*. This is performed for getting bumper catches and also for preventing the evil eye.

Thus, the Hindu fishing communities of Kerala observe most of the calendar festivals. They also observe some of the rites and ceremonies associated with these festivals in the most befitting manner. They have in recent times started paying visits to the temples belonging to Sanskritic deities like Lord Ayyappan of Sabarimala, Palani Subramanian and Guruvayur Krishnan. They send images of fish to these temples for curing illnesses and for getting bumper catches. They go, once a year, at least, on pilgrimage to these temples. They also observe fast and grow beard before they go on pilgrimage to the Sabarimala temple.

The Hindu fishermen in general, and the Bovis-Mogeyar in particular, believe in many superstitions which have been conceived arising out of their utter dependence on the marine environment. This is particularly to overcome their fear and cover the uncertainties. As such, it is seen that the Hindu fishermen across the state are steeped in religious beliefs and practices, and a sizeable part of their income is set apart to perpetuate these practices.

In the past the fishing nets were taken to the village temple at the beginning of every season, after they were repaired, treated with preservatives and sun dried. The Bovis-Mogeyar go to village temple on the first day of the fishing season with the fishing nets over their shoulders to seek the blessings of their patron goddess and other deities. Before pushing out their boats into the sea on the first day of the season, the owner of the fishing unit sprinkle sea water over the boats and nets. Coconuts are broken over the hull of the boat and incense sticks lit. In the northern parts of Malabar the deity Babrian is also propitiated on the first day of the season with *ada*, for calm seas and bumper catches. In this connection Zachariah (1996) has said that "in many places, offerings were made to the sea god Babrian and in some places like Kasargod, Kumbla, Uppala and Kunjathur to the deity Gulikan also, for *Chakara* to appear. It was told in Shriya, Mogral-Puthur, Kunjathur and Manjeshwar that these mudbanks were attributed with supernatural beliefs and characters, and prayers were observed to be offered before these formations. Fishermen everywhere here believe that *Chakara* has life of its own and moves unpredictably, and that the sea cannot upheave because *Chakara* is so strong. It forms and disappears without any indication. The removal or artificial spreading of *Chakara* it seems does not affect the behaviour of the sea."

The Bovis-Mogeyar still have their own priests from their own community, which is different from what is happening among other Hindu fishing communities in other areas of Kerala where there are Brahmin priests employed by most of their temples. This is observed in Alappat Panchayat of Quilon district. There are as many as a dozen temples dedicated to Bagavathy, Subrahmanya and Siva and other deities managed by the Arayans where mostly the Brahmins are the priests. They are also invited annually for *Shudhikalasam* on festival days. It is observed that at most of the temples belonging to the other Hindu fishing communities, the Namboodiri Brahmins officiate as priests and they are also invited for conducting special *Pujas* for a particular purpose.

In this connection, Zachariah (1996) has observed that "the change in this trend now appears to be in the nature of the process of Sanskritization. Most of the younger generation claim that this employment of higher caste members is only because the Brahmin members are well versed and more familiar with matters pertaining to religion, especially the Hindu *Shloka,* and not because of the elevated caste status of the individuals themselves. Some of the older generation however claim that their own youngsters these days have no patience to get themselves familiarised with the elaborate and difficult rituals, and also that they lack the will power to abstain from things like meat, sex and liquor during some days of the rituals, which demand strict abstinence and observation of rituals. Moreover, there are a number of elders who feel that it is the Brahmin who is ordained to perform religious ceremonies, especially at important functions and festivals".

"The Mogeyar are highly superstitious and indulge a lot in the arts of sorcery, witchcraft (*Manthravaadham*) etc. They consult mediums and occultists when they have doubts or fear about inexplicable happenings. Many of their older *Velichapad* are reportedly experts in such activities. The communities here, including the Bovis-Mogeyar, consult him whenever necessary … . The Mogeyar have elaborate ceremonies before every event, and especially before a new boat is launched or beginning to be constructed or when a new net is used for the first time. The first fishing venture of a season is also a day beginning with a series of rituals. Elaborate preparations precede these some days before, after determining the most auspicious timings with the astrologer (Kanniyan or Podhuval). These ceremonies are reportedly getting more elaborate and expensive. The offerings previously were confined to mainly coconut, jaggery, plantains and such products, but that which are made now extends to comprise many articles that were not previously included, like silver articles, gold braid bordered linen clothes, assortment of fruits, nuts, etc." (Zachariah, 1996).

"The slightest calamity or incidents believed to be ominous are sufficient cause for the Bovis-Mogeyar to consult their traditional astrologers and, or local Brahmin priests, and to arrange for the conduct of numerous and elaborate rites, considered necessary as atonement for any displeasure believed to have been caused to the supernatural beings. Most of the Bovis-Mogeyar fishing villages are notable by the presence of numerous shrines which are said to have been made as offerings of gratitude or to appease the wrath and anger of the supernatural beings. All their festivals and rituals are conducted very meticulously and with no expenses and efforts spared. In Uppala and Kunjathur it was told that from quite an early period specialised rituals have been performed mainly by Brahmin priests together with their own religious specialists" (Zachariah, 1996).

According to our informant, Jayan, Kasaba beach, Kasargod, the Hindu fishermen send their votive offerings such as coconuts, jaggery, plantains and oil to the local Sri Kurumba Temple for getting bumper catches. Apart from this, during the monsoon period they send golden images of *Chemmin* to the Sri Kurumba Temple for getting bumper catches. It is reported there is a new ritual called *Samudrapuja* initiated by the Bharatiya Janata Party workers which is being performed at the temple. They also strongly believe that the chicken-pox is caused by the wrath of the mother Goddess. Each of the owners of a fishing unit sends Rs 10 daily to the temple fund for its maintenance.

Sorcery and magic are practiced by the Mukkuvas in Blangad for getting bumper catches and also for curing diseases caused by these nefarious deeds. The Mukkuvas also propitiate Nagaraja and Nagayakshi by sending offerings of *Pal* (milk) and *Noor* (turmeric powder) for their *Abishekha,* in addition to tender coconuts. They also send 5 to 10 per cent of the catch as offerings to the local Bagavathy temple. There is another Vishnu temple at Blangad where *Palpayasam* is offered for getting bumper catches.

8

General Observations and Conclusions

In the preceding chapters we have made an attempt to describe and analyse the interrelationship that exists between ecology, technology and economy among the Mappila, Hindu and Latin Christian fisherfolk. In this chapter we shall give an overall summary and general observations. An attempt is also made to discuss the general principles underlining technological, economic, ecological and cultural changes that have occurred among the fisherfolk of Kerala since the introduction of innovations. While some aspects of the social structure, economic condition and technology of the fisherfolk of Kerala have changed, several other factors have remained intact. This uneven change among the Mappila, Latin Christian and Hindu fisherfolk cannot be explained without delineating the different types of structural relations. In the previous chapters we have tried to examine the social changes in the context of ecology, technology, economy and rituals in accordance with the functional criterion.

These aspects of culture were selected because we consider them as a very important factors that determine the social behaviour of the fisherfolk of Kerala. We are not taking into consideration all aspects of structural relations due to certain constraints.

The Mappilas of Tanur are converts to Islam and adhere blindly to their religious faith. Yet these converts have retained some of the Hindu religious rites and belief systems, although they embraced

Islam long ago. The pattern of persistence of folk beliefs among the Mappila fisherfolk was also a subject of our enquiry.

Ethno-science of the Mappilas

The marine resources of Kerala are vast and offer immense potential for the economic advancement of the fisherfolk of Kerala. Kurien (1985) observes: "The sea off the south-west coast of India, comprising the maritime states of Goa, Karnataka and Kerala, forms a relatively homogeneous aquatic eco-zone. The inshore or coastal waters (upto a depth of 50 metres) of this region measure 23,400 square kilometres and have a maximum sustainable yield (MSY) of 7,00,000 tonnes (George et al., 1977). The average fishery productivity potential of these waters works out to 30 tonnes per square kilometre (or 300 kg for every hectare) compared to 12.5 tonnes for all-India making it the most productive fishing zone in India. Kerala accounts for 12,570 sq. km of this coastal sea area which has an estimated MSY of 4,00,000 tonnes. The fishery resources in the tropical seas off Kerala are marked by a multitude of species attaining varying sizes at age of maturity."

The Mappilas' fishing ground is located in the area where maximum rainfall occurs. Boats and nets are liable to capsize by the swelling waters during the monsoon. The muddy walls and wattled huts of the Mappilas crumble under the swift waters. Epidemics like cholera and small-pox break out during the monsoon causing death to the fishermen. In fact, the life-giving and the life-sustaining south west monsoon is also often a season of misery, starvation, devastation and death to the Mappilas and Hindu fisherfolk.

The Kerala coastline facilitated easy water communication before the development of other means especially roads and railways. The river route from Trivandrum to Badagara is still in use. The navigability of the rivers have facilitated the fish trade. Road, rail, and water communication with important fish markets like Cochin, Trivandrum, Changanacherry, Quilon, Cannanore, Calicut and Tellicherry have facilitated the quick transport of the catch to these centres in recent years. Tanur, Blangad, Ponnani, Puthiyappa, Beypore, Mappila Bay, Munambu and Neendakara have been sheltered by promontories at different points along its coast. These fishing villages have made the landing stations ideally suited to fishing operations.

The most remarkable change in the history of the Mappilas is their evolution from the subsistence fishing economy to one of commercial enterprise. The advent of roads and railways in the 19th century reduced the distance of the villages from the wholesale fish markets at Calicut, Cochin, Trivandrum, Bangalore and Madras. Telecommunication gave a further fillip to commercial transactions. Many of the Mappila fishermen have got telephone connections now. STD and ISD booths have been functioning in Kerala for the past one decade or so. These fishermen can now easily contact their counterparts working in the Arabian Gulf countries. The development of transport and modern communication have offered facilities to the Mappilas to have constant contacts with the neighbouring towns, cities, and states, not only in India but also abroad, although their villages are far away from the urban and industrial centres.

Thus, the primary factors that gave an impetus to the technological, economic and ecological change among the Mappila fisherfolk of Tanur are derived from geographical and natural environment.

In Chapter II, we have a short description of the marine environment and the fishermen's traditional knowledge of the relevant eco-system including the behaviour of the fish, and different kinds of winds, currents and waves. We also tried to correlate this traditional knowledge of the Mappilas with those of the Hindu fishermen. We have also tried in this chapter to show the relevance of the specific technology in the context of the marine environment. It has been demonstrated that the Mappilas and also some of the Hindu fishermen, particularly the Mukkuvas of Puthiyappa, Bovis-Mogeyar and Mogavirar of Kasargod, have a remarkable capacity to adjust themselves with ecology and nautical cycles and seasons.

The Mappilas' knowledge of the various fishing grounds and the species inhabiting them is impressive. They adjust their fishing techniques according to the depth of each fishing ground. The Mappila and Hindu fishermen and marine biologist share the view that the *Puramkadal* (continental shelf) that lies beyond 10 fathoms is the habitat of all larger fish, which are caught with hook-and-lines and modern gears.

The Mappilas and the Hindu fishermen employ their techniques of fishing according to the habitat of the species. The habitat of any species of fish is related to the availability of food. The fisherfolk, particularly the Mukkuvas of Blangad and Puthiyappa, and the Bovis-Mogeyar of Kasargod have, over the years, stored enormous knowledge of the kinds of plankton which serve as the food of different species of fish. They have detailed knowledge of the migration and spawning period of almost all species of fish and they correlate the catch with these phenomena. Long before modern marine biologists did, the Mappilas classified the planktons into phytoplankton and zooplankton. The Hindu fishermen, especially the Arayans of Cochin, and the Mukkuvas of Blangad and Puthiyappa share the knowledge of the Mappilas. The Mappila and the Hindu fisherfolk adjust their fishing techniques according to the availability of a particular kind of plankton.

The Mappila and the Hindu fishermen possess abundant data on the behaviour of the various species of fish. They, like marine biologists, classify fish on the basis of their habitat, viz., (1) *Adiyilnilkkunna Min* (demersal), (2) *Polappu Min* (pelagic) and (3) *Edanirrilnilkkunna Min* (mid-water fish). The Mappilas, Mukkuvas as well as Bovis-Mogeyar manoeuvre their crafts and tackles according to the behaviour of the fish.

The need for efficient exploitation of marine products in order to improve the economic conditions of the fishermen led to the introduction of nylon nets and other innovations among the fisherfolk of Kerala. The adoption of nylon nets, mechanised boats, trawlers fitted with GPS, echosounder and wireless, and *Valiya Vallam* or *Valiya Vanchi* and Valiya glass Vallam fitted with inboard engines, helped better catch of fish, besides extending the range of fishing grounds. The Mappilas' knowledge of the secrets of the ecosystem including the behaviour of the fish helped them in reaping the benefits of new tools and technological innovations.

We now turn to the sphere of the more clearly formulated knowledge concerned with the relations of Mappilas' cosmology to the techno-economy. As has been demonstrated earlier, the Mappilas employ various techniques according to the occurrence of winds, currents, cross-currents, waves, tides and phases of the moon. They have acquired, through their age-old experience, enormous knowledge of cosmology and climatic conditions which guide them greatly in their fishing operations. This is also true in the case of the

Bovis-Mogeyar of Kasargod and the Mukkuvas of Puthiyappa and Beypore of Calicut district and Blangad and Puthan Kadappuram of Thrissur district.

We have already shown how climatic conditions influence the Mappilas' catch and technology. This is also true in the case of the Hindu fisherfolk and Latin Christians. The fishermen organise the labour according to seasons. For instance, in summer, individual units go out for fishing while during the monsoon joint operation of two to ten units is undertaken. In a joint expedition, only the minimum number of gears and crafts are employed. Able-bodied fishermen only are allowed to participate in joint expeditions. However, all members of the fishing units get their due shares from the catch.

The traditional fishermen, particularly those belonging to the Hindus and the Mappilas employ the maximum number of floats for their nets and lighter sinkers during the monsoon. On the contrary, only 50 per cent of the floats are used in the summer in addition to heavier sinkers. There is a marked difference in the length of float ropes used in various seasons. Shorter float ropes are employed in the rainy season and longer ones in the summer. The Mappila and the Mukkuva fishermen increase the length of the float ropes and weight of the sinkers so that the net could reach the bottom of the sea where the fish take shelter in the summer. Thus, both the Mappila and the Hindu fishermen adjust their technology according to seasons and habitat of the fish.

We have seen that the Mappilas have knowledge of eleven types of waves. They manoeuvre their fishing crafts according to the nature of the waves. Sometimes waves are taken as their guide during fishing expeditions. The Hindu fisherfolk, particularly the Mukkuvas of Blangad and Puthiyappa, share the traditional knowledge of waves with the Mappilas. Their knowledge of waves like *Karakadal Tiramala* and *Vayyekadal Tiramala* which are very dangerous to life and fishing craft, help them in their fishing expeditions. They say that the *Padikadal Tirmala* is the coldest in the sea and if a boat is capsized in it, the crew seldom escapes from a chilly death. *Coyala* is the most violent of all the waves and grows very roughly. The Mappilas believe that the *Coyala* is engendered by the malevolent spirit, *Gulikan*, and another sea sprite. If a fisherman falls into the

Coyala he courts a frozen death as this wave is as cold as the *Padikadal*. The Mukkuvas of Blangad and Puthiyappa have the same knowledge of *Coyala*.

Another factor that influences the catch is the occurrence of winds. The Mappila and the Hindu fishermen have classified winds into eight kinds and possess detailed knowledge about them like their occurrence on a particular day, month and season. They correlate the appearance of a particular species of fish with a particular wind. This knowledge of the wind helps the Mappila and the Hindu fishermen to adjust their nets and fishing techniques in catching particular species of fish.

The Mappilas, Mukkuvas and even Arayans of Alleppey have classified the currents into eight kinds. We have tried to show how the occurrence of different kinds of currents is related to winds, catch and fishing techniques. We have also discussed Mappilas' knowledge of different currents and winds and their influence on the technique of fishing and catches. The catch of certain species of fish is related to the *Tuyi* (light in the current) and *Nilavu* (moonlight).

The sea is compared by the Mappilas to a man and the current to the blood circulation. When different currents occur indiscriminately the sea becomes sick and the fishermen are unable to undertake fishing operations. When the *Kedar Nir* occurs the sea becomes inhospitable for fishing and it is then compared to a menstruating woman.

"The Mappilas chart their course during the fishing expeditions by the rising and setting of important stars, viz., *Kuttu Naksatram, Telinna Naksatram* or *Onnamyamakkaran*, *Randamyamakkaran* and *Munnamyamakkaran*. When these stars are obscured at night owing to dim light or storm, the Mappilas take the direction of winds and waves as their guide. The Mappilas have received their training in the customary techniques of navigation from their ancestors who gained this skill through long years of experience. Both Mappila and the Hindu fishermen believe that the *Telinna Naksatram* rises in the west and moves in the direction of east. When the *Telinna Naksatram* rises in the west after sunset, the Mappilas set out their fishing expedition keeping the prow of their boat facing this heavenly body. This star is taken as a guide for locating the east-west direction and for launching the gill-net to catch the seer fish. The Mappilas calculate the actual time spent in fishing by taking the course of the rising and setting of the *Randamyamakkaran*" (Mathur, 1978).

There is no doubt that the Mappilas are quite familiar with their environment. Their cosmological knowledge is supplemented by a detailed knowledge of fish and their behaviour. Environment plays an important role in the cultural behaviour of the Mappila, Hindu and Latin Christian fishermen. The Mappilas particularly, possess a vast cosmological knowledge which is their light-house in fishing expeditions. The fishing technology and economy of the fisherfolk of Kerala are related to their cosmological knowledge. The catch is influenced by both cosmological and ecological factors. Although the cognitive mapping of the natural environment and cosmology of the Mappila and the Hindu fisherfolk are often associated with myths, supernaturalism, and magico-religious notions, they do not stand in the way of detailed objective assessment and classification of the environment based on inductive reasoning.

Technology and Change

Our analysis is based on an examination of the impact of mechanisation of fishing equipment and modernisation of the commercial organisation. We have described the different types of country crafts and gears employed by the Mappilas. We have also compared and contrasted the fishing technology of the Mappilas with their Hindu counterparts in the light of response to new economic opportunities and technological innovations. We have tried to conclude with an evaluation of the impact of modern technology and innovations on the economy of the Mappila fishermen.

From the appearance and techniques of manufacturing of different types of country crafts, mechanised boats with outboard engine and big canoes with inboard engine and gears, it may be said that the traditional fishing technology appears to have developed in Kerala as a result of prolonged foreign contacts.

Nair (1971: 35) observes: "In the local fishery one can see an interesting fusion of many crafts and techniques. The widely used dug-out canoe made by stitching planks to the dugout is a result of Arab influence and the *Catamaran* used along the southern coast of Kerala is of Egyptian origin while the shore seines are of Portuguese import. The influence of Spanish is evident in the boat seines and stake nets which remind us of the stow nets of Belgium and German tidal floats. These were introduced into this country by the Dutch. The well-known Chinese dip nets and lure fishing have been incorporated into our fishing methods from the Chinese".

The dug-out canoe is most commonly used along the Kerala coast. The plank-built boats are generally larger than the dug-outs and the former are employed during the monsoon when the sea is rough. The important gears employed in fishing are shore seines, boat seines, gill-nets, cast nets and also hooks and lines. The range of operation of these traditional crafts and gears is generally confined to the inshore waters of 10 to 15 kms. The regions beyond this and the deep sea remained outside the purview of traditional technology. Exploitation of this vast expanse has recently been initiated with the help of mechanised boats, trawlers and big canoes, particularly those covered with fibre, as well as glass fibre canoes.

In our description and analysis of the technology of fishing we have also tried to show the social implications of different types of boats and canoes. We have shown how the economic organisation is based on two important factors, viz., (1) sociological differentiation of functions, and (2) the ritual and magical regulation of the work.

The *Toni* (plank-built boat) or the *Beppu Toni* (dug-out canoe) in Tanur was constructed by carpenters belonging to both Hindu and Muslim communities. A *Toni* or *Beppu Toni* was owned and engaged either individually or collectively, and this was done in accordance with certain norms. There are certain principles underlying the building, owning and sailing of a *Toni* or a *Beppu Toni*.

In the analysis of the sociological differentiation of functions, we have shown how the owner of the boat shoulders the entire responsibility for its construction. He meets the cost of the timber, pays the wages of carpenters, engages master-craftsmen and supervises the workmanship.

Earnings from a boat and net in a fishing unit are not confined to the *Karanavan* (head of the fishing unit). He used to get only 12 per cent and 15 per cent of the catch respectively for the boat and net. Other members of the crew get a fixed percentage of the catch. Nowadays, the owners of the boats and nets get 40 to 50 per cent of the catch and fishing labourers get 50 to 60 per cent of the catch. The owners of the boats and nets meet the expenses for fuel and food. The

Karanavan has definite social functions and privileges and exercises them in the running of a fishing unit. He recruits the members for running of a fishing unit. Sometimes he is required to clear the debts and look after their interests. The Karanavan is responsible for conducting the fishing rituals and magical rites associated with the construction of a boat and making of a net and their launching. The seaman, next in rank to the *Karanavan* is the *Chukkankaran* (steersman). He is technically qualified and knows the difference in speed, buoyancy and stability of a boat. He is a seaman in the real sense of the word.

It is evident from the description of the ceremonies connected with the technology of fishing that belief in the efficacy of the performance of rituals and magic is supreme among the Mappilas of Tanur. Rituals and magical rites are interwoven into the fabric of the Mappila society, as is natural in any pursuit where either danger or uncertainty is inherent.

The sociological weight of archaic tradition, that is, the degree to which the behaviour of the Mappila society is affected by the Hindu fishermen's customs and norms is indeed very heavy. Building a boat in conformity with ritual and magical practices is done without the slightest hesitation. Not a single boat or canoe in Kerala has been built without performing the prescribed fishing rituals and magical practices. Similarly, not a single net has been manufactured and launched without observing the prescribed fishing rituals. They have a strong conviction that if the traditional rites were not observed, evil would ensue. The boat built without performing the rituals, would not be seaworthy but would be slow in sailing and unlucky.

From the sociological point of view, the fundamental question that one may pose is, what is the economic function of rituals and magical practices in the process of boat-building and manufacture of nets? We shall answer this question by quoting the observation made by Malinowski (1953: 16): "magic puts order and sequence into the various activities, and that it and its associated ceremonials are instrumental in securing the cooperation of the community, and the organisation of communal labour. It inspires the builders with great confidence in the efficacy of their work, a mental state essential in any enterprise of complicated and difficult characters". Thus, magic and rituals keep the Mappilas confident about the success of the construction of a boat, making of a net and subsequent expeditions.

The Mappila fisherfolk of Tanur today operate trawl nets with the help of over 20 mechanised boats and bag substantial quantities of fish, particularly prawns, from the rich fisheries lying between 35 and 75 fathoms on the continental shelf.

Writing on the economic organisation of Mappila, Mathur (1978) had observed that "mechanisation and modernisation of the technology of fishing have made it profitable and consequently the Mappila fishermen today continue to concentrate all their energies on their traditional calling of fishing. The hereditary relation that exists between the owners of fishing units and the labour has not only received a new lease of life among the *Valakkar* but also among the *Beppukkar*. The persistence of traditional relations have, in their turn, put certain obstacles in the path of economic growth. Progressive-minded *Beppukkar* have been prevented from adopting modern technology in fishing."

What Epstein has observed holds well in the case of the *Beppukkar*. In her opinion, " a customary system of rewards impedes economic growth by tying people to traditionally accepted ways of production" (1967: 44) "Whenever such a system exists, we can expect a reluctance on the part of indigenous labour force to implement improved methods of production with the traditional crop until a method of payment has been devised" (Epstein, 1971: 3). In contrast, the *Valakkar* have responded remarkably to modern technology and innovations. They took to innovations like modern nylon nets, cooperative societies and mechanised boats. They also produced a number of engine drivers and technicians for manning mechanised boats. The speed with which the *Valakkar* have responded to changing technological innovations is rather a tribute to the genius of their community.

Nylon nets and mechanised boats covered with fibre were introduced to Tanur by the *Valakkar* because of the obvious economic advantages they have over smaller country crafts, cotton, jute and hemp nets, and the prestige they conferred. Mechanisation of the boats and the use of nylon nets, trawl nets, etc., have helped expeditions to go to fishing ground beyond 12 kms.

Non-availability of experienced seamen among the members of the cooperative society at Tanur, has been circumvented by the addition of experienced non-member fishermen who are able to handle the boat on longer trips. Mechanised boats operated by a highly experienced crew in order to land better catches. Accommodation of this new social element has, in fact, facilitated the Mappilas to recruit the members of the crew from their own kinship groups though they may not be members of the cooperative society. Subsequently they enroll themselves as members. It may be noted here again that majority of the mechanised boats are owned cooperatively by the members of the village who are kins.

In accordance with Linton's outline of the process of change the acceptance of nylon nets and mechanised boats involved a small group of *Valakkar* and was determined by economic change, perhaps increased prestige, and possibly the novelty of innovation. The *Valakkar* who welcomed the modern techniques were respected fishermen, and innovations like nylon nets and mechanised boats were adopted by the orthodox members of the community on account of their economic advantage. Adjustive modifications were carried out to alter the form of innovations by the addition of nylon nets, and they proved superior economic tools and prestige symbols. In the case of mechanised boats, however, "the important adjustive modification was the adjustment of social system in response to organisational requirements for efficient utilisation of the innovation" (Frazer, 1960: 240).

For the same reason the mechanised boats were introduced through the fishermen's cooperative societies in the village by a group of *Valakkar* in the initial stage. These individuals are comparatively rich fishermen having contacts with the leaders of the Muslim League and government officials. The prestige value of innovation was of particular significance as they brought nylon nets, paddy fields, coconut gardens and tiled houses to the innovators. Thus, before full economic advantages of the innovation were proved, six mechanised boats were introduced through cooperative societies within a year (1969). The Mappilas have adapted not only the *Valiya Vallam* with inboard engines but also the fibre covered boats in Tanur. But they failed to adapt trawlers due to the lack of harbour facilities in Tanur. But the Mappilas of Mappila Bay, Cannanore district have adapted trawlers fitted with echosounder, GPS, wireless and big canoes for fishing after building that harbour. The Mappila with the help of GPS can locate the sea bottom, rocky places with fish shoal up to 35 maarus and identify the occurance of winds and currents. The GPS also help the fishermen to locate their hamlet from distant fishing grounds.

Economic Organisation and Change

It has already been pointed out that the *Valakkar* and the *Beppukar* have responded in different degrees to the induced technological opportunities. Similarly the Mappila, Hindu and Latin Christian fishermen have responded to modern innovations in varying degrees. The economic, socio-cultural and ecological context of this differential response is discussed below.

By economic change we mean a change in economic roles and relations (Epestein, 1962: 314). Both the *Valakkar* and *Beppukar* have achieved considerable economic development during the past five decade or so. In other words, among both the social groups the output of goods and services have increased with the same or greater inputs. Both communities have changed from subsistence to cash economy, but the resultant economic change is unequal. The *Beppukkar's* economy has remained wholly fishing, whereas the *Valakkar's* diversification led to their close integration with the regional economy. Many *Valakkar* purchased paddy lands and coconut and arecanut gardens. A few of them have bought trucks. A considerable number of them took up fish trade though not without success, while yet others bought mechanised boats. Above all, a few have become partners in establishing ice plants in the village. Thus, the *Valakkar's* economic system has hinged on the wider commercial system. This is also true in the case of Latin Christians of Quilon district. But the Hindu fisherfolk, particularly the Arayans of Alleppey, are still very slow in adapting modern technological innovations compared to the Latin Christians inhabiting the same marine environment. Our informants have said that as the range of relations increases, the degree of reliance upon neighbours and contemporaries reduces. Thus, the Mappilas of Tanur (*Valakkars*) increased their range of economic relations, and the interdependence between the *Karanavans* and *Ottakkanmar* diminished, resulting in the severance of hereditary patron-client relationship. This did not mean that the traditional bond between them has been completely snapped. As has been already mentioned, the fishing labourers in a number of units have got ownership varying from 1/16th to 7/8th share but they are not able to free themselves from exploitation on account of their dependence upon their patrons for loans during off seasons and life crises. Moreover, they do not want to terminate their existing economic relationship with their patrons. As a matter of fact, in a large number of fishing units, the members of the crew continue to give the usual shares in the guise of commission to their patrons even after getting ownership. Although the fishing labourers are not bonded labourers, once they are employed under an owner of a fishing unit and have availed of loans from their masters, they cannot leave them until they clear off their debts. This is also true in the case of Hindu fishermen and traditional Latin Christians.

There are significant differences in the response to modern technology and innovations among the *Valakkar* and *Beppukar* among the Mappilas. Similarly, it was found that the Arayans of Quilon and Alleppey and Nullayans of Trivandrum are very slow in adopting innovations as compared to the Mappilas. Latin Christians have support from their churches. The Mappilas are interested in democratising the political institutions and in taking active part in the activities of the Muslim League and Latin christions. The reason for the Hindu fisherfolk's rejection of innovations can be explained by the social system, socio-economic organisation and natural environment.

Social System

The Hindu fisherfolk of erstwhile Travancore, like other groups, is composed of various castes. Caste membership is acquired by birth, and caste differentiation is sanctioned by the Hindu notion of Karma, i.e. the deeds done by an individual in his life-time. "The idea of Karma teaches a Hindu that he is born into a particular caste because of certain actions he performed in a previous life (*Janma*): Dharmasastras mention that if a man does evil deeds he will be born in a low caste, or even as an animal, a pig or a donkey" (Srinivas, 1962: 10). This religious belief is noticeable among the Arayans, particularly those of Alleppey and Quilon and Nulayans of Trivandrum, in a marked degree. On the other hand, the Mappilas of Tanur are divided into two endogamous groups, the *Valakkar* and the *Beppukkar*. They do not believe that their caste differentiation is sanctioned by the notion of Karma. To them it is a matter of occupational specialisation. However, even among the Mappilas, the different sects are hierarchically structured. The social system of the Arayans and Nulayans in the southern districts is very rigid. The relation between the master and the fishing labour among them has remained unaltered. However the Bovis-Mogeyar of Kasargod with the help of the temple organisation, have responded to modern technological innovations to a great extent. By contrast, the Mappilas, social system is highly flexible. Unlike the Hindu fishermen of Travancore, the Mappilas accorded little authority to social prescriptions. Men achieved renown or became rich by virtue of their entrepreneurship. It thus became possible for any Mappila with the necessary qualities of leadership to achieve recognition and also gather a following to become a 'big man'. This is also true in the case of Latin Christians. These positions are reserved to hereditary claimants among the Hindu fishermen of Travancore. The basic philosophy underlying the Mappila 'big man' system is one of *primus interpares*. The political and social institutions of Arayans of Travancore and some Mukkuvas of Malabar are even today filled with hereditary succession. A Mappila has to show outstanding personal qualities before he can acquire prestige and social status, as the society is egalitarian.

Socio-Economic Organisation

Ownership of fishing equipment among the Arayans of Travancore and Mukkuvas of Kasargod and Cannanore is mostly in the hands of rich joint families. Joint family is the predominant unit of production and consumption. Labour is pooled and the catch is shared. The Arayans are patrilineal and sons have equal rights in the paternal property. The Mukkuvas of North Malabar however are matrilineal. The disintegration of the joint family system under the impact of western influence has led to the fragmentation of fishing equipment.

The owners of fishing units among the Arayans and Nulayans of southern districts of Kerala have hereditary links with fishing labourers' households. They have almost the same customary relationship with the craftsmen and serving castes, whose services they regularly require. These hereditary services get customary rewards in kind. Fishing labourers turn to their patrons for the solution of any socio-political problem and the latter are responsible for their welfare.

The Mappilas, particularly the *Valakkar*, who took to innovations like a duck to water, accumulated wealth by both fair and foul means. The shrewd among the Mappilas often got fishing labourers by advancing loans in times of need. Affluent patrons made presents in kind such as ornaments and clothes, in addition to cash, to their clients, on ceremonial occasions like circumcision, *Kudikalyanam* (feast for mobilisation of credit), *Cayakalyanam* (tea party for mobilisation of credit) and marriage.

Variation in Natural Environment

How can we explain the mystery of simultaneous adoption of innovations by one group and rejection of anything new by another group, both of whom live in the same natural environment? This is the question one is prompted to ask about the Latin Christian fishermen and the Arayans who live in the same fishing village of Puthantura. The Arayans contend that they are too poor to afford costly nylon nets. They also do not have rich financiers to provide them with loans as and when needed. Alongside, their fishing grounds are full of rocks. Thus they feel that it is unwise to use these costly nets in the rocky seas. On the other hand, the Christians are comparatively economically better off and there are traders among them who readily help the fishermen in times of crisis and emergency. It is interesting to point out here that the mechanised boats issued to the Arayans of Puthantura under the Norwegian Scheme, were not welcomed on the pretext that the cost of operation was very high and repair charges prohibitive. So all the Arayans mortgaged the boats. This resulted in the seizure of these mechanised boats. Thus, we find that apart from the differences in the social systems and socio-economic organisations among the Mappilas and Arayans, psychological factors do play a decisive role in discouraging modern innovations. It is difficult, if not impossible, to determine the course of economic growth if certain social factors were different from what they actually are. For instance, it is extremely difficult to say with any certainty what Tanur's economic growth would have been had there been a very rigid caste system among the Mappila fisherfolk.

The Arayans of Puthantura, Quilon district, owned only three inboard *Vallams,* 55 outboard fibre *Vallams* and 15 trawlers in 2006 while the Latin Christians in the neighbouring Neendakara fishing village owned over 2,000 trawlers. So, the Latin Christians have adapted modern technological innovations very speedily compared to the Arayans inhabiting the same ecosystem.

Thus, the Mappila and Latin Christian fisherfolk of Kerala responded to economic change quickly whereas the Hindu fishermen of Puthantura lagged behind. It is true that the Mappilas' keenness in buying trucks, mechanised boats, establishment of ice plants etc. is the outcome of a flexible social system in which they operated. By contrast, the caste structure of the Arayans of Alleppey and Quilon discouraged innovations, coupled with their economic backwardness.

We must answer the reason why the *Beppukkar* are slow in adopting innovations although they have the same flexible social system as that of the *Valakkar* in Tanur. Some of the *Beppukkar* sought the assistance of the *Valakkar* in learning the new technique in the 1970s but the *Valakkar* were not only unhelpful to them but were also antagonistic to their innovations. For instance, one of the *Beppukkar* tried a nylon mackerel net in 1971 with the help of a *Valakkar*, but he was soon put out of business by the local *Valakkar.* Despite the hostility of the *Valakkar,* however, six out of 22 fishing units among the *Beppukar* have adopted nylon mackerel nets.

The introduction of nylon nets and mechanised boats and trawlers coincided with the demand for prawns in the USA. This invited traders from Trivandrum, Quilon, Kottayam, Alleppey, Cochin, Cannanore and Calicut who were willing to buy fish at higher prices for export. Thus, the Mappilas got windfall gains from the sale of their prawns. The advent of new traders at Tanur helped the *Beppukar* also to realise better prices for larger fish which they caught by their traditional hooks-and-lines. This changed the economy of the *Beppukar* to a limited extent from subsistence level to cash.

The Mappilas' ties with the outside world did not affect the customary relations between the owners of fishing units and fishing labourers. The services of a large number of fishing labourers were in great demand as the demand for fish increased by leaps and bounds. In fact, the patrons were only too eager to retain their labour force. Thus, the traditional relationship between patrons and clients received a fillip in the changed economic environment.

An important question that may be asked here is, to what extent is the caste system breaking in the modern fishing enterprise? It is evident from our above discussion that the *Valakkar* and *Beppukar* still follow their traditional specialisation of fishing. Yet it has come to our notice that many castes of Kerala in recent years have taken up fishing as their subsidiary source of income. For instance, we have seen that at Mappila Bay in Cannanore district, a Nair purchased a smaller trawler boat and seven fishing labourers are attached to his unit. Similarly, a number of Thiyyas in Azhikkode of Cannanore district have adapted fishing for their livelihood. A number of Iravas in Cochin, particularly at the Elamkunnapula village, worked as fishing labourers under their Hindu and Christian patrons. It is interesting to note that some of the Iravas owned fishing units and two of them have adopted mechanised boats during 1970s.

Organisational Change

Organisational change refers to a change in the principles of social organisation. In Tanur hereditary office is still basic to the social organisation. The hereditary principle is quite in conformity with the Mappilas' economic system. Status in the village is measured in terms of fishing units and other movable and immovable property one owns. Innovations have also enhanced economic capabilities. However, this does not mean that the hereditary position has been undermined by upstarts under the new economic environment. Epstein (1962: 324) observes: "Another aspect of the hereditary principle is the personal element in the relationship between the members of the social group. This personal element creates emotional attachments between parties to such a hereditary relationship". This observation holds good in the case of the relationship that exists between the patron owners of fishing units and their fishing labourers. Some of the owners of the unit take personal interest in the welfare of their fishing labourers. Conversely, the fishing labourers develop an emotional attachment to their patrons and take a personal interest in the fishing unit of their masters.

However, the economic relations have extended beyond the limits of the Mappila and Christian fishermen. This has resulted in decreasing the reliance on fellow fishermen to a limited extent; as a matter of fact, the personal element in the hereditary relationship has given way to an impersonal one. Thus, the Mappilas, and Latin Christians' participation in the regional economy has significantly increased their mobility. Personal drive, rather than membership in fishing units, has become the decisive factor in attaining economic status. This, in turn, has given birth to the competitive spirit which is seen in the eagerness to acquire ownership of shares by a number of fishing labourers.

A number of conclusions may be drawn from the discussions from the preceding pages on 'Capital, Saving and Credit'. The most obvious is that the majority of the Mappila fishermen of Tanur continued to be poor and do not have appropriate savings to meet their consumptional and occupational expenses. They mobilise most of the working capital by borrowing funds against the security of their catch. Financiers stipulate that the catch should be sold to, or marketed, through them and thus they get a monopoly over the fish trade. Although no interest, as such, is charged, it is implied that the financiers shall be paid 5 to 10 per cent commission on the sale proceeds. As a rule, the commission is not fixed and this results in the exploitation of the fishermen. This is also true in the case of the Hindu fisherfolk.

Ownership of a fishing unit with nylon nets and trawl nets, mechanised boats, big canoes, fibre *Vallams* and fibre glass *Vallams* and trawlers, confers a certain amount of prestige to the fisherfolk of Kerala. The highest status among the Mappila and Hindu fishermen is generally enjoyed by a fisherman who is the owner of more than one fishing unit and does not go out for fishing for his livelihood.

One of the important problems the Mappila economy faces today is the low level of income and the zero capital formation. Are the Mappilas interested in capital formation? An answer to this question is called for, because this will explain the Mappilas' failure to mobilisation of savings even after 30 years since our first field work in the 1970s. It is common knowledge that the Mappilas are not very keen to save but only like to maintain their current level of consumption. They have no concern for trawlers and trawl nets although they have adapted nylon nets and mechanised boats and fibre glass *Vallams*. This is partly true in the case of *Beppukar* because of their poor economic background. However, the *Valakkar* are interested in investing their savings in nylon nets, mechanised boats and *Valiya Vallams* and in agricultural lands and buildings.

One of the difficulties in capital formation in fishing is the narrow margins within which most of the Mappilas keep their assets. They are mostly kept in liquid form and so the Beppukkar find it difficult to accumulate capital. Thus they live at a subsistence level.

The Mappilas' economic activities are based partly upon their ritual relationship and kinship. This is evident from the rate of interest charged on loans which varies according to the social relationship of the debtor and creditor. Uterine and affinal kinsmen are preferred as creditors among the Mappilas even though fishing equipment are given as securities. The same is true in the case of ritual friendship. Social status is a relevant factor in obtaining credit. The *Valakkar* borrow from non-fishing Muslims as well as rich *Valakkar*, but *Beppukar*, for reasons of prestige, prefer to take loans from the *Angadikkar* (non-fishing Muslim traders) only.

Among the Mappilas, it is the husband who earns the livelihood, but it is the wife who manages the household affairs. Decisions about the mode of spending or saving are taken in consultation with her but the actual form of investment is left to the discretion of the head of the family. Group investment decisions are by and large taken with the consent of either member, including the head of the household. As the traditional economic system has been affected by innovations during the last four decades, decisions in respect of investment are taken individually. Previously the head of the kin group was able to mobilise the saved up resources of his group for investing in a new boat or net. Nowadays the head of the household takes the initiative in mobilising the credit for investing in fishing equipment by conducting feasts and tea parties. The organiser of the *Kudikalyanam* or *Cayakalyanam* receives gifts from kins and guests. If there is any surplus left after meeting the expenses, he uses this amount for investing in a fishing boat or net. Entrepreneurship coupled with the transformation of the economy of the Mappilas from subsistence to cash has helped them to mop up the savings of their kinsfolk as well as for investment in paddy lands, coconut plantations and nylon nets, all of which require substantial amounts. Investment in fishing equipment and land is considered the best form of savings by the Mappilas. This also enhances their prestige in the society.

Gold, in the form of simple jewellery, is favoured by the Mappilas as a kind of saving. They show a marked preference for weight and quality of the gold but never regard expensive workmanship as a waste. Jewellery is generally pawned with private banks in order to meet emergencies.

Some of the Mappilas hoard the cash but the wise ones among them have opened accounts with private banks. Rich fishermen always hold a sizeable amount of cash in order to invest in paddy fields, coconut gardens and arecanut plantations if an opportunity dawns upon them. Young fishermen invest their savings in luxury articles such as smuggled radio nets and wrist watches. They also invest money for getting visas to work as fishermen in foreign countries.

Wealth is not considered the end of, all but as a means to enhance social status. The Mappilas say that if a wealthy man does not use his riches for the welfare of the poor, it will not only bring unhappiness to him but also incurable diseases. If a fisherman buys paddy land, his neighbours often would comment that "he wants to own everything".

In our discussion on the pattern of borrowal we saw that the concept of credit exists among the Mappilas. "It involves an obligation by the borrower to make a return and confidence by the lender in the borrower's good faith and ability to repay" (Firth, 1964: 29). The fishermen borrow both for the purchase of materials for consumption, and for production. The former includes "normal consumption borrowings to tide over seasonal shortage and special consumption-borrowing of a social order to meet the expense of a marriage or funeral" (Firth, 1964: 29). Borrowal for meeting the expense of weddings, circumcisions and funerals is a normal part of the Mappila society in the special area of economy and a person is looked down upon if he does not conform. Grain and cash are also not infrequently borrowed by the poor Mappila families to tide over difficulties created by poor catch. Such borrowals permit them to carry on with productive activities such as manufacture and repair of nets, building of boats, etc., Such loans are repaid when there is a bumper catch. The debtors are under an obligation to supply the catch to their creditors at 10 to 20 per cent less than the market value.

Various forms of term credit exists in the Mappila economy. Mappila traders who make weekly settlements in fact operate on the capital of the sellers. There is no dearth of agencies to provide long term loans to Mappilas. In our analysis we have focused attention on the general difficulties in their repayments, including the role of money-lenders. Sometimes the fishermen borrow money from many small money-lenders for short periods to make part payment to the initial creditor. These short term loans are advanced against personal security. Sometimes the catch is also given as security.

It is evident from our discussion that credits are even today mobilised through organising feasts and tea parties by the Mappilas. The fact to be borne in mind here is that the fund is raised by feeding friends, kins and neighbours for the expenses of which the invitees make fixed contributions in cash. The ingenious methods of mobilising credit provide a ready-made solution to long term investment in fishing equipment. But it is very doubtful if such mobilisation of credit can really contribute substantially to capital formation for economic growth.

Now the question may be posed as to whether failure to repay the credit will lead to a drying up of the sources of credit or the debtor is liable to some legal action. Not a single instance of the creditor taking recourse to legal action to realise the amount advanced came to our notice. However, we came across with a few instances of cutting of the future supply of credit when the debtor failed to repay the loans for a long time. "But the whole question of seeking a credit and repayment involves a theory of obligation not comprised merely in terms of economic and legal sanctions" (Firth, 1964: 32). What Firth has observed is quite true in the case of Mappilas even today. Besides fear of supernatural powers makes the fishermen repay the loan without fail. Interest is realised under the guise of commission from the sale proceeds despite religious prescription against it. Thus, interest is generally realised in a disguised way for the loans advanced. Credit is, thus, an integral part of the Mappila economy. This is also true in the case of Latin Christians and some Hindu fishing communities.

In the section on "Trade and Market", we have tried to discuss certain economic aspects of marketing of the Mappilas in the wider social context. We have shown in our analysis that the cause of shortage of capital is not merely economic but also stems from the existing social patterns. Wholesale fish trade is handled and financed mostly by the non-fishing Muslims (*Angadikkar*). The capital resources at their disposal and their business acumen are largely due to their social organisation. Thus, in this chapter we have focused attention on this and other inter-related problems in the light of both economic and the social forces.

The retail fish trade is done by hawkers as consumers are very widely distributed in the fishing villages of Blangad, Puthiyappa and Beypore. The Mappila hawkers working as small traders compensate for their limited capital by using their own (unpaid) labour and local knowledge. The wholesalers have to maintain a large establishment spread over several fishing villages and towns in the marketing region. This is difficult and expensive, particularly in places with poor communication and transport facilities. There are three distinct spheres and types of operation in the marketing system. They are: (1) large-scale trade using (relatively) "capital and intensive techniques" (Deway, 1964) which is dominated by the *Angadikkar* (both iced, fresh, cured and iced fish merchants), (2) small scale trade, which is mostly operated by the Mappila traders "working with a minimum of capital and depending on labour intensive methods, local knowledge and individual initiative" (Deway, 1963: 233), and (3) hawkers who are relatively small traders turned into headload workers.

We have shown in our discussion that there are no restrictions on the entry into the fish market. A number of fishermen depend on trading to support themselves in the absence of contractual or paid employment in the fishing units. There is a great proliferation of individual fish traders who need to invest only their limited capital. This proliferation seems to be the result of unemployment and under-employment among the Mappilas despite technological innovations.

The development of modern communications and transport since the beginning of the 20th century, particularly after Independence, has solved the communication gap to a great extent and fish is marketed to distant towns and cities as soon as the catch is landed. This has enabled the fish traders, particularly the *Angadikkar*, to centralise the trade. In olden times the communication difficulties did not allow circulation of market intelligence. Information was in fact transmitted from one middleman to another as an accidental part of business. This limited the area over which any trader could operate successfully, because it was difficult, if not impossible, to maintain face to face contact with people in more markets than one. This was the situation in 1970s. But the introduction of STD, ISD and mobile phones have enabled the Mappilas to know the market intelligence in every second.

Banking, which was neglected by the Mappila traders, has developed with the advent of modern technology and innovations. Banking habit has been cultivated by the wealthy traders including a few from the Mappila fishermen, in order to cope up with the increasing volume of business. These influential fish traders of Tanur, Puthiyappa, Beypore and Blangad fishing villages have kinship relationship with the poor fishermen who are tied to them.

Mathur (1978) observed that "the Mappilas do appreciate the value of time in their trade transactions. They know that the time is vital in fishing, processing, and in marketing, where a few hours' delay may mean a considerable adverse change in prices."

The social structure of the Mappilas influences the economic organisation in several ways. The Mappila society has kingroups, lineages, age grades, and other groupings of the kin which bind a large number of persons together. The kinship system is seen as a collection of interpersonal relationship between different kinds of kins which involve a distinct pattern of social behaviour, of rights and obligations, and of expectations of beliefs and values. Thus, blood ties among the Mappila fishermen link members with one another and with groups. Although the population in Tanur has increased ten fold since our previous study, neighbourhood groups based on a combination of kinship and economic interdependence are of considerable importance. In this connection, Mathur (1978) observed that "within these groups informal social and economic sanctions are very effective. The fishermen depend on each other for help, irrespective of the fact that they are traders or owners of fishing units, both in day-to-day life and in times of crises. The poor fishermen exchange labour for payment in cash or in kind. They in turn give their labour to the wealthier fishermen for cash wages and access to their fishing units. The patrons including the traders who are wealthy have enormous capital but need labour which the clients can supply. The interdependence of fishing labour on the one hand and the traders on the other provides an economic basis for strong informal sanctions in the fishing economy. It is also seen that the kinsmen in Tanur clustered in the local Kadappurams, have common fishing and trading interests. The big and small traders in Tanur are the residents of the same village and have social and economic commitments in their dealing with one another. As a matter of fact, there is perfect competition and free play of market forces since capital is fairly distributed, but the bases for a patron and client relationship are not lacking in fish marketing. In most of the trading activities carried on by the Mappila fishermen, small scale labour-intensive methods are found to be more advantageous than capital-intensive methods and so most traders in Tanur, particularly the *Valakkar*, dealing with small scale traders, prefer to operate independently and in competition with each other. But they have to contend with poor transport and communication facilities. Sometimes fish pass from hand to hand through a chain of intermediaries. On the one hand, the small retailers bring the fish to market and sell them to middlemen, who, on the other hand, operate almost entirely within the market place itself. The middlemen may then sell the fish to a housewife, petty trader or to an intermarket middleman who will take the fish to another market. It is interesting to point out that generally there are no social ties between the people involved in different parts of the chain of intermediaries. Thus, each small trader sells fish to small retailers who have the choice of many markets. There are a large number of traders who may buy the fish and in turn, have a wide variety of customers. The above observation is very much true in the case of Mappilas and Mukkuvas inhabiting the fishing villages of Puthiyappa and Beypore of Calicut district and Blangad of Thrissur district and Mappila Bay of Cannanore district.

The *Angadikkar* have established trade relations over wider geographical areas without unduly increasing the number of individuals involved. For instance, we have seen how some wholesale traders send fish for marketing to Trivandrum, Kottayam, Changanacherry, Quilon and Alleppey and also to Bangalore and Madras. The *Angadikkar* have monopolised the fish trade. The Mappila fishing community, particularly the traders, is sometimes hostile to, and often discriminates against them. The non-Mappila fishermen still monopolise the fish trade in the Malabar area. But in the southern districts of Kerala the Latin Christians are the monopolists in fish trade.

The striking characteristic of the marketing system in the Malabar area is the adherence to accepted business ethics. A trader who has promised to deliver certain species of fish on a particular date tries his level best to fulfill his commitment even if it means a financial loss, for it is essential that he keeps up his word. If the trader is found to be dishonest or undependable, both the Hindu and Mappila fishing communities sometimes bring strong informal sanction, social and economic, to bear on him. Another interesting feature of the marketing system is the grant of credit for the mere asking, and a trader who gives it knows he is assured of future catches. The advent of modern technology and innovations has made the rich fishermen wise and they sometimes send some of their members to Cochin, Trivandrum and other wholesale markets to negotiate large purchases directly from individual fishermen so that they have a good supply of fish of uniform quality which they can dispose of in one transaction at a reasonable price. In other words, if the wealthy fishermen are not satisfied with the price offered by the wholesalers in the fishing villages and are under no obligation to sell the catch to them, they may get market intelligence by sending a member to wholesale markets outside the village to find out the possibilities of sale there. But nowadays they get the market intelligence though ISD, STD and mobile communications. If the group succeeds in striking a deal, they hire trucks to transport the fish and send one or two of their members to receive the payment.

Sale on credit is a common phenomenon in the fish marketing system in Kerala. The traders contact a fisherman and offer him a tentative price per basket or *Kalli*, as soon as the catch is landed, based on the estimate of the prevailing market price. If the buyer forces down the traders, the price paid eventually to the fishermen falls accordingly. In this way the traders shift their risks on to the fishermen. The traders, however, bear the cost of handling and thus undertake all consequences.

"As has been shown in our discussion most of the capital comes from the earnings of the traders themselves. Sometimes the traders also take loans from private banks, rich fishermen and money-lenders. Each trader keeps enough out of his profits each day to finance all or part of the next day's trading. Cash is retained only for a few days before investment" (Mathur, 1978).

Another point worth mentioning in connection with the flow of capital in fishing villages of Kerala is the absence of formalities. Except for loans from the Fishery Department, Matsyafed and nationalised and private banks, formal documents and terms of transactions are rarely drawn up. This is because the Mappila, Hindu, and even Latin Christian fishermen are, on the whole, unaccustomed to using documents in their daily life. The low level of literacy makes it difficult to introduce formal credit instruments. This has hampered free flow of institutionalised finance.

As has been shown in our discussion in this chapter the cost of marketing is related to the transport and communication facilities, the availability of ice, packing and loading charges, commission to the intermediaries and curing and peeling charges in the case of cured fish, besides the quality and size of the species. Thus, the cost of marketing varies from one species of fish to another. As the wholesale trade is monopolised by the *Angadikkar*, Latin Christians, the Mappila and Hindu fishermen have little control over the marketing, processing and exporting of their own catch. Majority of them continue to be at the mercy of the traders.

Ritual and Supernatural Beliefs: Synthesis, Co-existence, Adaptation and Persistence

Those Mukkuvas who embraced Islam are called Mappila fisherfolk in Malabar. The Mappila society today presents the spectacle of a remarkable process of adaptation from the elements of Islam civilisation and retention of some of the original Hindu cultural traits. As a consequence of conversion, there are some modifications in the ritual idioms, beliefs, customs and practices among the Mappilas and pronounced changes in the complexion of rituals and magic.

As has been stated earlier, the break in the customary economic relations between the owners of fishing units and their fishing labourers in Malabar in general and Tanur in particular, did not lead to a severance of ritual relations that involved the individual households. Even today the Mappilas perform the rituals connected with the technology of fishing, calendar festivals, and life crises, and practice sorcery and magic by inviting their religious leaders and sorcerers. Hindu sorcerers are still required to render their services for which customary dues are paid. Moulavis and Qazis officiate at rituals in connection with the making and launching of new nets and boats. They are asked to perform these ritual duties as the fishermen consider them sacred. Clashes have never occurred between the activities and demands of the religious leaders (who are mostly non-fishing Muslims) and the fishermen. The advent of technological innovations has only buttressed the ritual ceremonies in Tanur and other fishing villages, thus giving rise to enhanced prestige to the religious leaders. Technology has thus cemented the ritual. Moreover many Mappila fishermen act as Khatib, cheifimans (Quadiyar) Mukri, and Moulavis in local mosques including the Jumat mosques. More than 1,000 young Muslim fishermen undertook religious education from the local Arabic college.

Mathur (1978) observes that "a portion of the day's catch is set apart for the performance of ritual thanks-giving to the Prophet, his predecessors and descendants. Besides observing the Muslim calendar festivals like Muharam, Id and Bakrid, the Prophet and his descendants are propitiated for getting bumper catches, curing incurable diseases and warding off dangers at sea. Ancestors are not neglected. Households propitiate ancestors by offering the favourite species of fish to the departed souls. No ritual is performed, when a fisherman receives his reward for doing any non-fishing work. Some of the neo-rich fishermen who have taken up cultivation as their subsidiary occupation perform a number of rituals for getting better harvests in honour of certain Tangals and other descendants of the Prophet. It cannot, however, be said that because farming activities have assumed importance in recent years in Tanur, the prominence of traditional rituals in economic activities (fishing) have decreased."

Thus, in Tanur and other fishing villages of Malabar, where the fishing economy is still rural, ritual occupies an important place in the economic activities. Here the ritual status is the dominant factor in the determination of the social status. There is, however, a tendency among the neo-rich fishermen among the Mappilas to gain ritual status through political influence. Sometimes they show more fanaticism in observing classical Islamic festivals than the non-fishing Muslim communities. Thus, we find an increase in the number of intra-Mappila relations with the advent of innovations, while ritual relations between the religious leaders (including Hindu magicians) and the Mappila functionaries remain unchanged.

There has, therefore, been some ritual change after the Mukkuva fishermen embraced Islam, but they have retained a good number of Hindu elements in the sphere of rituals and magic. Sometimes the Mappilas have tried to reinterpret their Hindu beliefs and practices in order to fit them into the universalistic Islam.

We have shown in the section on "Healing, Sorcery and Magic" how beliefs and practices centring around the supernatural play a significant role in the life of the Mappilas. Most of these are shared with not only non-fishing Muslims of the village but also with the Hindus, particularly the Mukkuvas, while some have been evolved and developed with reference to their economic pursuit.

Mathur (1978) said that "religious activities associated with diseases are in the form of magical rites and vows. Most of the diseases are attributed to evil spirits, sorcery and black magic and invariably a cure is sought by invoking saints. Hindu sorcerers are invited to perform the magical rites to ward off serious diseases caused by sorcery. This is more common in the case of the Mappila fishermen, unlike the local non-fishing Muslims. Vows in the names of various saints are another method of curing diseases. As a matter of fact it is very common for a Mappila to resort to vows in times of crisis, for the success of fishing expeditions and also for fulfilment of particular wishes such as offsprings. Generally, vows involve promises to offer money, gold, silver, pisiforms such as an eye, nose, leg or hand made of silver, oil, rice, sugar, food, cloth, cow, goat, etc., to the mosque which is associated with that particular saint."

"Where the local Hindu character of magic is still maintained, the magician should belong to a Hindu community. In all other cases the local Muslim character of magic is still recognised. The traditional character of Hindu magic and magical filiations of the performer find expression in several significant features of the spells. In some of them, as we have pointed out, names of mythical ancestors are uttered" (Mathur, 1978).

"A distinction between what Malinowski called 'systematic' and 'independent' magical rites and formulae has been made in our discussion. Some of the rites and spells are 'independently, employed by the Mappilas themselves whenever the need arises'. For example, when a fisherman paddles his boat at sea and an unfavourable wind blows, he utters a spell to make it abate and change. Thus, the incantation is a free individual act which is generally performed in any of the circumstances which require it. On the other hand "systematic magic" consists of numerous formulae, forming one consecutive series. Such series among the Mappila fishermen are associated with the building of a boat, making of a net and launching of new boats and nets. In fact, most of the rites and spells described in the foregoing pages belong to "systematic magic". We also saw that the magic bridges the gap between the supernatural world of Mappilas and the day-to-day activities. Thus magic partakes of the supernatural beliefs of the Mappilas. The Mappilas know fully well that the speed and buoyancy of a boat are owing to good workmanship, properties of good material and finally to the expertise of the carpenter. Yet the Mappilas believe that the performance of magical rites at the time of launching a new boat or a net adds something supernatural. Thus, the force of magic has been regarded by the Mappilas as something independent of surpassing even, the power of all other personal charms (Mathur, 1978).

"The characteristics of magic have been described with reference to the techno-economy of fishing and curing incurable ailments. Magic is a hereditary profession and the secrets are passed on by father to son. The magicians' services are in great demand for curing illness, warding off evil spirits and propitiating enraged spirits. The magician receives payment in kind and cash for his services. There are several kinds of payment: some are given periodically by a person for particular acts of magic, as in the case of sorcery or counter magic; others paid at different times for curing illness caused by sorcery. In some cases the payments made are substantial, as in the case of very serious illnesses and failure of catches for a long period. Sometimes they amount to little more than a formal reward" (Mathur, 1978).

Thus, from the discussion on rituals and supernaturalism, we notice that the Mappilas have developed elaborate rituals and supernaturalism to insure the risk of uncertainly. In this process, the Mappila fishermen have incorporated a number of original Hindu traits.

As has been stated earlier, the Mappila fishermen observe minutely all the Muslim festivals and ceremonies. This indicates that although that the Mappila form one of the endogamous groups of the larger Muslim community, they have integrated with the other Islamic castes to a great extent. In the course of this integration they have reinterpreted the Hindu elements in order to fit them into the Koranic tradition, while a number of local Hindu practices remain unchanged. Even today the Mappila fishermen are looked down upon by the non-fishing Muslims and are branded 'neophytes'.

Although the Mappilas observe the rituals connected with the Muslim great tradition and life crises, majority of them have little basic Islamic theological training. No Mappila fisherman has become an Imam, Khatib or even a teacher at the local Madrassa during 1970's. However, the Mappila fishermen have trained their own Moulavis who are invited to perform the rituals, but they are paid only nominal rewards.

Persistence of Non-Koranic Tradition

"The important trait that struck me is the Mappilas' belief in the spirit of the departed. The Mappilas confirm that the *Roohu* (soul) live in the brain of a man. The *Roohu* is taken to Allah by the angel Israel, when a fisherman dies. When the soul leaves the body it is called *Ruhan* (ghost). The *Ruhan* of the dead takes its abode among the living, particularly in the house of the deceased for 40 days and returns to the grave and lives on the *Meesankallu* (grave-stone). The *Ruhan* visits the house of the deceased on the morning of Monday and Friday. During the period of ritual mourning the Mappilas conduct the recitation of the Quran and the Confession of Faith as an act of appeasement of the *Ruhan*. The Hindus of Kerala, particularly the Mukkuva fishermen, propitiate the departed soul for 41 days, though the rites are concluded on the 16th day by offering rice balls etc. The Mappilas belief in *Ruhan* seems to be a vestige of their original Hindu faith. Dutiful sons annually honour the memory of their departed parents by conducting memorial services and by offering dishes which they relished. The Hindu fishermen also conduct the *Sradham* (ancestral worship) annually by offering the dishes which the departed liked best" (Mathur, 1978).

The very fact that the Mappila fishermen propitiate the non-Koranic spirits like *Karinkutty, Kutticcattan, Ottamulakkaratty, Akasa Gandharvan, Cetta, Kalladimuttan*, etc., makes it clear that the

Mappilas have not abandoned the magico-religious ideas of the local Hindus, particularly those of the Hindu fishermen, the Mukkuvas. There are some magicians who are reputed as 'possessed' curers. It is interesting to point out that the Mappila fishermen often consult not only Muslim diviners but also Hindu astrologers to find out the suitability of a particular sorcerer for a given patient.

"*Homam* is one of the curing techniques employed by both Hindu and Muslim sorcerers when the disease is caused by spirits like *Karinkutty, Kutticcattan, Apasmaramurti,* and *Gulikan. Ulinnuvannal* (waving off) is another important technique of treatment. The performance of *Homam* and *Ulinnuvannal* is also the favourite method of caste Hindus as well as Hindu fishermen for curing diseases caused by spirits and sorcery" (Mathur, 1978).

"The Mappila fishermen of Malabar like their Hindu counterparts believe that smallpox and cholera are caused by the spirit called *Cetta*. A coconut or a coin is waved over the head of the victim after invoking the local goddess *Bhagavaty* and it is hung inside the house. As soon as the patient recovers from the illness, the coconut or the coin, as the case may be, is sent to the temple of the *Bhagavaty*" (Mathur, 1978).

"Again the Mappilas' belief in 18 spirits that cause fits, epilepsy and convulsion is similar to that of the local Hindus. Both communities make the patient smell a leaf of Aruta plant (a medicinal herb). Sometimes wicks made of the Aruta and *Karinocci* (Nergundo) leaves, Tulasi (*ocimum sanctum*) and Ramaccam (*cuscus grass*) are burnt for inhaling their medicated vapours. This method of curing is widely prevalent among both the Hindus and the Mappilas. At times the blood of a fowl is given to the patient to drink. If the epilepsy is very violent in nature both *Ulinnuvannal* and *Homam* are conducted" (Mathur, 1978).

"There are some interesting methods of black magic practiced by both Hindu and Mappila fishermen. The sorcerer draws some cabbalistic figures on the skull of an oxen, snake or a human being. After hammering a number of nails into it the skull is buried beneath the enemy's house. This is believed to cause his death" (Mathur, 1978).

"We have outlined here the successive application of black magic by the sorcerers among the Mappilas and the local Hindus in curing illness caused by sorcery and for causing ailments and death to the enemies. There is co-existence and integration of the traditions of Hinduism and Islam. Thus, the local Hindu tradition is yoked to the universalistic Islamic tradition" (Mathur, 1978).

The results of conversion from Hinduism to Islam are very complex. The Mappilas have abandoned a large number of religious observances but few rituals. They have not completely abandoned the traditional ideas about certain gods and goddesses like Ganapaty, Bhagavaty and a large number of non-Koranic spirits.

The Mappila fisherfolk, particularly the *Valakkar*, have participated in the market operations of the Malabar region for a number of years and their economic orientation has come to vary widely from the cooperative, communal and kinship-based orientations of the Hindu fishermen. In this connection what Frazer has said is true in the case of the Mappilas. He observes: "This economic orientation can be summed up as stressing individualism, competition within limits, and acquisition and utilisation of capital goods" (Frazer, 1960: 219). Individualism, as a value, is constantly displayed by the Mappilas, particularly the *Valakkar*. Stress is given on diligence and productive enterprise by the individual than on group activity. Each Mappila excels himself in fishing. When he is called upon to do some services to others, it is strictly on reciprocal basis. Emphasis is placed upon the individual share in fishing units. The economic motives predominate, although some of the values connected with fishing appear to be beyond economic realms. Although fishing expeditions generally require the cooperation and coordination of all the members of the crew for efficient netting, the avowed orientation of the fishermen is individualistic. The steersmen of the fishing units, through their cosmological and ecological knowledge, are individually responsible for the success of the expedition. Crew members are recruited and remain with the fishing units on the basis of their skill but not exclusively on the basis of kinship or common ownership. This is true even today after the adaptation of modern technological innovations like GPS, wireless and echosounder.

The operation of individualistic and ecocompetitive values in the economic activities of the Mappilas is supported and complemented by an orientation towards investment in capital goods, paddy lands and arecanut orchards. The introduction of nylon nets and mechanised boats, trawlers and canoes illustrate the positive values attached to investment in fishing tools. This is also true in the case of the Bovis-Mogeyar of Kasargod, Mukkuvas of Puthiyappa, Beypore and Mappila Bay, but not in the case of Mukkuvas of Kasargod.

Continuity and Change

From the accounts presented in the preceding pages it is apparent that the Mappila fishermen of Malabar have shown remarkable resilience in accepting modern technological innovations. By and large, this has not only displaced their traditional structure and cultural values but in some ways it has even reinforced their cultural focus. They are even more committed to fishing now than was the case before. This is evident from the fact that 300 to 500 Mappila fishermen migrated to Arabian Gulf countries to work as fishing labourers from Tanur, Beypore, Puthiyappa and Blangad. Also, it has been recorded that the traditional social structure and status hierarchy has been fairly adaptive to the demands of the new technology. Both Hindu and Muslim rituals connected with fishing continued to be applied to the new techniques.

We have discussed at some length that while the Mappila *Valakkar* went ahead in adopting more of modern technology the *Beppukar* lagged behind. The lag is not mainly on account of differences in value systems but due to differences in initial technology and economic base. This is even today true in the case of Tanur village.

We have described in detail how through deep and intimate association with rituals, mythology, social structure and social sentiment, fishing is essentially a moral activity. In spite of its connection with the market mechanism, there is an abundance of traditional lore that the Hindu and Mappila fisherfolk possess, not only in the technology of fishing, but also in the marine ecology. Climate is indeed very impressive. The superstructure of supernaturalism and rituals do not distort the inductive rational care of their ethno-science.

However, we do not want to convey an impression that nothing has changed on account of incorporation of modern technological innovations. We find that the traditional spirit of individualistic competition has been exaggerated in the face of the new opportunity.

Also the lower strata of fishermen belonging to Mappila and Hindu fishermen are getting more opportunities to own shares in the new fishing units. However, the overall picture is that the bulk of the fisherfolk of Kerala engaged in the primary activities of fishing are under the stranglehold of the enterprising capitalistic upper strata. It appears that this stranglehold cannot be overcome through indigenous efforts in Tanur and other fishing villages of Malabar but some kind of systematic government intervention is required.

The important fishing communities that have been included in the discussions here are the Bovis-Mogeyar, Mogavirar, Arayan, Nullayan, Mukkuva (all Hindu communities), and non-Hindu fishing communities professing the Latin Christian and Muslim faiths. While the Bovis-Mogeyar, Mogavirar and Mukkuva are found in the northern region of the state sharing similar cultural ecology and environment along with the Muslim fishermen, the others are living in the south. The Mappila fisherfolk are numerically a dominant Muslim fishing community inhabiting the Malabar area, while the *Marakkar* (Muslim fishermen) are found in most of the fishing villages in the south, particularly in the Trivandrum district. The Latin Christian fisherfolk are found in localised fishing villages in the north, and are concentrated mainly in the southern districts. They are called variously as *Mukkuvar* (Latin Christians) and *Marakkan* (Muslim fisherman) in Trivandrum, which names seems to be generic, and not caste-based.

The fact that gains predominance in our study is that some fishing communities like Bovis-Mogeyar among the Hindu fisherfolk have been able to successfully adopt modern technologies introduced in marine fishing, while the other Hindu fishermen communities particularly the Mukkuvas occupying the same marine environment failed very obviously and for reasons that were generally unknown. It must be also noted that the Bovis-Mogeyar have also adapted trawlers and trawl nets. It may be stated here that it was not that the community that succeeded or failed belonged to any particular caste or religious group as described and concluded by earlier scholars like Klaussen (1968), Platteau (1982), Abraham (1982, 1985), Kurien and Mathew (1982), Kuruvilla et al. (1982) and Gulati (1984). But it could perhaps be that the fishing communities studied by earlier researchers were too small or too localised.

"A scrutiny of the different fishing communities would show that their successes and failures depend on the traditional organisational structures or networks that existed within these communities, and the abilities of these social groups to adapt to the technological changes and provide supports to its members. The attitudes to change were also very significant. It is observed that the successful fishing communities have the willingness to try and to innovate new techniques. They are therefore very quick in adapting the technological changes despite having retained their traditional cultural heritage for centuries" (Zachariah, 1996).

The religious organisations of the Latin Christians, Mappila fisherfolk and Hindu fishermen were also the traditional structures that dispensed the mechanisms of social control within their respective communities. However the establishments of the religious, political, ritual and economic bodies of the successful fishing communities have an additional feature that runs on similar lines. This is that they all have provisions for group related development programme, mechanisms to mobilise funds for these, and to use the same for the different needs of their members in acceptable manners. The main variations in this between the Bovis-Mogeyar (and the Mogavirar), and the Latin Christian and Muslim fisherfolk is that, the former mobilises funds from their caste members who were predominantly traditional fishermen, while the latter comprises of members of different religions with fishing as the main occupation.

"The fund mobilisation that had origins in religious activities and charity works among their own members, like help for social needs and education, were extended in these cases, to help poor fishermen purchase or replace fishing equipment and gears. The religious organisations either on their own or through representative committees, discussed these matters and took necessary steps to get their objectives fulfilled. In the case of the Bovis-Mogeyar fishing community, initially the temple organisation, or the *sabha*, mobilised funds through membership fees alone, mainly for the upkeep and maintenance of their temples, for religious activities, as they had no other means to do this. They realised the need to provide financial assistance for their poorer members, and were able to collect money for the same, mostly from their better off members. These aids to begin with, was to help deserving and poor students among them, to construct or repair dilapidated houses, to assist families during life crisis ceremonies, and later to facilitate poor fishermen replace fishing equipment like boats, oars, nets, etc., that were lost or damaged in calamities. Therefore, the *sabha* (caste temple organisation) to extend this kind of activity to the purchase of more sophisticated gears was not too deviant or unfamiliar a step. A similar trend can also be seen among the other successful groups too" (Zachariah, 1996).

"At the advent of the modern technological innovations that were introduced in the 1960s and 1970s, besides the above mentioned facts, the Mogeyar fishing community here in Kerala had the advantage of the development programmes that were already being introduced among their kinsmen in the adjoining State of Karnataka. The positive gains of that situation for these members were in numerous ways. Salient among these were that some of the Mogeyar fishermen were able to earn better wages by joining their kinsmen's boat units, they were able to learn from their own and, or from the other fishermen's experiences. They had also the advantage of the supports of a well managed and informed official machinery in the Karnataka State. Their temple organisation was also fortunately able to step in during this time to provide help for acquiring and establishing some essential infrastructures like nylon nets, trucks, ice plants and freezing units because of their inherent capacity to mobilise funds" (Zachariah, 1996).

The *sabha* use their influences, acquaintances and knowledge to secure loans for their members from banks, and ensure that these are repaid by the borrowers. They help their fishermen organise themselves so as to be able to capitalise on the new innovations, and help with advices, planning, etc. The financial inputs of the *sabha,* although very modest in comparison to that of the government agencies, are timely and helpful to their members. All these, and especially the supports of the *sabha* are important contributions that enable the economic successes for this community. The changes the form of technological advances in fishing that were coming up in the areas described can be compared to a kind of crisis and the responses of the Bovis-Mogeyar community to these changes as being responses to the crises. They had to respond to these changes probatively, failing which they would have been doomed to an economic death.

In case they failed to do so, as did the other Hindu fishing communities, then they would have also been overtaken and superseded by other groups, especially the 'encroachers' into this field, thus being thrown into worse situations. Hamblin (1958) has portrayed how groups behave in response to crisis situations and the behaviour of the Bovis-Mogeyar community is similar and comparable to that. If the fishing communities do not accept progress they will not remain stagnant but will retrogress instead, as had happened among the 'failed' groups.

Among the Bovis-Mogeyar, the beneficiaries are expected to repay the *sabha* for most of the assistance provided from their improved earnings or when they start to earn. This reinforces a feeling of worth and value among the beneficiaries for the assets and help provided, and the efforts made by the *sabha* on their behalf. This also recoups for the *sabha* the money they invest, to be reused for other similarly needy causes. A feeling of complete dependence and abusing sensitivity to charity as such is also thus avoided.

Any assistance provided should have therefore provisions also for repayment they said, and what the *sabha* did. Moreover this strategy ensured that no one took unwarranted liberties with the help provided. They also felt that beneficiaries valued such assistance better than an all-free assistance. Such consideration for the feelings and values of their members even if poor, is seen to be conscientious and just. Dittes (1959) had pointed out how the self-esteem of a group members functions as an attraction for the members to remain in the group. Assistance as against charity, contributes to the feelings of self-esteem among the Bovis-Mogeyar group too.

"Psychologically the Bovis-Mogeyar and Mogavirar communities had experienced lesser of the demeaning feelings that the other Hindu fishing communities had suffered on account of the latter's inferior caste status. These other communities, like the Arayan, Nullayan and Valan were in a psychological lockup, where their sense of inevitability created in them feelings of hopelessness. Quite often they felt that no matter what they achieved or did, they could never hope to shed the stigma of their socially low caste ranking. Some of the fishing community members professing other faiths too expressed, and wallowed in such sentiments. The Mogeyar community which enjoyed a relatively higher social status, on the other hand had lesser degrees of such negative emotions to handle or overcome, although they too had suffered untouchability practices in the past at the hands of caste Hindus like the Brahmin and Bant" (Zachariah, 1996).

The *sabha* paid little attention to such negative outlooks and instead placed high importance and values on other achievements. These age-old feelings and frustrations of dejection and hopelessness on account of low (caste) status are however becoming things of the past. Nevertheless, as mentioned above, even till recently it had played an important role in hindering the processes of development, especially among the Hindu Arayan, Nullayan and Mukkuva fishing communities.

In undertaking this detailed work of an ethnographic analysis, we have sought to address central issues in Indian Anthropology and culture, as well as more general issues of social theory. In analysing the changes in the lives of the Mappila, Latin Christian and Bovis-Mogeyar fishermen, we have tried to show a gulf between two major issues of social theory, the inadequacies of each, and the possibility of going beyond them, in a way that can begin to span the gulf.

"The first body of social theory is the Marxist tradition. Despite all attempts at reassigning a greater role to the 'superstructure' in Marxist theory (such as the Althusserian attempt to delay economic determination 'till the last instance'), the crushing weight of 'economism', and even more fundamentally, of an attempt to separate culture from material practice, allows the residual power of old Western dualisms to continue to haunt the philosophical premises of Marxist theory and political practice" (Ram, 1991).

For several Marxist scholars (Mencher, 1974; Meillassoux, 1973; Gough, 1971) what is distinctive about caste is the organisation of power around a moral hierarchy of purity and pollution – that is seen as epiphenomenal, reducible to the exercise of economic power in the appropriation and exploitation of labour. A similar logic is played out in explanations of women's subordination: women become subsumed under a broader social category, that of the class to which they belong among the fisherfolk of Kerala.

According to Sahlins (1976: x), anthropological concept of culture as 'meaningful orders of persons and things' has the potential to 'leave behind just such antique dualisms as mind and matter, idealism and materialism' (p. ix). Yet nevertheless the analysis of social systems and historical processes in terms of culture and symbolic structures is not necessarily more productive of insight than a materialist approach. History and culture may disappear from sight under the weight of static, quasi-eternal, intellectually elaborated structures of explication which bear the all-too-visible mark of the privileged standpoint of the western scholars.

"Economistic studies of caste have their 'culturalist' counterparts: the work of Dumont, for all its suggestive insights, stresses the moral and ideational elements in the organisation of social structure virtually at the expense of materialist concerns, for example, with labour and livelihood. A more striking example is the understanding of caste, kinship and gender in India, fostered from the 1970s by the Chicago brand of cultural anthropology. These categories have been interpreted in terms of culturally-defined transactions in food, bodily fluids and blood (Marriott and Inden, 1977; Inden and Nicholas, 1977). Daniel (1984), in exploring how Tamils construct their view of the person, has carried further the notion of 'ranked substances' as the basis for negotiating everyday life in a Tamil village. Studies of Tamil women by Egnor (1978), and Wadley (1980) again exhibit the strengths and weakness of what I am calling a purely culturalist account, conveying a vivid sense of symbolic constructions of meaning, located in the context of kinship and ritual, but ignoring or minimising historical processes, relations of power, and what in the Marxist account would be called 'relations of production'" (Ram, 1991: 231).

In the report on "Ecology, Technology and Economy: Continuity and Change" we have explored areas of the life and culture of Mappila fisherfolk, Hindu fishermen and Latin Christians which are quite intractable if viewed in terms of these European polarities of thought. The fishermen's interaction with the physical environment cannot be explained wholly in terms of economic relations. Ecological data and the question of fishing people's location in the geographical landscape have long been culturally overlined by the traditional Kerala culture. The geographical zones of Malabar, Cochin and Travancore are inscribed with cultural meanings reflecting a sense of hierarchical ordering which plays just as important a part in the fisherfolk's of Kerala self-definition as do habitation or occupation. The system of ranking social groups according to criteria of purity and pollution is a later historical accretion among the Hindu fisherfolk of Kerala, adding another layer of cultural determination to the issue of fishermen's relationship with their ecology and environment. The agrarian orientation of the ancient Kerala world-view, combined with caste society's labeling of fishing communities as polluted and inferior, has intimately shaped the historical evolution of the Kerala fishermen's definition of their work, their community, and of caste itself.

At the same time these cultural determinations have been radically shaped by the specificities of fishing as a sub-economy and labour process, which is not directly dependent on the dominant land-holding groups in Kerala society. The fact that the relations of the Hindu, Mappila and Latin Christian fishermen with caste society are based on trade and monetary exchange rather than direct servitude, has had enormous implications in tempering the lowly status ascribed to them. It means that fisherfolk of Kerala, particularly the men who are directly engaged in fishing, are able to construct an alternative self-image based on their relation to the sea rather than to the land. But here again, their relation to the sea is not a material relation as such: it is dominated by a religious conceptualisation of the sea and of nature in general. Such conceptualisations are remote from the mind-body dualism of Western philosophy.

In our presentation of these cultural constructions (of caste, the landscape, the labour process), we have tried to avoid discussing them as static, unified, internally consistent cultural products. Rather, we have argued for culture as a field of conflict, internally split by relations of power and contestation between dominant and subordinated groups. The Hindu fishermen contest the upper caste Hindu view of them by emphasising the independence of their existence out at sea, the internal homogeneity and tight-knit nature of their community, their orthodox Hindu belief system and, above all, the positive valence of their sense of difference from agrarian society. The contestation is often a latent one, full of implicit possibilities which only partially come to the foreground in specific historical settings.

As the objects of these complex and contradictory attributes, Bovis-Mogeyar women and Latin Christian women enter into relationships with their physical environment, with the economic resources available to the community, and with the labour process, not simply as versions of the 'economic man' posited by materialism, but as bearers of a (contradictory) gender-specific subjectivity. The battle between the Church, Mosque and Temple can be seen as a metaphor, as well as a very real site for the cultural tensions which feed into all the activities and relationships into which women enter. Their means of access to the resources of nature are not the same as that of the men in their community; they cannot go out to the sea and fish, and indeed, they cannot even occupy the spaces contiguous to the sea, for fear of upsetting the balance between men and their environment.

A long as women view maternal nurturance (of children and men) as their prime area of self-definition and keep their dimensions under restraint, they are accepted. Indeed, they may even succeed, as we have suggested, in covertly enlarging the actual physical spaces allotted to them. But if women step outside the charmed circle or the Lakshman Rekha (to use an image from Hindu mythology), of domesticity, legitimised by the Christian, Hindu and Mappila model, they run a profound risk. The fish-traders in the Mappila, Latin Christian and Bovis-Mogeyar communities are a living example of what awaits the fisherwomen who compete directly with men for trade and cultural space. The fisherwomen themselves are culturally disowned and their femininity subject to hostile rejection. Their success or failure as traders cannot be judged or even adequately described in purely economic terms.

In this connection, Ram (1991) held the view we have argued for, which is placing culture and sexuality at the centre of capitalist transformation and the new division of labour between men and women renders problematic one of the key distinctions between pre-capitalist and capitalist society. This distinction has been inherited from Marx. "In Marx's formulation, the distinction revolved around this very question of determination. In his characterisation of 'Forms which precede capitalist production' in the *Grundrisse* [(1857–58), 1973: 471–79], he describes pastoral society as one where 'the communality of blood, language, customs' is the 'first presupposition ... for the appropriation of the objective conditions of their life, and of their life's reproducing and objectifying activity ...'

(1973: 472). Similarly, in Asiatic landforms, the relation of the individual to the conditions of labour is mediated by 'the total community' (1973: 473). In sharp contrast, the situation under capitalism ruptures the mediating role of culture and the community, between the individual and the conditions of labour: the economy comes into its own as the ultimate determinant of all human activity, once 'production appears as the aim of all mankind and wealth ... the aim of production' (1973: 487, 488). There are a number of problematic features in Marx's formulation, not least of all his association of pre-capitalist culture with the sphere of nature, a stage where man is no more than a 'species-being' (1973: 496). The inadequacies of such a position from the perspective of contemporary anthropology have been clearly articulated by Sahlins (1976: 49)".

"The thrust of my critique of Marx's position lies elsewhere: it is aimed at the model of determination he posits. I have shown that culture does not cease to play a crucially formative and shaping role with the advent of capitalism. Despite the loss of control over the organisation of production and domestic life in the community, culture is mediating the very process of capitalist transformation among the fisherfolk of Kerala, there is by now a vast literature demonstrating that capitalist forms of production, new technology and market relations of exchange do not simply displace pre-existing social formations. Instead, as writers such as Frank (1978), Laclau (1977), Wallerstein (1974) and Wolf (1982) tried to show, capitalism typically enters into complicit relations with pre-capitalist relations of labour utilisation and exploitation, propping them up and even accentuating them in the process. The debate has largely been confined to a strictly economic and political terrain. Ironically, the closest one gets to an appreciation of cultural factors is in the 'modernisation' theory of the 1960s, where 'culture' was singled out as one of the key obstacles to progress in the Third World" (Foster, 1967; Huntington, 1968; Ram, 1991: 235). Modernisation theory is now mainly discredited for its intellectually and politically unacceptable premises, entailing an uncritical universalisation of the western model of modernisation. However, the alternative model which has replaced modernisation theory is a mechanistic one. Systematic forces such as modes of production operated without any reference to the cultural construction of meaning, let alone the capacity of social subjects to negotiate and reconstruct the social meanings of their world (Banaji, 1977; Hindess and Hirst, 1977; and Wolf, 1982).

In the report we have made attempts to show the importance of cultural constructions of sexuality, particularly fisherwomen's sexuality, in rendering capitalism itself a gender-specific historical process. We have also presented this culture as a relatively fluid system of meanings, such that even those who are most constrained by the system can also selectively modify and re-interpret its hegemonic interpretations. These re-interpretations range from the unconscious, 'hysterical' resistance of ill and possessed fisherwomen who rupture the smooth reproduction of daily life, to the older women who travel far and wide in the name of their religious and familial obligations and extend the spaces of domesticity well beyond the confinement of the domestic dwelling; and finally to the select group of educated young girls who have succeeded in manipulating the cultural capital available to them to pursue upward mobility without sacrificing their acceptability within the community.

Is this, then, an interpretation which leans towards the 'culturalist' side of the old culture/materialism dichotomy? We have tried throughout the report to anchor our account of culture firmly in recognition of the independent transformative power of capitalism. We have traced the effects of capitalism not only on work and the labour process, but an important aspect of daily life which are of equal importance to fisherwomen. For example, marriage payments play a fundamentally different role in today's fishing economy. Women's dowry now functions increasingly as a source of capital, to be invested in male ventures either into wage labour or into small-scale capitalist entrepreneurship in ownership of trawlers, paddy land and coir factories. Similarly, in household formation, the pattern favouring female solidarity and management of household finances is slowly altering. Men's incorporation into wage labour is taking them further away from the village, whereas women's wage labour ties them more firmly to the four walls of the home than was the case even with domestic work. Given such divergent modes of work, the patterns of living and residence are drastically affected. Households become environments of women, children and the very old for several months of the year. Where separation from the male wage-earner lasts even longer, as where men have gone to the

Arabian Gulf countries, the woman may find herself pressured to live with in-laws or parents, thus conforming to the pattern of the joint family, close to the Brahmanic patrilineal ideal, rather than to the pattern of matrifocal, nuclear families typical of the Arayans, Bovis-Mogeyar and Mukkuvas of Kerala.

However, rather than collate culture into social relations of power, we have kept it distinct, at least analytically, from the conceptual fields of discourse, social relations and practice. Culturally constituted subjects are not formed by any one of these in isolation, nor do the three fields reinforce a common within each field and between them, which allow us to avoid the determinism of a widely prevalent extreme sociologism which reduces the problem of the construction of the subject to an effect of social relations of power. The dissonance or lack of a neat fit between discourse, practices and social relations allows room in which to posit a degree of agency for the subject. Thus, the fisherwomen successfully evade the full force of the religious discourse on the domesticated and maternal 'feminine', and carve out for themselves a much wider sphere of legitimate action in their practice. Such practice in turn becomes possible only because the discursive field is itself internally given. The hegemony of the religious discourse is in perpetual tension with popular religion, thanks to the ambiguous figure of goddess Bagavathy who occupies a position in Hindu fisherfolk's consciousness analogous to the figure of the repressed, disowned, but always reappearing. The counter-hegemonic discursive field occupied by the figure of Hindu fisherwomen is in turn reinforced by social relations which assign to women a position of authority in kin and economic matters.

"There is nothing immutable about this balance of forces between the different fields. New kinds of social relations, involving male migration and a tendency to female dependency, erode key aspects of women's ability to negotiate the more restrictive interpretations of femininity. Representations of women such as one finds in the journal of the young, ambitious migrant fisherman John – in which women, while remaining maternal, are relied on to supply romantic intimacies missing in life away from home – may already register a discursive shift, one which women may be in a weaker position to negotiate in practice" (Ram 1991: 237).

Our report has explored the lives of Latin Christians, Mappilas and Hindu fisherfolk as social subjects who are situated at the nodal point of heterogeneous and distinct relations of power. The fisherwomen have to be considered as members of an occupationally polluted community, which is also both a majority and minority community, as much as in terms of the relations of power based on gender and class. Attention to the multiple determinations of fisherwomen's lives may take us a little further in challenging western representations of Indian women, which often emerge untransformed in ethnography. The fisherwomen of Kerala, rich in spiritual strength but oppressed materially, as she emerges from the cultural anthropology of the 1970s, is not so different from the woman who inhabited the discourse constructed in the nineteenth century in the dialogue between colonisers and male elites. The polarisation between the confined, spiritual upper caste woman and the lower caste woman who is free from the burdens of caste purity and sexual disciplines (read: free of culture itself), but exploited simply as a member of her class, partakes of a complex history. It has been our aim to show that in giving an adequate and meaningful account of the lives of Kerala fisherfolk in general and fisherwomen in particular; one is necessarily engaged in the wider project of transgressing and subverting such artificial polarities.

Concluding Remarks

We have seen from the forgoing pages how the marine environment exerts causal force on fisherfolk and their thoughts, feelings and actions. Furthermore we looked at their economy, technology and the institutions related to marine ecosystem.

Robert B. Edgerton (1971: ix) observes: "That man is not a passive recipient of his environment but is instead an active reactor against it may reduce our ability to predict how his behaviour will relate to specific environmental conditions ... of course, other environments are far less coercive in that they permit a wide range of economic adaptation even within simple technological systems. Similarly, sociocultural environments differ in their coercive force. Some, like many religious orders, may be highly intolerant of ambiguity or innovation; others, like some academic institutions, may be quite tolerant. However, as yet we know far too little about the relationship between the limits of physical environments for economic exploitation and the development of flexible or restrictive sociocultural control systems" (p. xi). The situation among the fisherfolk of Kerala is no different from the above observations. For instance the Hindu fisherfolk are highly intolerant of innovations whereas Latin Christians fisherfolk and Mappila are very quick in adapting the modern innovations. The differences in adaption of sociocultural environments and technological systems are due to the ecological differences between these communities.

Each of the fishing communities in Kerala had long been subject to modern innovations. It is not unreasonable to assume that the three religious groups should have been uniformly affected by the external influences. But the Mappila and Latin Christians were more affected by the external forces. It was found that the behaviour of the Hindu fisherfolk was conservative, that there was resistance to change and evidence of irrational and non-adaptive behaviour. On the other hand, the Mappila and Latin Christians recognised: (1) militaristic virtues such as bravery, readiness to take direct action, capacity to endure hardship and pain; (2) decision making ability; and (3) sexual prowess, in adapting modern innovations.

We observed that Mappila and Latin Christians were more independent, reflected more direct aggression, and were generally able to express their emotions. By contrast, the Hindu fisherfolk suppressed their emotions, were more hostile in interpersonal relationship but were unable to express such sentiments through direct action. We have also seen that the Hindu fisherfolk were more devious and more given to witchcraft as a means of handling their aggressive feelings. We were not sure whether the greater restraint among the Hindu fisherfolk would lead to greater impulsivity of action, but the findings clearly indicate that such impulsivity is associated with constraints.

In this connection UNESCO holds (1995): "In the face of such examples, it must be acknowledged that the expansion of Western technology is mainly driven by a powerful economic incentive, the desire to make profits. Accordingly, it calls into a question – the local know-how, distribution of wealth, and social organisation of the community. The participation of the population is only rarely effective and the transfer of technology is generally only partially successful".

Traditional communities like the Hindu fisherfolk were reluctant to adapt modern innovations in fishing unlike the modern societies like the Latin Christians and the Mappilas. UNESCO rightly observes "Furthermore, this trend raises the question of the attitude of the population to innovation. Traditional societies, which are supposed necessarily reject innovation, and modern societies supposedly in constant evolution, are too often considered to be mutually exclusive. In fact, every type of society is capable of innovation. However, the application of new technologies, endogenous or exogenous, depends on numerous parameters and, in particular, on the risk factor".

Thus, we note that over the past 40 years numerous innovations could be successfully introduced among the Latin Christians of southern district of Kerala, Mappila fisherfolk of Malabar and Hindu fisherfolk. Their adaptation to the modern innovations depends on the climatic change, national policies, international economic environment and the risk in which the fisherfolk of Kerala are exposed. Among the Latin Christians of southern districts of Kerala several innovations were introduced; such as trawler, mechanised boats, and trawl nets. But at the same time the Arayans of these districts lag behind in utilising the modern technological innovations.

Certain development projects introduced for the welfare of the fisherfolk have not been successful due to cultural and ecological factors. In this connection UNESCO (1995) observes: "Considering the frequency of failure, requiring projects to be completely redesigned, and, conversely, the possibility of innovation under certain conditions, the questions that need to be asked are how one should manage the dissemination of technology having regard to the cultural dimensions linked to a specific ecosystem and how traditional skills and technologies can be combined with modern knowledge and know-how".

Appendices

Appendix I
Impact of Gulf Boom Among the Fisherfolk

Many of the Mappila fishermen migrated to Saudi Arabia, Kuwait, Oman, Qatar, Dubai, and so on, and worked as fishing labourers and fish traders. According to our informants more than 500 Mappila fishermen of Tanur have migrated to Gulf countries since 1980 where they are working as fishing labourers and fish traders. The following are the few Mappila fishermen who were able to construct houses in Tanur with their savings from Gulf countries.

	Name of the fishermen	*Approx. cost of the house*	*Nature of the employment*
1.	C.M. Hasan Bava, S/o Mammuhaji, Cheriya Moidenteakathu.	4 lakhs	Fishing labourer
2.	C.M. Kunjibaba Haji and his two sons	10 lakhs	Fishing labourer and Contractors.
3.	A.K. Koakutty Haji, Angadikadavathu and his six sons.	20 lakhs	Fishing labourer and fish traders.
4.	E.P. Cheriyabava Haji, Iswarante purakkal and his four sons.	20 lakhs	Fishing labourer
5.	C. Umar, Cheriya Akathu and his five sons.	20 lakhs	Fishing labourer
6.	Saithalvi, Jarkante Purakkal and his four sons.	4 lakhs	Fishing labourer and employed in a Construction Company
7.	Muhammad Haji, Cheriya Moithente Akathu and his three sons.	20 lakh	Fishing labourer
8.	Kunjam Bava, S/o Valiya Karanavar, Jarkante Purakkal and his three sons.	20 lakhs	Fishing labourer and fish traders.
9.	Kunjumuhammad, Asante Purakkal and his three sons.	10 lakhs	Fishing labourer
10.	Moithin Bava Haji, Kuttiali Kadavathu, Idakadappuram	15 lakhs	Fishing labourer

Viswambaran, a Mukkuva fisherman of Blangad, was in a Gulf country for ten years. His two sons-in-law were also in Gulf countries. He returned to his native village and started working as a fishing labourer. He invested his income in fishing crafts and gears and along with five other fishermen, owns one fibre boat, *Cheriya Kettuvallam* and an *Ozhukkuvala* (gill-net). The cost of the equipment are as follows:

Cheriya Kettuvallan	=	Rs. 1 lakh.
Fibre boat with a gill net	=	Rs. 1½ lakhs.
Big fibre boat with 30 men crew	=	Rs. 8 lakhs.

Viswambaran's son is also working in a Gulf country for the past several years. His two daughters are also working in a Gulf country. Viswambaran built a house in 2003, incurring an expenditure of Rs 3 lakhs. But he is proud of his traditional occupation of fishing. His brother Rajan has also been working in a Gulf country as a fishing labourer for 10 years. His younger brother Sudheer is working in a super market in a Gulf country. Viswambaran's elder son Haridas who has studied upto SSLC, is employed in the Gulf. His eldest daughter Kavitha, who has studied upto B.Com is a lecturer ina Parallel College. His younger daughter's husband is also working in a Gulf country as fishing labourer. Viswambaran died when his boat capsized at sea and his family got Rs 1 lakh as compensation.

Anil Kumar, a migrant from Pavarty has two sisters. Their husbands are working in Gulf countries. One of them is employed in a construction company as a driver while the other is working as a fishing labourer.

Ayyappan originally belongs to Pavarty village. His eldest daughter Sindhu has studied upto B.A and she is married in the Pavarty village. Her husband is working as a driver in the Gulf. According to Ayyappan, a large number of fishermen of Pavarty are working in various Gulf countries as employees in construction companies. Pavarty is only 10 kms from Blangad. Ayyapan migrated to Blangad from Pavarty ten years ago. He is still working as a fishing labourer under his wife's brother's fishing unit. He does not own any house in Blangad.

Appendix II
Kshemanidhi Board

Under the Kshemanidhi Board, group insurance programme and other welfare schemes are implemented for the welfare of the fisherfolk. The following are the benefits paid:

(a)	Death due to accident in the sea	=	Rs. 1 lakh
(b)	Permanent and total disability	=	Rs. 1 lakh
(c)	Partial disability	=	Rs. 50000
(d)	Medical expenditure	=	Rs. 2000

Other benefits

1.	If a death takes place before and after the fishing expedition	=	Rs. 20000
2.	For the marriage of the daughter	=	Rs. 1500
3.	Financial assistance for the death of a dependent	=	Rs. 600
4.	Old age pension	=	Rs. 120
5.	Temporary disability due to accident	=	Rs. 500
6.	Financial assistance to the dependents of a disease-stricken pensioner	=	Rs. 5000
7.	Cash award for the students who pass SSLC with highest marks.	=	Rs. 5000, Rs. 3000 and Rs. 2000
8.	Continuing education per month	=	Rs. 100
9.	For the partition of the family	=	Rs. 500
10.	Treatment of cancer and dreadful diseases.	=	Rs. 50,000
11.	Chairman's relief fund	=	500–2500
12.	First aid for delivery	=	Rs. 750
13.	Award for the great fishermen	=	Rs. 10001
14.	Widow pension per month.	=	Rs. 120

Appendix III
List of Fishing Villages Surveyed

Name of Fishing Village (Kadappuram)	*Name of District*
Blangad	Thrissur (=Trichur)
Puthankadapuram	Thrissur
Azhikkode harbour	Thrissur
Puthiyappa	Kozhikode
Beypore harbour	Calicut
Puthiyangadi	Calicut
Tanur	Malappuram
Elaram Kadapuram	Malappuram
Mappila Bay	Cannanore (=Kannur)
Theyil Kadappuram	Cannanore
Ayikkara	Cannanore
Kasabha beach	Kasargod
Kunjathur	Kasargod
Uppala	Kasargod
Kumbla	Kasargod
Tumpoli kadappuram	Alleppey (=Alappuzha)
Vembandu Kayal	Alleppey
Kanhiranchira	Alleppey
Ambalapuzha	Alleppey
Arathuppuzha	Alleppey

Name of Fishing Village (Kadappuram)	*Name of District*
Azheekkal, Karunagappaly Taluk	Quilon (=Kollam)
Sraikkadu	Quilon
Cheriyazheekkal, Parayakadavu	Quilon
Puthanthura	Quilon
Ayiram Tengu	Quilon
Vadi	Quilon
Needakara harbour	Quilon
Shaktikulagara	Quilon
Panathura	Trivandrum (=Thiruvananthapuram)
Punthara	Trivandrum
Virinjam harbour	Trivandrum
Beemapally Kadapuram	Trivandrum
Munambu	Ernakulam
Kalamakku harbour	Ernakulam
Vypin	Ernakulam
Murikkinpadam	Ernakulam

Plates

Plate 1. Mohammad Kutty, a happy hook-and-line fisherman of Tanur (Malappuram district, Kerala) with his family members.

Plate 2. Mohammad Kutty of Tanur with his drift-line in hand.

Plate 3. Interaction between hook-and-line fishermen and the researcher at Tanur.

Plate 4. Trawlers with their catches arriving at Puthiyappa harbour.

Plate 5. Removal of the catch from trawlers for retail sale at Puthiyappa harbour, Calicut district.

Plate 6. The researcher talking to the owner of a trawler at Puthiyappa harbour.

Plate 7. *Chemmin* or prawn is the prized catch of Kerala. The researcher having an interaction with the buyers and owner of the trawler at Puthiyappa harbour.

Plate 8. Separating the *Naran Chemmin* from other species of *Chemmin* at the Puthiyappa harbour.

Plate 9. Landing the *Puthiyapplakili* catch from the trawler at Puthiyappa harbour.

Plate 10. Various species of fish from the catch are separated at Puthiyappa harbour before they are retailed.

Plate 11. Fish traders, fish vendors, commission agents and moneylenders at the auctioning of the catch at Puthiyappa harbour.

Plate 12. Glistening iced sardines, ready for marketing at Puthiyappa harbour.

Plate 13. A headload worker carrying iced sardine to the truck at Puthiyappa.

Plate 14. Fibre boats at the Blangad fishing village with fibre boats with coconut palms as onlookers.

Plate 15. Traditional wooden boats at Blangad, showing their scenic beauty.

Plate 16. The traditional boats in Blangad showing their different *Kallis* (compartments).

Plate 17. Sri Nagayakshi, a *Valiya Vanji* (big canoe) at Blangad.

Plate 18. A fibre boat with facilities for storing ice and catch at Blangad.

Plate 19. Nets kept in coconut groves, for drying before launching at the Blangad.

Plate 20. Launching of a traditional wooden boat at Blangad fishing village.

Plate 21. Shore seine net (*Kamba Vala*) in its operation at Blangad beach.

Plate 22. Hauling the fish caught in the shore seine net at Blangad.

Plate 23. Fishermen launching a fibre boat at Blangad fishing village.

Plate 24. An inside view of a *Valiya Vanji* at Blangad fishing village.

Plate 25. A group of Mukkuvas of Blangad posing for a photograph with a homestead and coconut groves.

Plate 26. An outboard motor being fitted into the boat at Blangad.

Plate 27. A traditional boat showing its different compartments at Blangad.

Plate 28. Fishmongers eagerly awaiting the arrival of a canoe with its catches at Blangad fishing beach.

Plate 29. Repairing of a Gill net is a laborious job. A fisherman is engaged in this job at Blangad beach.

Plate 30. A *Tangu Vala* taken for launching at Blangad.

Plate 31. Repairing of a *Tangu Vala* by three fishermen at Blangad.

Plate 32. Two research scientists discuss the merits and demerits of modern technology at Kappad (Kozhikode) where Vasco de Gama landed in his sailing ship.

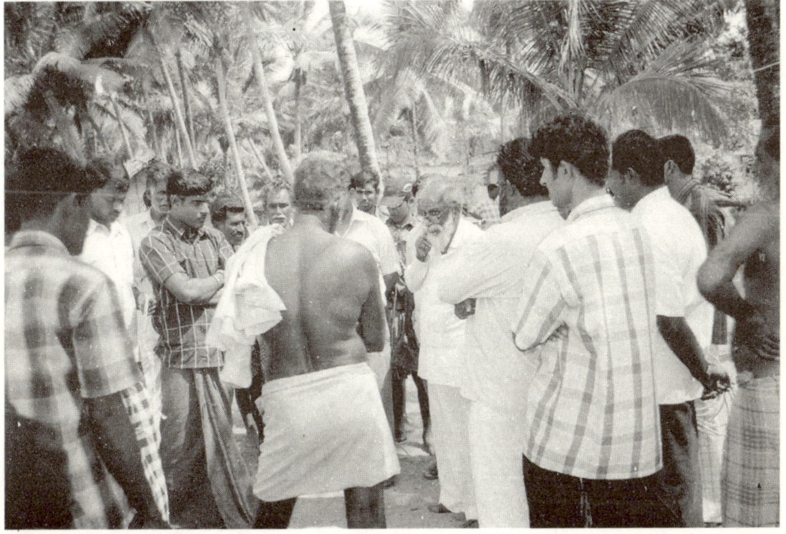

Plate 33. The Mukkuva fishermen discussing their problems with two researchers at Kappad.

Plate 34. The Kappad fishermen joining the piece nets (*Tangu Vala*).

Plate 35. A traditional fibre boat at Kappad, ready to be launched. Two research scientists watch the operation.

Plate 36. The fishermen at Kappad engaged in tying the ropes to the nets. Their tiled and walled houses in the midst of coconut grove gives an appropriate setting to this laborious job.

Plate 37. A fibre boat with its crew members ready for going out for fishing at the Mappila Bay fishing harbour, Cannanore.

Plate 38. A trawler in operation at Mappila Bay fishing harbour, Cannanore.

Plate 39. A *Tangu Vala* is being placed in the boat before going out for fishing at the Mappila Bay fishing harbour.

Plate 40. A trawler getting ready for going out for fishing at the Mappila Bay.

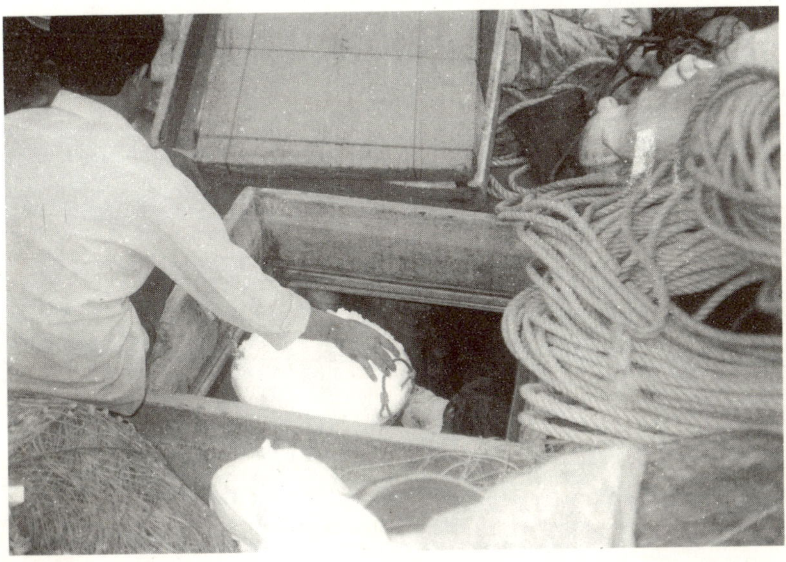

Plate 41. Storing ice in a fibre boat before going out for fishing at Kunjathur, Kasargod.

Plate 42. Three Mogavirar fishermen at Kunjathur, Kasargod, explaining the various parts of their traditional boat to the research worker.

Plate 43. Women fish-vendors at Kasaba village, Kasargod, rushing to the consumers on foot.

Plate 44. A traditional boat of *Pathi*, at Kunjathur, Kasargod, with its specific compartments.

Plate 45. Covering fibre on a boat at Kunjathur, Kasargod, by an expert builder.

Plate 46. Udayawara Bagavathy temple which is the centre of the fishermen at Kunjathur, Kasargod. Devotees offer their prayers here.

Plate 47. Veeraputhran, Nagan, Goddess Bagavathy, Vishnu, and Gulikan at the Udayawara Bagavathy temple, Kunjathur, Kasargod.

Plate 48. A traditional fishing boat at Tumboli Kadappuram, Alleppey district.

Plate 49. A head-load worker transporting the fish to the landing station at Tumboli, Alleppey.

Plate 50. Unloading the catch from *Vakkil Mayavi,* a canoe, at Tumboli fishing village, Alleppey.

Plate 51. Fisherwomen unloading and carrying the catch from the canoe at Tumboli village, Alleppey.

Plate 52. Storing the ice for curing fish at Tumboli fishing village, Alleppey.

Plate 53. Fisherwomen wading through the sea for a share in the catch at Tumboli village, Alleppey.

Plate 54. A busy scene of auctioning and marketing of the catch by the traders and commission agents at Tumboli, Alleppey.

Plate 55. Fisherwomen waiting for the catch at Tumboli, Alleppey, for retail sale in the villages.

Plate 56. A scene of traditional boats ready for going out to fishing at Azheekkal harbour, Quilon.

Plate 57. Bagavathy Temple at Azheekkal, which is the ritual centre of the local Arayans.

Plate 58. Canoes ready for going out to fish at Azheekkal harbour, Quilon.

Plate 59. A view of the Subramanya Temple at Azheekkal fishing village, Quilon.

Plate 60. A trawler arriving with its catch at Neendakara port, Quilon.

Plate 61. A panorama of the Neendakara port with its fishing craft and fisherfolk.

Plate 62. A scene at the auctioning of the catch at Neendakara harbour. Men and women joining to the auction centre to buy the fish for retailing.

Plate 63. Araya women selling the fish at Neendakara harbour after buying the catch at auctions.

Plate 64. After the auctioning at Sakthikulangara harbour, fish is ready for transportation to the consumer centres.

Plate 65. Two fishermen with their traditional boat at Sakthikulangara harbour, Quilon district.

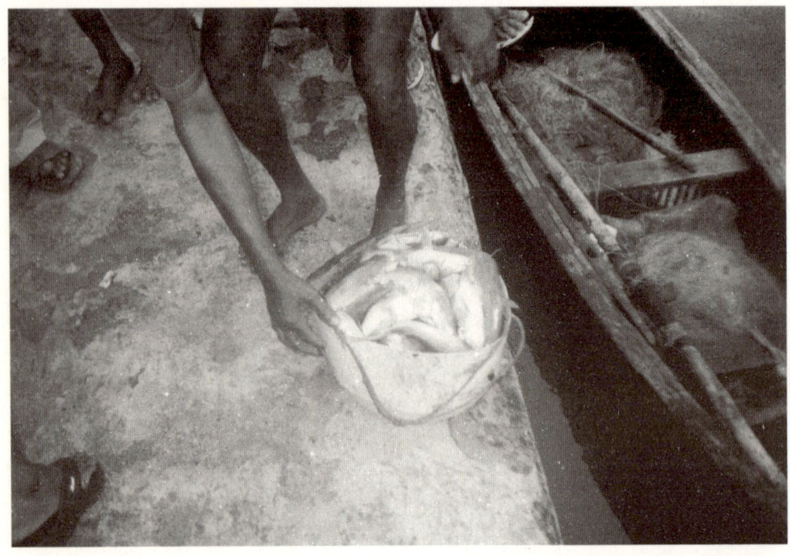

Plate 66. A traditional boat with a catch at Sakthikulangara harbour, Quilon.

Plate 67. Traditional boats in operation at Sakthikulangara harbour, Quilon.

Plate 68. A trawler arrives at Sakthikulangara harbour, Quilon.

Bibliography

ABRAHAM, A. 1982: *A Study of Household Incomes among Marine Fishermen of South Kerala*. Trivandrum: PCO Centre.

— 1985: "Subsistence Credit: Survival Strategies among Traditional Fishermen". *Economic and Political Weekly,* XX (6): 247–252.

ACHESON, J.M. 2006: "Lobster and Groundfish Management in the Gulf of Maine: A Rational Choice Perspective". *Human Organization,* Vol. 65, No. 3.

ALEXANDER, P. 1975: "Innovation in a Cultural Vacuum: The Mechanisation of Sri Lanka Fisheries". *Human Organization*, 34: 333–344.

ALLEN, S.D. et al. 2006: "Monitoring Environmental Justice Impacts: Vietnamese-American Longline Fishermen Adapt to the Hawaii Swordfish Fishery Closure". *Human Organization*, 65, No. 3.

ARNOFF, J. 1967: *Psychological Needs and Cultural Systems*. Princeton: D.Van Nostrand.

BAKER, Paul, T. 1961: "The Application of Ecological Theory to Anthropology". *American Anthropologist*, 64, No. 1, Part 1.

BAKS, C. 1975: "Fishing Communities in the Scottish East Coast: Traditions in a Modern Setting". *Man and Life,* 1 & 2, January-December 1975.

BANAJI, 1977: "Modes of Production in a Materialist Conception of History". *Capital and Class,* 3: 1–44.

BEEBE, J. 2001: "Rapid Assessment Process". Walnut Creek, CA: Alta Mira Press.

— 2005: "Rapid Assessment Process". In *Encyclopaedia of Social Measurement*. Kimberley Kempf-Leonard, (ed.), pp. 285–291. San Diego: Elsevier Inc.

BERKES, F. 1977: "Fishing Resource Use in a Subartic Indian Community". *Human Ecology* 5: 289–307.

BETEILLE, A. 1966: *Caste, Class and Power: Changing Patterns of Stratification in a Tanjore Village*. Bombay: Oxford University Press.

BORT, J. 1978: "Technological Innovation and Organizational Implications: The Case of Pair Trawling in New England's Fishing Industry". Paper Presented at the Annual Meeting of the American Anthropological Association, Los Angeles, California, 1978.

BRANDT, A. Von. 1964: *Fishing Catching Method of the World.* London: Fishing News Books Ltd.

BROSIUS, J.P. 1999: "Analyses and Interventions: Anthropological Engagements with Environmentalism". *Current Anthropology*, 40: 277–309.

BROX, Ottar 1972: *Newfoundland Fishermen in the Age of Industry: A Sociology of Economic Dualism*. Toronto: University of Toronto Press.

BUCHANAR, F.H. 1807, *A Journey from Madras through the Countries of Mysore, Canara and Malabar*. 3 Vol., London: Cadel and Davis.

COLBURN, L.L. et al. 2006: "Anthropological Applications in the Management of Federally Managed Fisheries: Context, Institutional History, and Prospectus". *Human Organization*, 65, No. 3.

D'SOUZA, Miriam 1978: "Fighting Over Fishing Rights and Pose Threat to Goa's Ranponkars". *Sunday Standard*. May 21, 1978.

DALTON, George 1961: "Economic Theory and Primitive Society". *American Anthropologist,* 63: 1–25.

DANIEL, V. 1984: *Fluid Signs: Being a Person the Tamil Way*. Berkeley and Los Angeles: University of California Press.

DEWEY, Alice G. 1962: *Peasant Marketing in Java.* New York: Free Press of Glencoe.

— 1964: "Capital, Credit and Saving in Javanese Marketing" in *Capital, Saving and Credit in Peasant Societies*. R. Firth and B.S. Yamey (ed.). London: Allen & Unwin.

DIGBY, M. 1973: *The Organization of Fishermen's Co-operatives.* Oxford: Plunkett Foundation for Co-operative Studies.

DIGBY, Margaret 1961: Fishermen Co-operatives. Rome: FAO.

DUBE, S.C. 1955: *Indian Village*. New York: Cornell University Press.

DUMONT, L. 1980: *Home Hierarchicus: The Caste System and its Implications.* (revised ed.) Chicago: The University of Chicago Press.

EGNOR, M. 1978: "The Sacred Spell and other Conceptions of Life in Tamil Culture". Ph.D Thesis, The University of Chicago.

EDGERTON, Robert B. 1971. *The Individual in Cultural Adaptation,* University of California Press.

EPSTEIN, A.L. 1967: *The Craft of Social Anthropology*. London: Tavistock Publication.

EPSTEIN, T. 1962: *Economic Development and Social Change in South India.* Manchester: Manchester University Press.

EPSTEIN, T.S. 1973: *South India: Yesterday, Today and Tomorrow*. London: Macmillan.

EPSTEIN, T. Scarlett 1971: *A Comparative Analysis of Economic Development and Social Change in South India and New Guinea Villages*. University of Mysore.

EVANS-PRITCHARD, E.E.,1937: *Witchcraft, Oracle and Magic Among the Azande.* Oxford: Clarendon Press.

FAO (United Nations Food and Agricultural Organization) 1971: Manual on Fishermen's Co-operatives. Rome: UNFAO Organization.

—, 2005: *The State of the World's Fisheries and Aquaculture.* Rome: FAO Fisheries Dept.

FERRIS, A.L. 1988: "The Uses of Social Indicators". *Social Forces* 66(3): 601–617.

FIRTH, Raymond, 1939: *Primitive Polynesian Economy* (1st edn.) London: George Routledge.

— 1946: *Malay Fishermen: Their Peasant Economy* (1st edn.) London: Hegan Paul, Trench, Trubner.

— 1967: *Tikopia Ritual and Belief.* George Allen & Unwin Ltd.

FIRTH, R. and B.S. YAMEY (eds.) 1964: *Capital, Saving and Credit in Peasant Societies:* London: Allen & Unwin.

FORDE, Drayil, 1934: *Habitat, Economy and Society*. London: Methuen & Co. Ltd.

— 1956: *Primitive Economies: Man, Culture, and Society*, Hary L. Shapiro, (ed.) New York: Oxford University Press.

FOSTER, G. 1973: *Traditional Societies and Technological Change*. Berkeley: University of California Press.

FOSTER, G. 1967: *Tzintzuntzan,* Boston: Little Brown.

FRAKE, Charles, O. 1962: "Culture, Ecology and Ethnography". *American Anthropologist,* 64 No. 1, Part 1.

FRANK G. 1978: *World Accumulation,* 1492–1789. New York: Monthly Review Press.

FRAZER, T.M. Jr. 1960: *Rusembolan: A Malay Fishing Village in Southern Thailand.* Cornell University Press.

— 1966: *Fishermen of South Thailand: The Malay Villagers.* Holt: Rinehart and Winston, Inc.

GAHIN, R. and PATERSON, C. 2001: "Community Indicators: Past, Present, and Future". *National Civic Review,* 90: 347–360.

GEERTZ, C., 1962: *The Religion of Jawa*. Chicago: The University of Chicago Press.

— 2000: *The Interpretation of Culture: Selected Essays*. New York: Basic Books.

GEORGE, A.I., 1971: "Fisheries Development Programmes of Kerala". *Sea Food Export Journal Annual,* 3 No. 1.

GEORGE, P.C. et al. 1977: Fishery Resources of the Indian Exclusive Economic Zone, in *Souvenir*. Integrated Fisheries Project, Cochin.

GEORGE, P.C., 1958: "Sole Fisheries". *Fisheries of West Coast of India* (ed.) Jones, S.

GERSUNY, C. and POGGIE, J.J. 1974: "A Fishermen's Co-operative: Open System Theory Applied". *Maritime Studies and Management,* 1: 215–222.

GLAZIER, Edward W. et al. 2006: 'Toward Mitigating Problems at the Fisheries-Oil Development Interface: The Case of the Salmon Drift Gillnet Fishery in Cook Inlet, Alaska". *Human Organization* 65, No. 3.

GOUGH, K. 1971: "Caste in a Tanjore Village" in E.R. Leach (ed.) *Aspects of Caste in South India, Ceylon and NW Pakistan.* London: Cambridge University Press, pp. 11–60.

GROVES, C.W. 1934: "Fishing Rites at Tabar". *Oceania, 4* (4): 432–457.

GULATI, L. 1984: "Technological Change and Women's Work: Participation and Demographic Behaviour. A Case Study of Three Fishing Villages". *Economic and Political Weekly,* XXIX (49): 2089–94.

HAJRE, D. 1967: Marine Fisherfolk of Gujarat, Preliminary Exploration. Seminar on Research Programme in Cultural Anthropology and Related Disciplines, Surajit Sinha (ed.). Calcutta: ASI: 47–64.

HINDESS, B. and HIRST, P. 1977: *Mode of Production and Social Formation.* London: Macmillan.

HORNEL, J. 1950: *Fishing in Many Waters*. Cambridge: University Press.

HUNTINGTON, S.P. 1968: *Political Order in Changing Societies.* New Haven: Yale University Press.

HUTTON, J.H. 1961: *Caste in India.* London: Oxford University Press.

INDEN, R.B. and NICHOLAS, R. 1977: *Kinship in Bengali Culture.* Chicago: The University of Chicago Press.

IYER, L.K.A. 1909: *The Tribes and Castes of Cochin.* Vols. 1–3. New Delhi: Cosmo Publications.

JHINGRAN, V.G. 1982: *Fish and Fisheries of India.* Delhi: Hindustan Publishing Corporation.

JOHNSON, J.C. 1990: "Selecting Ethnographic Informants". Newbury Park, CA: Sage Publications.

JULIA, Olson 2006: "Changing Property, Spatializing Difference: The Sea Scallop Fishery in New Bedford, Massachusetts". *Human Organization* 65, No. 3.

KITNER, R. Kathi 2006: "Beeliners, Pinkies, and Kitties: Mobility and Marginalization in the South Atlantic Snapper Grouper Fishery". *Human Organization*, 65, No. 3.

KLAUSSEN, A.M. 1968: *Kerala Fishermen and the Indo-Norwegian Pilot Project.* London: Allen & Unwin.

KOVAKANANDY, R. 1984, "Purse Seine Fishing in Kerala: Its Economics and Politics". *Economic and Political Weekly,* XIX, No. 13: 566–572.

KURIEN, J. and MATHEW, S. 1982, *Technological Change in Fishing: Its Impact on Fishermen*. Trivandrum: Centre for Development Studies.

KURIEN, J. and THANKAPPAN ACHARI, 1998: Overfishing the Coastal Commons: Causes and Consequences in *Social Ecology*. Ramachandra Guha (ed.), New Delhi: Oxford University Press.

KURIEN, J. 1985: "Technical Assistance Projects and Socio-economic Change: Norwegian Intervention in Kerala's Fisheries Development". *Economic and Political Weekly*, XX (25, 26) A: 70–87.

KURUVILLA, A.O. et al. 1982: "Fishermen's Perceptions of Problems and Prospects of Kerala Fisheries". Trivandrum: PCO Centre.

LACLAU, E. 1977: *Politics and Ideology in Marxist Theory*. London: New Left Books.

LAMUG, C.B. 1994: "A Guide for Participatory Rural Appraisal of Coastal Communities". Laguna, Philippines: Environment and Resource Management Project, Institute of Environmental Science and Management.

MALINOWSKI, B. 1912: *Argnonauts of the Western Pacific*. New York: E.P. Dutton & Co.

— 1918: "Fishing in the Trobriand Islands". *Man* 18 (58): 87–92.

— 1948: *Magic, Science and Religion and Other Essays.* Glencoe: The Free Press.

MARRIOTT, M. and INDEN, R.B. 1977: "Towards an Ethnosociology of South Asian Caste Systems" in K. David (ed.) *The New Wind: Changing Identities in South Asia.*The Hague: Mouton.

MARX, K. 1973, 1857–58: *Grundrisse*. Harmondsworth: Penguin.

MATHUR, P.R.G. and ZACHARIAH E., 1983: The Hindu Fisherfolk of Kerala: Why Were They Not Classified as Scheduled Castes. Kozhikode: Unpublished Report of the Department of KIRTADS.

MATHUR, P.R.G. 1978: *The Mappila Fisherfolk of Kerala.* Trivandrum: Kerala Historical Society.

— 1979: "The Marine Hindu Fisherfolk: Some Ethnographic Notes". *Journal of Kerala Studies,* Vol. VI. Parts I & II.

— 2005: "Ecological and Cultural Heritage of Hindu Fishing Communities in India" (Unpublished).

McCRACKEN, J. 1990: Rapid Rural Appraisal for Fishereis Development: Testing the Approach in a Fisherfolk Community. Tamil Nadu, India: Report of a Consultancy from the Bay of Bengal Fisheries Development Programme.

MEILLASOUX, C. 1973: "Are There Castes in India?" *Economy and Society*, 2, pp. 89–111.

MENCHER, J. 1974: "The Caste System Upside Down: Or the Not So Mysterious East". *Current Anthropology*, 15(4): 469–494.

MICHAEL PAOLISSO et al. 2006: "Restoration of the Chesapeake Bay Using a Non-Native Oyster: Ecological and Fishery Considerations". *Human Organization*, 65, No. 3.

MILTON, K. 1996: *Environmentalism and Cultural Theory: Exploring the Role of Anthropology in Environmental Discourse*. New York: Routledge.

MOSES, S.T. 1929: "The Valayans of Pamban". *Man in India*, 9(1): 149–156.

MURICKAN, J., S.J. 1991: *Religion and Power Structure in Rural India.* Jaipur: Rawat Publications.

NADEL, S.F. 1960: "Malinowski on Magic and Religion" in *Man and Culture: An Evaluation of the Works of Malinowski*. Raymond Firth (ed.). New York: Humanities Press.

NAIR, B. 1971: Water Wealth of Kerala in *Sea Food Export Journal.* Vol. III. No. 1.

NICHOLSON, F. 1909: "The Preservation and Curing of Fish". *Madras Fish Bulletin* No. 3.

NOOR, K. and NOOR, J. 1974: Environmental and Technical Factors Influencing Power in Work Organizations: Ocean Fishing in Peasant Societies. *Sociology of Work and Occupations*. No. 1: 219–251.

NORBECK, E. 1974: *Takashima: A Japanese Fishing Village*. Salt Lake City: University of Utah Press.

PANICKER, N.K. 1958: "Indian Fisheries: An Overall View" in *Fisheries of the West Coast of India*. Jones, S. (ed.), Mandapam Camp, Central Marine Fisheries Research Station.

PIDO, M.D. et al. 1996: *A Handbook for Rapid Appraisal of Fisheries Management Systems*. Manila: International Center for Living Aquatic Resource Management.

PILLAI, V.K. et al. 1958: "Fishing Curing and Fishing Techniques". *Fisheries of West Coast of India*, Jones, S (ed.), Mandapam Camp, Central Marine Fisheries Research Station.

PLATTEAU, J. 1982: *A Study on Credit and Indebtedness among Marine Fishermen of Southern Kerala: A Summary*. Trivandrum: PCO Centre.

POGGIE, J.J. and GERSUNY, C. 1974: *Fishermen of Galilee: The Human Ecology of a New England Coastal Community*. Kingston: University of Rhode Island.

POGGIE, J.J. 1978: Ritual Adaptation to Risk and Technological Development in Ocean Fisheries: Extrapolations from New England. Paper Presented at the Society for Applied Anthropology Annual Meetings, Merida, Mexico.

POGGIE, J.J. Jr 1980: "Maritime Anthropology: Socio-cultural Analysis of Small-Scale Fishermen's Co-operatives". *Anthropological Quarterly*, 53 No. 1: 1–3.

— 1980: "Small-scale Fishermen's Psychocultural Characteristics and Cooperative Formation". *Anthropological Quarterly*, 53 No.1: 20–28.

POGGIE, J.J., POLLNAC, R.B. and GERSUNY, C. 1976: "Risk as a Basis for Taboos among Fishermen Southern New England". *Journal for the Scientific Study of Religion,* 15: 257–262.

POLLNAC, R.B and POGGIE, J.J 2006: "Job Statisfaction in the Fishery in Two Southeast Alaskan Towns". *Human Organization*, 65, No. 3.

POLLNAC, R.B. and CARMO, F. 1980: "Attitudes toward Co-operation among Small-Scale Fishermen and Farmers in the Azores". *Anthropological Quarterly*, 53 No. 1: 12–19.

POLLNAC, R.B. and POGGIE, J.J. 1978: "Economic Gratification Orientations among Small-Scale Fishermen in Panama and Puerto Rico". *Human Organization*, 37: 355–367.

— 1979: The Structure of Job Satisfaction Among New England Fishermen. *Anthropology* Working Paper No. 31. Kingston R.I: University of Rhode Island.

POLLNAC, R.B. and RUIZ STOUT, R. 1977: Small-Scale Fishermen's Attitudes Towards the Occupation of Fishing in the Republic of Panama. Marine Technical Report No. 44. Kingston R.I: University of Rhode Island.

POLLNAC, R.B. (ed.) 1977: Panamanian Small-Scale Fishermen: Society, Culture and Change. Marine Technical Report No. 44. Kingston R.I: University of Rhode Island.

POLLNAC, R.B. 1976: Continuity and Change in Marine Fishing Communities. Anthropology Working Paper No. 10. Kingston R.I: University of Rhode Island.

POLLNAC, R.B. 1977(a): Investment Orientations Among Small-Scale Fishermen in the Gulf of Nicoya, Costa Rica Anthropology Working Paper No. 19 Kingston R.I: University of Rhode Island.

— 1977(b): "Illusion Susceptibility and Adaptation to the Marine Environment: Is the Carpentered World Hypothesis Seaworthy?" *Journal of Cross-Cultural Psychology* 8: 425–434.

— 1978: Socio-cultural Aspects of Technological and Institutional Change Among Small-Scale Fishermen. Anthropology Working Paper No. 22. Kingston R.I: University of Rhode Island.

POLLNAC, R.B., GERSUNY, C. and POGGIE, J.J. 1975: "Economic Gratification Patterns of Fishermen and Millworkers in New England". *Human Organization* 34: 1–7.

POSTEL-COSTER, EIs and HEIJMERINK, J.J.M. 1976: *Fishing Communities on the Scottish East Coast*. Leiden: Press and Journal.

PRAMANIK, S.K. 1993: *Fishermen Community of Coastal Villages in West Bengal*. Jaipur: Rawat Publications.

PRASAD, R.R. 1958: "The Arabian Sea Marine Biology of the West Coast". *Fisheries of the West Coast of India*. Jones, S. (ed.). Mandapam Camp, Central Marine Fisheries Research Station.

RADCLIFFE-BROWN, A.R. 1948: *The Andaman Islanders*. Illinois: The Free Press.

RAM, K. 1991: *Mukkuvar Women,* London: Allen & Unwin Private Limited.

RAO, D.L.P. and KUMAR, R.Y. 1984: "Impact of Mechanisation on a Traditional Economy: Case Study of Fishermen in Visakapatnam". *Man and Life* 10 (1 & 2) January–June: 31–42.

RAO, K.M 1984: "Fisherman Castes of Coastal Andhra". *Man and Life* 7 (1 & 2) January–June: 11–16.

RAO, Naga Raja, 1964: *An Introduction to Fisheries.* Hyderabad.

RAO, U. 1981: *Deprived Castes in India*. Allahabad: Chugh Publications.

RAPPAPORT, R. 1984: *Pigs for the Ancestors: Ritual in the Ecology of a New Guinea People*. New Haven: Yale University Press.

RAYCHAUDHURI, B. 1980: *The Moon and Net: Study of a Transient Community of Fishermen at Jambudwip*. Calcutta: ASI.

ROBBINS, M., Robbins, L. and POLLNAC, R.B. 1977: Co-operative Membership and Optimism Among Small-Scale Fishermen in Panama. Marine Technical Report No. 44. Kingston R.I: University of Rhode Island.

SAHLINS, M. 1976: *Culture and Practical Reason,* Chicago: The University of Chicago Press.

SAMUEL, C.T. 1968: *Marine Fisheries of Kerala*. Ernakulam: S.D. Reddiar & Sons.

SANYAL, S. 1984: "Some Demographic Aspects of the Fishermen Communities of Orissa". *Man in India,* 64 (3): 309–319.

SEPEZ, J. et al. 2006: "Fish Scales: Scale and Method in Social Science Research for North Pacific and West Coast Fishing Communities". *Human Organization,* 65, No. 3.

SINHA, S. (ed.) 1970: *Science, Technology and Culture* "Introduction", India International Centre.

SREEKRISHNA, Y. and SHENOY, L. 2001: "Fishing Gear and Craft Technology". Indian Council of Agricultural Research, New Delhi.

SRINIVAS, M.N. 1952: *Religion and Society Among the Coorgs of South India.* New Delhi: Oxford University Press.

— 1962: *Caste in Modern India and Other Essays*. Bombay: Asia Publishing House.

— 1966: *Social Change in Modern India*. Bombay: Allied Publishers.

— 1976: *The Remembered Village*. Delhi: Oxford University Press.

STEWARD, J.H. 1955: *Theory of Culture Change.* Urbana: University of Illinois Press.

SUBBA RAO, N. 1978: "Development of Fisheries in Srikakulam". *Yojana* 22 (3), February 16.

SURYANARAYANA, M. 1977: *Marine Fisherfolk of North East Coastal Andhra Pradesh*. Calcutta: ASI.

THURSTON, E. 1900: *Sea Fisheries in India*, Ernakulam: S.D. Reddiar & Sons.

— 1909: *Caste and Tribes of Southern India*. Vol. I–VII Delhi: Cosmo Publications.

TUNSTALL, J. 1962: *The Fishermen*. London: Routledge.

UNESCO 1995: "Cultural Dimension of Development: Towards Practical Approach". Paris: UNESCO Report.

United States Department of Agriculture (USDA), 2003: National Report of Sustainable Forests–2003. United States Department of Agriculture, Forest Service FS–766, p. 139.

WADLEY, S. (ed.) 1980: *The Powers of Tamil Women*, Syracuse: Syracuse University Press.

WALLERSTEIN, E. 1974: *The Modern World System: Capitalist Agriculture and the Origins of the European World System.* New York: Academic Press.

WILSON, D. et al. 1998: "A Social and Cultural Impact Assessment of the Highly Migratory Species Fisheries Management Plan and the Amendment of the Atlantic Billinfhs Fisheries Management Plan". The Ecopolicy Center for Agriculture, Environmental, and Resource Issues, Rutgers University and NMFS. New Brunswick, NJ.

ZACHARIAH, E. 1996: Mogeyar of Malabar: A Study in Interrelationship between Environment, Technology, Economy, Polity, Society and Culture. Unpublished Ph.D Thesis.

Index